Sequence Space Theory with Applications

The book features original chapters on sequence spaces involving the idea of ideal convergence, modulus function, multiplier sequences, Riesz mean, Fibonacci difference matrix etc., and illustrates their involvement in various applications. The preliminaries have been presented in the beginning of each chapter and then the advanced discussion takes place; so it is useful for both expert and nonexpert on aforesaid topics. The book consists of original thirteen research chapters contributed by the well-recognized researchers in the field of sequence spaces with associated applications.

Features

- Discusses the Fibonacci and vector valued difference sequence spaces
- Presents the solution of Volterra integral equation in Banach algebra
- Discusses some sequence spaces involving invariant mean and related to the domain of Jordan totient matrix
- Presents the Tauberian theorems of double sequences
- Discusses the paranormed Riesz difference sequence space of fractional order
- Includes a technique for studying the existence of solutions of infinite system of functional integro-differential equations in Banach sequence spaces

The subject of book is an active area of research of present time internationally and would serve as a good source for researcher and educators involved with the topic of sequence spaces.

Sequence Space Theory with Applications

Edited by

S. A. Mohiuddine

Department of General Required Courses, Mathematics
Faculty of Applied Studies
King Abdulaziz University, Jeddah, Saudi Arabia

Operator Theory and Applications Research Group,
Department of Mathematics, Faculty of Science,
King Abdulaziz University, Jeddah, Saudi Arabia

Bipan Hazarika

Department of Mathematics
Gauhati University
Guwahati, Assam

CRC Press
Taylor & Francis Group
Boca Raton London New York

CRC Press is an imprint of the
Taylor & Francis Group, an **informa** business
A CHAPMAN & HALL BOOK

First edition published 2023
by CRC Press
6000 Broken Sound Parkway NW, Suite 300, Boca Raton, FL 33487-2742

and by CRC Press
4 Park Square, Milton Park, Abingdon, Oxon, OX14 4RN

CRC Press is an imprint of Taylor & Francis Group, LLC

ISBN: 978-1-032-01325-1 (hbk)
ISBN: 978-1-032-01331-2 (pbk)
ISBN: 978-1-003-17820-0 (ebk)

DOI: 10.1201/9781003178200

Typeset in CMR10 font
by KnowledgeWorks Global Ltd.

Publisher's note: This book has been prepared from camera-ready copy provided by the authors.

Contents

Preface

This book features original chapters on the theory of sequence spaces and illustrate their involvement in various applications. Each chapter describes the problem of current importance, and to summarize ways of their solution and possible applications, and to improve the current understanding pertaining to sequence spaces. The book consists of thirteen original research chapters contributed by the well-recognized researchers in the field of sequence spaces with associated applications. The list of chapters is arranged alphabetically by the last names of the first author of each chapter.

Chapter 1 is devoted to prove the Hahn-Banach, sandwich and extension theorems, the Fenchel duality theorems, the Moreau-Rockafellar formula, the Farkas theorem, the Karush-Kuhn-Tucker conditions, and separation by means of convex sets, for vector lattice-valued operators, invariant or equivariant with respect to a given (semi)group G of homomorphisms, are always true if and only if G is amenable, and are equivalent each other. Some applications to extensions of finitely additive invariant set functions, related to a Tarski-type theorem and a concept of "event" viewed as a suitable product set, indexed by the time variable are given.

Chapter 2 makes use of the techniques associated with measures of noncompactness to obtain the existence of solution of infinite system of functional integro-differential equations with integral initial conditions in Banach spaces and results are illustrated with the help of examples.

Chapter 3 is devoted to defining the notions of strongly λ-summable of order α and statistically λ-convergent of order α for the sequence of interval numbers, and proves some inclusion relations between aforesaid notions. The notion of interval valued sequence space $\bar{l}(p)$ is introduced and shows that this is solid, monotone and not convergence free in general.

Chapter 4 obtains necessary and sufficient conditions for $(A^{r,\delta}, p, q; 1, 1)$ summability under which the converse of the implication "if $\lim_{m,n\to\infty} x_{mn} = l$ and (x_{mn}) is bounded, then $\lim_{m,n\to\infty} \sigma_{m,n}^{r,\delta,p,q}(x) = l$" holds. This chapter also discusses $(A^{r,*}, p, *; 1, 0)$ and $(A^{*,\delta}, *, q; 0, 1)$ summability methods and obtains Tauberian theorems for these methods.

Chapter 5 makes use of newly introduced regular matrix to define matrix domains of it in the classical sequence spaces c_0, c, ℓ_∞ and computes α-, β- and γ-duals of the resulting spaces as well as presents the characterization of certain matrix mappings on these spaces.

Chapter 6 discusses the construction of new sequence spaces by means of the matrix domain of a particular limitation method, more precisely, using the Fibonacci difference matrix $\hat{F}$ defined by Fibonacci sequence (f_n) for $n \in \{0, 1, ...\}$ and the notion of ideal convergence, introduces some new sequence spaces related to the matrix domain of $\hat{F}$. Some topological and algebraic properties, and inclusion relations of these spaces are studied.

Chapter 7 discusses some operators, namely, strong intuitionstic fuzzy continuous operators and strong intuitionistic fuzzy bounded operators, in complete intuitionistic fuzzy normed linear spaces in order to obtain the approximation property, as well as discusses the existence of approximation properties with the help of these operators in the framework of complete intuitionistic fuzzy normed linear spaces.

Chapter 8 introduces a new notion of (F, χ)-set contraction mappings for class of functions involving measure of noncompactness in Banach space, and establishes some basic Darbo type fixed and coupled fixed point results. This chapter also discusses the applicability of derived results to the Volterra integral equations in Banach algebras.

Chapter 9 discusses the sufficient conditions that ensure the existence of a unique common positive definite solution of the pair of matrix equations, for this, establishes new common fixed point results for mappings in the frame work of closed subsets of Banach spaces. Some examples are presented to validate the fixed-point results and the convergence analysis of non-linear matrix equations.

Chapter 10 defines the σ-regularity for a series method and discusses the necessary and sufficient conditions for a series method to be σ-regular, σ-regularity of dual summability methods and some sequence spaces by using invariant mean and sequence of matrices. Further, these sequence spaces are extended with the help of modulus function, and examines some matrix transformations.

Chapter 11 is devoted to studying the Carathéodory solutions of dynamic equations on time scales. This chapter also deals with the solutions of Carathéodory dynamic problem based on the concept of Henstock-Kurzweil delta integral as well as some other related and interesting results.

Chapter 12 introduces the vector valued generalized difference ideal convergent sequence spaces associated with the multiplier sequence and investigates some algebraic and topological properties such as solid, symmetry, convergence free, monotone, completeness and compactness. The multiplier problem is characterized, and suitable examples are discussed to justify some some failure cases and for the definitions.

Chapter 13 introduces the paranormed Riesz difference sequence space of fractional order obtained by the domain of generalized difference operator in Maddox's space $\ell(p)$, and obtains some topological and geometric properties, and Schauder basis of this sequence space. The α-, β- and γ-duals and characterization of some matrix classes on the new space are presented.

Editors

S. A. Mohiuddine is a full professor of Mathematics at King Abdulaziz University, Jeddah, Saudi Arabia. An active researcher, he has coauthored three books, Convergence Methods for Double Sequences and Applications (Springer, 2014), Advances in Summability and Approximation Theory (Springer, 2018) and Soft Computing Techniques in Engineering, Health, Mathematical and Social Sciences (CRC Press, Taylor & Francis Group, 2021), and a number of chapters and has contributed over 140 research papers to various leading journals. He is the referee of many scientific journals and member of the editorial board of various scientific journals, international scientific bodies and organizing committees. He has visited several international universities including Imperial College London, UK. He was a guest editor of a number of special issues for Abstract and Applied Analysis, Journal of Function Spaces and Scientific World Journal. His research interests are in the fields of sequence spaces, statistical convergence, matrix transformation, measures of noncompactness and approximation theory. His name was in the list of Worlds Top 2% Scientists (2020) prepared by Stanford University, California.

Bipan Hazarika is presently a professor in the Department of Mathematics at Gauhati University, Guwahati, India. He has worked at Rajiv Gandhi University, Rono Hills, Doimukh, Arunachal Pradesh, India from 2005 to 2017. He was professor at Rajiv Gandhi University upto 10-08-2017. He received his Ph.D. degree from Gauhati University and his main research interests are in the field of sequences spaces, summability theory, applications of fixed point theory, fuzzy analysis and function spaces of non absolute integrable functions. He has published over 150 research papers in several international journals. He is an editorial board member of more than 5 international journals and a regular reviewer of more than 50 different journals published from Springer, Elsevier, Taylor & Francis, Wiley, IOS Press, World Scientific, American Mathematical Society, De Gruyter. He has published books on Differential Equations, Differential Calculus and Integral Calculus. He was the guest editor of the special issue “Sequence spaces, Function spaces and Approximation Theory”, in Journal of Function Spaces..

Contributors

Antonio Boccuto
Department of Mathematics and Computer Sciences, University of Perugia
Perugia, Italy

Anupam Das
Department of Mathematics, Cotton University
Guwahati, Assam, India

Ayhan Esi
Department of Basic Sciences of Engineering, Engineering Faculty, Malatya Turgut Ozal University
Malatya, Turkey

Ayten Esi
Department of Mathematics, Adiyaman University
Adiyaman, Turkey

Çağla Kambak
Department of Mathematics, Ege University
İzmir, Turkey

İbrahim Çanak
Department of Mathematics, Ege University
İzmir, Turkey

Emrah Evren Kara
Department of Mathematics, Düzce University
Düzce, Turkey

Necip Şimşek
Department of Mathematics, İstanbul Commerce University
İstanbul, Turkey

Merve İlkhan Kara
Department of Mathematics, Düzce University
Düzce, Turkey

Vakeel A. Khan
Department of Mathematics, Aligarh Muslim University
Aligarh, India

Kamal M. A. S. Alshlool
Department of Mathematics, Aligarh Muslim University
Aligarh, India

Sameera A. A. Abdullah
Department of Mathematics, Sultan Qaboos University
Muscat, Oman

Nabanita Konwar
Department of Mathematics, Birjhora Mahavidyalaya
Bongaigaon, Assam, India

Pradip Debnath
Department of Applied Science and Humanities, Assam University
Silchar, Cachar, India

Hemant Kumar Nashine
Department of Mathematics, School of Advanced Sciences, Vellore Institute of Technology
Vellore, TN, India

Sourav Shil
Department of Mathematics, School of Advanced Sciences, Vellore Institute of Technology
Vellore, TN, India

Ekrem Savaş
Department of Mathematics, Uşak University
Uşak, Turkey

Sanket Tikare
Department of Mathematics, Ramniranjan Jhunjhunwala College
Ghatkopar (West), Mumbai, India

Binod Chandra Tripathy
Department of Mathematics, Tripura University
Agartala, Tripura, India

Taja Yaying
Department of Mathematics, Dera Natung Government College
Itanagar, Arunachal Pradesh, India

Bipan Hazarika
Department of Mathematics, Gauhati University
Guwahati, Assam

S. A. Mohiuddine
Department of General Required Courses, Mathematics, Faculty of Applied Studies
and
Operator Theory and Applications Research Group Department of Mathematics, Faculty of Science King Abdulaziz University
Jeddah, Saudi Arabia

Chapter 1

Hahn-Banach and Duality Type Theorems for Vector Lattice-Valued Operators and Applications to Subdifferential Calculus and Optimization

Antonio Boccuto

University of Perugia, antonio.boccuto@unipg.it

1.1 Introduction

The Hahn-Banach theorem is one of the most important and investigated results in Functional Analysis. It has several developments and applications in various fields of Mathematics, for example Optimization, Convex Analysis, Numerical Analysis, Differential Equations and Calculus of Variations (see, e.g., [3, 21, 23, 25, 50, 67, 74]). By means of this theorem, it is possible to prove extension and sandwich-type theorems for operators and measures, results on separation of convex sets by means of hyperplanes and several other related theorems. For a related survey, see, e.g., [26, 61].

Among the consequences we recall, for instance, the Fenchel-type duality theorem, which has several applications in minimax theorems, in particular

DOI: 10.1201/9781003178200-1

in Economics and stochastic programming (see, e.g., [7, 28, 72, 75, 76, 80, 81]) and image and/or document reconstruction, and is related to the problem of minimizing the dual energy, which is in general easier to handle than the corresponding one concerning the primal energy (see, e.g., [4, 5, 10, 17–19, 30, 31, 35, 36]).

Some other consequences deal with Subdifferential Calculus (see, e.g., [22, 41, 58, 59, 82] and the references therein), Operations Research and optimization problems with constraints, for example Farkas and Kuhn-Tucker-type theorems, related to optimization problems with suitable constraints (see, e.g., [3, 8, 9, 38, 42, 49, 57, 65, 84, 91]).

Another related topic is the study of invariant and equivariant linear functionals, invariant measures and amenable (semi)groups. Note that every abelian semigroup is amenable, while the group of all permutations $\phi : \mathbb{N} \to \mathbb{N}$ which keep fixed all but finitely many numbers is amenable but not abelian. Moreover, the group $SO(2, \mathbb{R})$ of all orthogonal 2×2-matrices whose determinant is equal to 1 (that is, the group of all rotations of $\mathbb{R}^2$) is amenable, while $SO(n, \mathbb{R})$ is not amenable for all $n \geq 3$ (see, e.g., [68]).

These topics have several applications also in Measure Theory, Probability and stochastic processes (see also [27, 51, 76]). In [51] it is dealt with different kinds of probabilistic symmetries, exchangeability and coding results, for random elements invariant under finite or compact groups (see also [24]). In many cases it is advisable to deal with ordered vector spaces, since we often consider operators or probability measures which can depend not only on the considered event, but also on the time and on the state of knowledge, as well as conditional expected values, which can be viewed as operators taking values in the space of all integrable functions. For a related literature, see, e.g., [11, 12, 14, 15, 20, 26, 27, 29, 33, 46, 47, 53, 68, 70, 77–79, 87, 88]. Moreover, in Probability and Statistics, there are several types of translation invariant or equivariant functionals (for example, deviation risk and acceptability functionals, respectively), invariant statistic models, invariant testing problems and equivariant estimators on the mean vector in Multivariate Analysis of Variance (see, e.g., [39, 70] and the references therein). There are some other recent studies about invariance and equivariance, for example in applications to Machine Learning (see, e.g., [32, 48, 55]).

In [12] (resp. [13]), we extended to linear and invariant or equivariant vector lattice-valued operators earlier results proved in [41, 66, 90–92] in the linear case (without invariance or equivariance) on Hahn-Banach, sandwich and extension theorems, separation theorems by means of affine "hyperplanes", Fenchel-type duality theorems, Moreau-Rockafellar formula, subdifferential calculus, Farkas- and Kuhn-Tucker-type theorems on convex optimization under constraints, and proved the equivalence of the given theorems with amenability. In this work we unify the invariance and equivariance approaches and present a comprehensive overview of such results, giving also some improvements and new proofs of some of them. We are inspired by a Hahn-Banach-type theorem given in [29] for invariant partially ordered vector space-valued operators with respect to amenable semigroups of transformations (here, the role of amenability is essential). Furthermore, we prove that our given theorems are equivalent to each other and equivalent to amenability of the involved (semi)group. Some Hahn-Banach and/or monotone extension theorems for invariant linear vector lattice-valued operators were given also in [29, 47, 78, 79]. Moreover, in [11] and [14] some Hahn-Banach, sandwich and monotone extension theorems and a characterization of amenability were given for linear invariant vector lattice-valued operators, by considering subadditive (resp. sublinear) operators defined on suitable monoids or groups (resp., cones or vector spaces). Here we deal with *convex* functions, defined on suitable *convex* sets, which are not necessarily subspaces or cones, and so we consider the *algebraic* relative interior of a set. Note that, in our context, no topological structure is required, and in the spaces of the type $L^0(\Omega, \mathcal{A}, \mu)$, where μ is a σ-finite and non-atomic countably additive extended real-valued positive measure (with identification up to μ-null sets), order convergence coincides with almost everywhere convergence, and is not topological (see, e.g., [69]). As an application, we give a Tarski-type theorem for finitely additive and invariant vector lattice-valued set functions, and an example of extensions of vector lattice-valued finitely additive "probability measures" on product spaces, in connection with a suitable concept of "event", which can be identified with a family of sets indexed by the time variable.

1.2 Basic Notions and Results

1.2.1 Relative interior points and convexity

Let X be a real vector space. An *affine combination* of elements x_1, $x_2, \ldots, x_n$ of X is any linear combination of the type $\sum_{i=1}^{n} \lambda_i x_i$, with λ_1, $\lambda_2, \ldots, \lambda_n \in \mathbb{R}$ and $\sum_{i=1}^{n} \lambda_i = 1$. An *affine manifold* of X is a nonempty subset of X, closed under affine combinations.

Let $\emptyset \neq D \subset X$. We denote by $\text{span}(D)$ (resp., $\text{span}_{\text{aff}}(D)$) the smallest linear subspace (resp., affine manifold) of X which contains D.

A point $x_0 \in D$ is an *algebraic relative interior point of* D iff for each $x \in \text{span}_{\text{aff}}(D)$ there is $\lambda_0 > 0$ with $(1-\lambda)x_0 + \lambda\, x \in D$ whenever $\lambda \in [-\lambda_0, \lambda_0]$. We denote by $int(D)$ the set of all algebraic relative interior points of D. It is not difficult to check the following property, which will be useful later.

Proposition 1.2.1. *It is* $0 \in int(D)$ *if and only if for every* $x \in span(D)$ *there exists a positive real number* λ_x *such that* $\lambda\, x \in D$ *for all* $\lambda \in [-\lambda_x, \lambda_x]$.

A nonempty subset D of a real vector space X is said to be *convex* iff $\lambda x_1 + (1-\lambda)x_2 \in D$ for every x_1, $x_2 \in D$ and $\lambda \in [0,1]$.

Given two real vector spaces X, Y, we say that a function $T \in Y^X$ is *affine* iff there exist a linear function $L \in Y^X$ and an element $y_0 \in Y$ with $T(x) = L(x) + y_0$ for every $x \in X$.

If Y is partially ordered and $D \subset X$ is convex, we say that a function $U \in Y^D$ is *convex* on D iff $U(\lambda x_1 + (1-\lambda)x_2) \leq \lambda U(x_1) + (1-\lambda)U(x_2)$ for every x_1, $x_2 \in D$ and $\lambda \in [0,1]$.

1.2.2 Dual spaces of vector lattices and representation as spaces of continuous functions

Here we recall some fundamental properties of vector lattices, which will be useful for the construction of the integrals we present later.

A partially ordered vector space R is called a *vector lattice* iff for each r_1, $r_2 \in R$ there is in R the supremum $r_1 \vee r_2$.

Let R be a vector lattice. A nonempty set $A \subset R$ is an *upper* (resp., *lower*) *order bounded* iff there is $y_0 \in R$ such that $y \leq y_0$ (resp., $y \geq y_0$) whenever $y \in A$. We say that A is an *order bounded* iff A is both upper and lower bounded.

A vector lattice R is said to be *Dedekind complete* iff every nonempty upper (resp., lower) order bounded subset of R has supremum (resp., infimum) in R.

Let R and S be two vector lattices, $R^+ = \{y \in R : y \geq 0\}$, $S^+ = \{y \in S : y \geq 0\}$. An *$S$-valued functional* (*on R*) is any function $\varphi : R \to S$. A *real functional* is an $\mathbb{R}$-valued functional.

An S-valued functional φ is said to be *positive* iff $\varphi(r) \in S^+$ whenever $r \in R^+$. A *positive order continuous S-valued functional* is a positive S-valued functional φ such that, for each upward directed increasing net $(r_\lambda)_{\lambda \in \Lambda}$ of elements of R with $R \ni r = \bigvee_\lambda r_\lambda$, it is $\varphi(r) = \bigvee_\lambda \varphi(r_\lambda)$.

A linear R-functional is an *order bounded* (resp., *order continuous*) iff it can be expressed as a difference of two positive (resp., positive order continuous) R-functionals. Observe that, if R is a Dedekind complete vector lattice, then an R-functional on X is order bounded if and only if it maps order bounded subsets of X into order bounded subsets of R (see, e.g., [2]).

Now we recall some notions and properties on vector lattices and their duals, which will be useful later (for a related literature see, e.g., [1, 2, 34, 44, 62, 63, 69, 73, 86, 89]).

The *algebraic dual* of R is the ordered vector space R^* whose elements are the linear real functionals on R, where the addition and the scalar multiplication are those inherited by $\mathbb{R}^R$, and whose order relation $\leq$ is defined by setting $\varphi_1 \leq \varphi_2$ iff $\varphi_2 - \varphi_1$ is a positive linear real functional on R. The *order dual* (resp., *order continuous dual*) of R is the ordered vector space $R^\sim$ (resp., $R^\times$) whose elements are the order bounded (resp., order continuous) linear real functionals on R, where the addition, the scalar multiplication and the order relation are those inherited by R^*. Note that a linear real functional y belongs to $R^\sim$ iff it maps order bounded subsets of R into order bounded subsets of $\mathbb{R}$.

Now, let $Y = R^\sim$ (resp., $R^\times$), and $c : R \to Y^*$ be the *evaluation map*, defined by $c(r)(y) = y(r)$, $r \in R$, $y \in Y$. Note that c is a lattice homomorphism from R into $Y^\times$. A vector lattice R is a $\sim$*-space* (resp., π*-space*) iff the evaluation map $c : R \to Y^\times$ is one-to-one.

A subset I of R is said to be *order dense* in R iff for each $r \in R^+$, $r \neq 0$, there is $u \in I$ with $0 \le u \le r$ and $u \neq 0$. A subspace I of R is called an *ideal* of R iff $r_1, r_2 \in I$, $r_3 \in R$ and $r_1 \le r_3 \le r_2$ imply $r_3 \in I$.

Let Υ be the class of all order dense ideals of R, and set $\Phi = \bigcup_{I \in \Upsilon} I^\times$. Then, a function $\varphi \in \mathbb{R}^R$ belongs to Φ iff there exists $I \in \Upsilon$ such that φ is an order continuous linear real functional on I. If $\varphi \in \Phi$, we denote by I_φ its domain. Given $\varphi \in \Phi$ and $r \in R^+$, set $|\varphi|(r) = \sup\{|\varphi(s)| : 0 \le s \le r\}$, and for $r \in R$, put $|\varphi|(r) = |\varphi|(r^+) - |\varphi|(r^-)$.

Let $D_\varphi = \{r \in R : |\varphi|(r) < +\infty\}$, then D_φ is the largest order dense ideal of R on which $|\varphi|$ can be extended finitely.

Given two elements φ_1, $\varphi_2 \in \Phi$, we say that $\varphi_1 \approx \varphi_2$ iff the set $\{r \in R : \varphi_1(r) = \varphi_2(r)\}$ contains an order dense ideal of R. Since Υ is closed with respect to finite intersections, we get that $\approx$ is an equivalence relation. We denote by the symbol $[\varphi]$ a generic class of equivalence with respect to $\approx$.

Let $R^\rho = \Phi/\approx$. Given $[\varphi_i] \in R^\rho$, $i = 1, 2, 3$, we say that $[\varphi_1] + [\varphi_2] = [\varphi_3]$ iff there exist $\varphi_i' \in [\varphi_i]$, $i = 1, 2, 3$, such that the set $\{r \in R : \varphi_1'(r) + \varphi_2'(r) = \varphi_3'(r)\}$ contains an order dense ideal. It is not difficult to see that this definition is independent of the choice of φ_i'. Similarly, it is possible to define $\alpha \cdot [\varphi]$, for each $[\varphi] \in R^\rho$ and $\alpha \in \mathbb{R}$. Moreover, we say that $[\varphi] \ge 0$ iff there is a $\varphi' \in [\varphi]$ such that the set $\{r \in R : \varphi'(r) \ge 0\}$ contains an order dense ideal of R, and $[\varphi_1] \le [\varphi_2]$ iff $[\varphi_2] - [\varphi_1] \ge 0$. Observe that for each element $[\varphi] \in R^\rho$ there exists $\varphi_* \in [\varphi]$ such that for every $\varphi' \in [\varphi]$ it is $I_{\varphi'} \subset I_{\varphi_*}$ and $\varphi' = \varphi_*$ on $I_{\varphi'}$. So, given $\varphi \in \Phi$, we say that φ_* is the *maximal element* determined by φ, and we will identify φ with φ_*. If $y = [\varphi] \in R^\rho$, then we put $\mathcal{I}_y = D_{\varphi_*}$. So, by means of this identification, y can be viewed as a linear real functional on $\mathcal{I}_y$. It is possible to see that R^ρ, endowed with these structures, is a vector lattice.

Note that, in the space R^ρ, the lattice supremum can be identified with the pointwise supremum on an order dense ideal of R. Indeed, we get the following

Proposition 1.2.2. (see, e.g., [1, Theorems 1.67 (b) and 1.73], [62, Theorem 1.5]) *The spaces $R^\sim$ and $R^\times$ are Dedekind complete vector lattices, and the lattice supremum of any upward directed net of positive elements coincides with its pointwise supremum along the set R^+.*

Moreover, if $([\varphi_\lambda])_{\lambda\in\Lambda}$ is an upward directed net of positive elements of R^ρ, upper order bounded by an element $[\varphi_0] \in R^\rho$, and $\varphi_\lambda' \in [\varphi_\lambda]$, $\lambda \in \Lambda$, $\varphi_0' \in [\varphi_0]$, *then there exists* $\varphi' \in (D_{\varphi'_0})^\times$ *with* $\bigvee_{\lambda\in\Lambda} \varphi'_\lambda = \varphi'$, *that is* $\sup_{\lambda\in\Lambda} \varphi'_\lambda(r) = \varphi'(r)$ *for each* $r \in D_{\varphi'_0} \cap R^+$. *If* φ_* *is the maximal element determined by* φ', *then* $[\varphi_*]$ *is the lattice supremum of the net* $([\varphi_\lambda])_{\lambda\in\Lambda}$. *Therefore, the space* R^ρ *is Dedekind complete. Analogous results hold for the infimum.*

Now, we define the evaluation map $c : R \to R^{\rho\rho} = (R^\rho)^\rho$ as follows. For each $r \in R^+$, set $I_r = \{y \in R^\rho : r \in \mathcal{I}_y\}$. For every $r \geq 0$, $y \in I_r$ and $\varphi \in y$, let $c(r)(y) = \varphi(r)$, and for each $r \in R$ set $c(r) = c(r^+) - c(r^-)$. Observe that I_r is an order dense ideal of R^ρ, the definition of c does not depend on the choice of $\varphi \in y$, $c(r)$ is an order continuous linear functional on I_r for every $r \in R^+$ and c is a lattice homomorphism. A vector lattice R is a *ρ-space* iff the evaluation map c is one-to-one.

Now we recall the following properties of ρ-spaces.

Proposition 1.2.3. (see also [62, Corollary 2.4]) *A ρ-space R is Dedekind complete if and only if R is an ideal of $R^{\rho\rho}$, and in this case R is order dense in $R^{\rho\rho}$.*

Proposition 1.2.4. (see also [62, Theorem 2.5]) *A vector lattice R is a ρ-space if and only if there exists an order dense ideal I of R such that $I^\times \neq \{0\}$.*

Example 1.2.5. Let $(\Omega, \mathcal{A}, \mu)$ be a measure space, where μ is any positive countably additive measure. We say that μ is *non-atomic* or *diffused* iff for every $A \in \mathcal{A}$ with $\mu(A) > 0$ there exists $B \in \mathcal{A}$ with $B \neq A$, $B \subset A$ and $\mu(B) > 0$. The measure μ is said to be *semi-finite* iff $\mu(\Omega) < +\infty$ or every element of $\mathcal{A}$ of infinite μ-measure contains an element of $\mathcal{A}$ of finite strictly positive μ-measure. We say that μ is *localizable* iff μ is semi-finite and for each class $\mathcal{M} \subset \mathcal{A}$ there exists $H \in \mathcal{A}$ such that

(1.2.5.1) $\mu(E \setminus H) = 0$ for all $E \in \mathcal{M}$, and

(1.2.5.2) if $A \in \mathcal{A}$ and $\mu(E \setminus A) = 0$ for all $E \in \mathcal{M}$, then $\mu(H \setminus A) = 0$.

Let $L^0(\Omega, \mathcal{A}, \mu)$ be the space of all real-valued μ-measurable functions on Ω, with the identification of μ-null sets, and $L^\infty(\Omega, \mathcal{A}, \mu)$ be the subspace of $L^0(\Omega, \mathcal{A}, \mu)$ consisting of all μ-essentially bounded functions (a function f is *μ-essentially bounded* iff there is a positive real number C with $|f(\omega)| \leq C$ for all $\omega \in \Omega$, with the exception of a μ-null set). Observe that $L^\infty(\Omega, \mathcal{A}, \mu)$ is an order dense ideal of $L^0(\Omega, \mathcal{A}, \mu)$, and $(L^\infty(\Omega, \mathcal{A}, \mu))^\times \neq \{0\}$ (see [34, Corollary 5.3.11]). Thus, by Proposition 4.27, $L^0(\Omega, \mathcal{A}, \mu)$ is a ρ-space. However, if μ is non-atomic, then $L^0(\Omega, \mathcal{A}, \mu)$ is not a $\sim$-space, and a fortiori neither a π-space (see, e.g., [34, 89]).

If μ is semi-finite, then $L^0(\Omega, \mathcal{A}, \mu)$ is Dedekind complete if and only if μ is localizable (see [45, 241G]). For example, the usual Lebesgue measure is non-atomic and localizable (see also [45, 211G, 211L, 211M]). Moreover, if $\mathcal{A}$ is the countable-cocountable σ-algebra of $\mathbb{R}$ and μ is the restriction of the counting measure to $\mathcal{A}$, then μ is semi-finite but not localizable (see [45, 211Y]). Thus, the associated space $L^0(\mathbb{R}, \mathcal{A}, \mu)$ is a ρ-space, but not Dedekind complete.

Now we recall the Maeda-Ogasawara-Vulikh theorem, which gives a representation of vector lattices as subsets of continuous extended real-valued functions defined on suitable topological spaces (see also [43, Theorem 2.1], [64, 86]). We denote by the symbols $\bigvee$ and $\bigwedge$ the supremum and infimum in the involved vector lattice and by sup and inf the pointwise supremum and infimum, respectively.

Theorem 1.2.6. *Given a Dedekind complete vector lattice R, there exists a compact extremely disconnected topological space S, unique up to homeomorphisms, such that R can be embedded isomorphically as an order dense ideal of $C_\infty(S) = \{f \in \widetilde{\mathbb{R}}^S : f$ is continuous, and $\{s : |f(s)| = +\infty\}$ is nowhere dense in $S\}$. Moreover, if we denote by $\widehat{a}$ an element of $C_\infty(S)$ which corresponds to $a \in R$ under the above isomorphism, then for any family $(a_\lambda)_{\lambda\in\Lambda}$ of elements of R with $R \ni a_0 = \bigvee_\lambda a_\lambda$ (where the supremum is taken with respect to R), we get $\widehat{a_0} = \bigvee_\lambda \widehat{a_\lambda}$ with respect to $C_\infty(S)$ and $\widehat{a_0}(s) = \sup_\lambda \widehat{a_\lambda}(s)$ in the complement of a meager subset of S. The same is true for $\bigwedge_\lambda a_\lambda$.*

1.2.3 The ρ-integral in vector lattice setting

Now we recall the construction of an integral for functions with values in ρ-spaces, with respect to finitely additive real-valued measures (see also [11]).

Let G be any infinite set, R be a vector lattice, $l_b(G, R)$ be the space of all bounded functions of R^G, and $f \in l_b(G, R)$. There are r_1, $r_2 \in R$ with $0 \le [f(g)]^+ \le r_1$, $0 \le [f(g)]^- \le r_2$ for all $g \in G$. By identifying r_i with $c(r_i)$, $i = 1, 2$, we find an order dense ideal I of R^ρ such that $r_1|_I$, $r_2|_I$ are two positive linear order continuous real-valued functionals.

Now, fix $g \in G$. There exists an order dense ideal K_g of R^ρ with $0 \le [f(g)]^+(y) \le r_1(y)$ for every $y \in K_g, y \ge 0$. Without loss of generality, we can suppose $K_g \subset I$. Let $\overline{r_1}(y) = \sup\{r_1(u)\colon 0 \le u \le y, u \in K_g\}$, $\overline{[f(g)]^+}(y) = \sup\{f(g)(u)\colon 0 \le u \le y, u \in K_g\}$, for every $y \in R^\rho$, $y \ge 0$. It is possible to see that $\overline{r_1}(y_1 + y_2) = \overline{r_1}(y_1) + \overline{r_1}(y_2)$, $\alpha\,\overline{r_1}(y_1) = \overline{r_1}\;\alpha(y_1)$, and $[y_1 \le y_2] \Rightarrow [\overline{r_1}(y_1) \le \overline{r_1}(y_2)]$ for every $y_1, y_2 \in R^\rho, y_1, y_2 \ge 0$, and for each $\alpha \in \mathbb{R}_0^+$; similar results hold for $\overline{[f(g)]^+}$. Moreover, it is $0 \le r_1(u) \le r_1(y)$ for all $y \in I$, $y \ge 0$, and for every $u \in K_g$, $0 \le u \le y$. Taking the supremum, we obtain $0 \le \overline{r_1}(y) \le r_1(y)$. As $0 \le [f(g)]^+(u) \le r_1(u)$ for each $u \in K_g$, $u \ge 0$, then $0 \le \overline{[f(g)]^+}(y) \le \overline{r_1}(y) \le r_1(y)$ for any $y \in R^\rho$, $y \ge 0$. Thus, $\overline{[f(g)]^+}$ is an additive, monotone, positively homogeneous real functional on the set of all positive elements of I, and then it can be extended linearly on I. We denote this extension by $\overline{[f(g)]^+}$ again. Since r_1 is positive order continuous on I, then so is $\overline{[f(g)]^+}$. Analogously, we can construct $\overline{[f(g)]^-}$. Note that $f(g)$ is equivalent (with respect to the relation $\approx$) to $\overline{[f(g)]^+} - \overline{[f(g)]^-}$. With this identification, $f(g)$ is an order continuous linear real-valued functional on I

for every $g \in G$, and $-r_2(y) \le f(g)(y) \le r_1(y)$ for every $y \in I$, $y \ge 0$ and $g \in G$. Now, let $\mu : \mathcal{P}(G) \to [0,1]$ be any finitely additive measure. For every $y \in I$, let

$$J_f(y) = (L) \int_G f(g)(y) \, d\mu(g) \tag{1.1}$$

be the classical Lebesgue integral of the function $g \mapsto f(g)(y)$ with respect to μ. Note that $J_f(y)$ is well-defined for all $f \in l_b(G, R)$, since μ is defined on *all* subsets of G. It is not difficult to check that J_f is a linear real-valued functional on I. Since $-r_2$ and r_1 are order continuous on I, then so is J_f. By identifying J_f with its class of equivalence with respect to the relation $\approx$, J_f will be an element of $R^{\rho\rho}$. It is not difficult to see that the map $f \mapsto J_f$ is a positive linear $R^{\rho\rho}$-valued functional. When R is Dedekind complete, by Proposition 4.26 R can be identified with an order dense ideal of $R^{\rho\rho}$, and thus $J_f \in c(R)$ for each $f \in l_b(G, R)$.

Set $\mathcal{J}_f = c^{-1}(J_f)$, then $\mathcal{J}_f$ will be called the *ρ-integral* of f (with respect to μ), and we write

$$(\rho) \int_G f(g) \, d\mu(g) = \mathcal{J}_f. \tag{1.2}$$

It is possible to see that $\mathcal{J}_f$ is well-defined, and the map $f \mapsto \mathcal{J}_f$ is a positive linear R-valued functional.

The next lemma will be useful in proving the existence of G-equivariant R-functionals.

Lemma 1.2.7. *Let $\mu : \mathcal{P}(G) \to [0,1]$ be a finitely additive measure with $\mu(G) = 1$, $f \in l_b(G,R)$, $h \in R^R$ be an affine order bounded R-valued functional, and $\mathcal{J}_f$ be as in* (1.2). *Set $(hf)(g) = h(f(g))$, $g \in G$. Then, $h(\mathcal{J}_f) = \mathcal{J}_{hf}$.*

Proof. Since h maps order bounded sets into order bounded sets, the union of the ranges of f and hf is order bounded. By proceeding analogously as in the construction of the integral is (1.1), we find an order dense ideal I_h of R^ρ such that $f(g)_{|I_h}$ and $(hf)(g)_{|I_h}$ are linear order continuous real functionals for every $g \in G$.

For every $r \in R$ and $y \in I_h$, define $h_y \in \mathbb{R}^{\mathbb{R}}$ by $h_y(y(r)) = h(r)(y)$. Since h is affine, then h_y is affine for all $y \in I_h$, and hence there are a_y and $b_y \in \mathbb{R}$ with $h_y(x) = a_y x + b_y$ for all $x \in \mathbb{R}$. If $J_f(y)$ and $\mathcal{J}_f$ are as in (1.1) and (1.2), respectively, since $\mu(G) = 1$ we obtain

$$\begin{aligned} h(\mathcal{J}_f)(y) &= h_y(J_f(y)) = a_y(J_f(y)) + b_y = J_{a_y f}(y) + \mu(G)\, b_y = \\ &= J_{a_y f}(y) + J_{\mathbf{b}_y}(y) = J_{a_y f + \mathbf{b}_y}(y) = \mathcal{J}_{hf}(y), \end{aligned} \tag{1.3}$$

where $\mathbf{b_y}$ denotes the constant function defined on G which assumes the value b_y. From (1.3) we deduce that $h(\mathcal{J}_f) = \mathcal{J}_{hf}$, that is the assertion. □

Remark 1.2.8. Observe that, when h is linear and positive, the assertion of Lemma 1.2.7 holds for any finitely additive measure $\mu : \mathcal{P}(G) \to [0, +\infty[$.

1.2.4 A Chojnacki-type integral for vector lattice-valued functions

We recall the construction of an integral for functions taking values in an *arbitrary* Dedekind complete vector lattice R, with respect to finitely additive positive real-valued set functions, which is inspired by a technique due to W. Chojnacki (see, e.g., [29]).

First, we consider the particular case $R = \mathcal{C}(S) = \{f \in \mathbb{R}^S, f \text{ continuous}\}$, where S is a compact and extremely disconnected topological space. By virtue of a classical result (see, e.g., [71]), there is a discrete set D such that S is homeomorphic to a retract of βD, where βD denotes the Stone-Čech compactification of D. For example, we can pick D as S itself, but endowed with the discrete topology. We denote by r the retraction and identify S with this retract. Given $f \in l_b(G, \mathcal{C}(S))$, define $f_r : G \to \mathcal{C}(\beta D)$ by $f_r(g)(\xi) = f(g)\ (r(\xi))$, $\xi \in \beta D$, $g \in G$. Note that, for every $g \in G$, the map $\xi \mapsto f(g)\ (r(\xi))$ is the unique continuous extension to βD of the function $d \mapsto f(g)\ (r(d))$, defined on D (see, e.g., [29, Théorème 2]). For each $d \in D$, let $I^*_f(d) = (L) \displaystyle\int_G f(g)(r(d))\, d\mu(g)$ be the classical Lebesgue integral. The function $d \mapsto I^*_f(d)$ has a unique continuous extension $\xi \mapsto \widehat{I}_f(\xi)$ to βD. Our integral, which we denote $I_f = (C) \displaystyle\int_G f(g)\, d\mu(g)$, will be the restriction of $\widehat{I}_f$ to S.

Now let $R = \mathcal{C}_\infty(S)$, where S is compact and extremely disconnected. For each positive function $f \in l_b(G, \mathcal{C}_\infty(S))$, put $\overline{I}_f = \displaystyle\bigvee_{m \in \mathbb{N}} I_{f \wedge \underline{m}}$, where $\underline{m}$ is the element of $\mathcal{C}(S)$ which associates the constant $m \in \mathbb{N}$ with each $s \in S$, and $f \wedge \underline{m} : G \to \mathcal{C}(S)$ is defined by $(f \wedge \underline{m})(g) = (f(g)) \wedge \underline{m}$, $g \in G$. It is not difficult to check that $f \mapsto \overline{I}_f$ is a monotone, additive and positively homogeneous map, and hence it can be uniquely extended linearly on $l_b(G, \mathcal{C}_\infty(S))$. We denote by I_f this extension. Then, $I_f = (C) \displaystyle\int_G f(g)\, d\mu(g)$ is our Chojnacki-type integral of f. It is not difficult to see that I_f is a positive linear R-valued functional.

In the general case, that is when R is embedded as an order dense ideal of $\mathcal{C}_\infty(S)$, this construction still makes sense, because I_f is an element of R.

1.2.5 Basic assumptions and properties

Let G be an infinite semigroup, $\mathcal{P}(G)$ be the family of all subsets of G, R be a vector lattice. A set function $\nu : \mathcal{P}(G) \to R$, such that $\nu(G) \in R^+$,

$\nu(G) \neq 0$ and $\nu(h^{-1}(E)) = \nu(\{hg : g \in E\})$ (resp., $\nu(\{gh : g \in E\})) = \nu(E)$ for every $E \subset G$ and $g \in G$, is called a *left* (resp., *right*) *G-invariant mean.* We say that G is *left* (resp., *right*) *amenable* iff there is a left (resp., right) G-invariant mean $\nu : \mathcal{P}(G) \to [0,1] \subset \mathbb{R}$ with $\nu(G) = 1$, and that G is *amenable* iff it is both left and right amenable. In general, left and right amenability do not coincide, but, if G a group, then they are equivalent (see, e.g., [37]).

A set $\emptyset \neq D \subset X$ is said to be *G-invariant* iff $hx \in D$ whenever $x \in D$ and $h \in G$. Note that, if G is a group, then $hx \in D$ if and only if $x \in D$.

A function $F : X \to R$ is *G-subinvariant* (resp. *G-superinvariant*, *G-invariant*) iff $F(hx) \leq F(x)$ (resp., $F(hx) \geq F(x)$, $F(hx) = F(x)$) for every $x \in X$ and $h \in G$.

A function $F \in R^X$ is *G-equivariant* iff $h^{-1}(F(hx)) = F(x)$ (or equivalently $h(F(x)) = F(hx)$) whenever $x \in X$ and $h \in G$.

From now on, when we consider the problem of finding linear invariant functionals, we assume that X is a real vector space, $G \subset X^X$ is a right amenable semigroup of X-valued linear functionals with $(gh)(x) = g(hx)$ for any g, $h \in G$ and $x \in X$, and R is a Dedekind complete vector lattice.

When we deal with the problem of finding equivariant functionals, we do the additional structural hypotheses that X is a Dedekind complete vector lattice, G is a group, the elements of G are positive, $R \subset X$ is a Dedekind complete ρ-space, R is G-invariant and R is contained in the domain of the involved convex functions. Note that finding equivariant functionals with values in *any* Dedekind complete vector lattice is still an open problem (for a related literature see, e.g., [77, 88]).

The following property will be useful later.

Proposition 1.2.9. *Let $G \subset X^X$ be a group of positive linear functionals. Then, for every $h \in G$ and for each family $(r_\xi)_{\xi \in \Xi}$ in X, both upper- and lower-order bounded, it is*

$$h\Big(\bigvee_\xi r_\xi\Big) = \bigvee_\xi h(r_\xi), \quad \textit{and} \tag{1.4}$$

$$h\Big(\bigwedge_\xi r_\xi\Big) = \bigwedge_\xi h(r_\xi). \tag{1.5}$$

Proof. We prove only (1.4), since the proof of (1.5) is analogous.

Pick arbitrarily $h \in G$. Since h is positive, it is $h\Big(\bigvee_\xi r_\xi\Big) \geq h(r_\xi)$ for each $\xi \in \Xi$. Taking the supremum, we obtain

$$h\Big(\bigvee_\xi r_\xi\Big) \geq \bigvee_\xi h(r_\xi). \tag{1.6}$$

Moreover, as h is chosen arbitrarily and G is a *group*, the inequality in (1.6) holds also when h is replaced by h^{-1}. Hence, we get

$$h^{-1}\Big(\bigvee_{\xi} h(r_\xi)\Big) \geq \bigvee_{\xi} h^{-1}(h(r_\xi)) = \bigvee_{\xi} r_\xi. \tag{1.7}$$

By applying h in (1.7), we obtain

$$\bigvee_{\xi} h(r_\xi) = h\Big(h^{-1}\Big(\bigvee_{\xi} h(r_\xi)\Big)\Big) \geq h\Big(\bigvee_{\xi} r_\xi\Big). \tag{1.8}$$

Thus, (1.4) follows from (1.6) and (1.8). □

We denote by $\mathcal{L}(X,R)$ the set of all linear R-functionals on X, by $\mathcal{L}_{\text{inv}}(X,R)$ (resp., $\mathcal{L}_{\text{equiv}}(X,R)$) the set of all G-invariant (resp., G-equivariant) R-functionals of $\mathcal{L}(X,R)$, and by $\mathcal{L}_{+,\text{inv}}(X,R)$ (resp., $\mathcal{L}_{+,\text{equiv}}(X,R)$) the set of all positive R-functionals belonging to $\mathcal{L}_{\text{inv}}(X,R)$ (resp., $\mathcal{L}_{\text{equiv}}(X,R)$).

Given $0 \neq L_* \in \mathcal{L}(X,R)$, $0 \neq L \in \mathcal{L}(R,R)$ and u_0 in R, set

$$H = \{(x,y) \in X \times R : L_*(x) + L(y) = u_0\}. \tag{1.9}$$

Observe that the set H defined in (1.9) is empty or is an affine manifold of $X \times R$ (see also [54]).

If A, B are two nonempty subsets of $X \times R$ and $H \neq \emptyset$ is as in (1.9), then we say that H *separates* A and B iff $A \subset H^+$ and $B \subset H^-$, where

$$\begin{aligned} H^+ &= \{(x,y) \in X \times R : L_*(x) + L(y) \leq u_0\}, \\ H^- &= \{(x,y) \in X \times R : L_*(x) + L(y) \geq u_0\}. \end{aligned}$$

Let $U : D(U) \to R$, $V : D(V) \to R$ be two convex and G-subinvariant (resp., G-equivariant) functions, where $D(U)$ and $D(V)$ are convex and G-invariant subsets of X. The *G-invariant conjugate* of U is the function U^c_{inv} defined by

$$\begin{aligned} &U^c_{\text{inv}}(L) = \bigvee\{L(x) - U(x) : x \in D(U)\}, \quad L \in D(U^c_{\text{inv}}), \quad \text{where} \\ &\emptyset \neq D(U^c_{\text{inv}}) = \{L \in \mathcal{L}_{\text{inv}}(X,R) : \bigvee\{L(x) - U(x) : x \in D(U)\} \\ &\text{exists in } R\}. \end{aligned}$$

The *G-equivariant conjugate* of U is the function U^c_{equiv} defined by

$$\begin{aligned} &U^c_{\text{equiv}}(L) = \bigvee\{L(x) - U(x) : x \in D(U)\}, \quad L \in D(U^c_{\text{equiv}}), \quad \text{where} \\ &\emptyset \neq D(U^c_{\text{equiv}}) = \{L \in \mathcal{L}_{\text{equiv}}(X,R) : \bigvee\{L(x) - U(x) : x \in D(U)\} \\ &\text{exists in } R\}. \end{aligned}$$

If $x_0 \in D(U)$, then we call *G-invariant subdifferential* (resp, *G-equivariant subdifferential*) *at* x_0 the set $\partial_{\text{inv}}U(x_0)$ (resp., $\partial_{\text{equiv}}U(x_0)$) defined as

$$\partial_{\text{inv}}U(x_0) = \{L \in \mathcal{L}_{\text{inv}}(X,R) : L(x) - L(x_0) \leq U(x) - U(x_0) \text{ for all } x \in D(U)\}$$

(resp.,

$$\partial_{\text{equiv}} U(x_0) = \{L \in \mathcal{L}_{\text{equiv}}(X, R) : L(x) - L(x_0) \leq U(x) - U(x_0) \text{ for any } x \in D(U)\}).$$

An element $L \in \partial_{\text{inv}} U(x_0)$ (resp., $\partial_{\text{equiv}} U(x_0)$) is called *G-invariant* (*G-equivariant*) *subgradient of* U *at* x_0.

Given $\emptyset \neq A \subset X$ and $x_0 \in X$, we call *G-invariant* (resp., *G-equivariant*) *polar of* A *at* x_0 the set

$$A^*_{\text{inv}}(x_0) = \{L \in \mathcal{L}_{\text{inv}}(X, R) : L(x) - L(x_0) \leq 0 \text{ for all } x \in A\}$$

(resp.,

$$A^*_{\text{equiv}}(x_0) = \{L \in \mathcal{L}_{\text{equiv}}(X, R) : L(x) - L(x_0) \leq 0 \text{ for any } x \in A\}).$$

In our Fenchel-type duality theorems, we will consider the following problems:

(1.2.5.1) Find $r = \bigwedge\{U(x) + V(x) : x \in D(U) \cap D(V)\}$ in R.

(1.2.5.2) Find $s = \bigvee\{-U^c(L) - V^c(-L) : L \in D(U^c) \cap D(V^c)\}$ in R (with $D(U^c) \cap D(V^c) \neq \emptyset$),

where U^c and V^c mean U^c_{inv} and V^c_{inv} (resp., U^c_{equiv} and V^c_{equiv}) according to the context (see, e.g., [90]).

1.3 The Main Results

We extend to the setting of invariance and equivariance some extension, sandwich, Hahn-Banach, Fenchel duality, Moreau-Rockafellar, Farkas and Kuhn-Tucker-type theorems. We begin with the following Hahn-Banach-type theorem, extending [92, Theorem 2.1]. Our technique is based on the existence of linear functionals, not necessarily invariant or equivariant, which exist thanks to Hahn-Banach classical-like theorems, and on the construction of suitable integrals, by means of which it is possible to have G-invariant or G-equivariant linear functionals.

Theorem 1.3.1. *Let* $U : D(U) \to R$ *be convex,* $D(U) \subset X$ *be convex and G-invariant,* $0 \in int(D(U))$ *and* $U(0) \geq 0$*. If* U *is G-subinvariant, then there exists* $T \in \mathcal{L}_{\text{inv}}(X, R)$*, with* $T(x) \leq U(x)$ *for all* $x \in D(U)$*. If* U *is G-equivariant, then there is* $L \in \mathcal{L}_{\text{equiv}}(X, R)$ *such that* $L(x) \leq U(x)$ *for any* $x \in D(U)$*.*

Proof. Let $X_1 = \text{span}(D(U))$. Observe that X_1 is G-invariant, because the elements of G are linear and $D(U)$ is G-invariant.

Pick arbitrarily $x \in X_1$. As $0 \in int(D(U))$, by Proposition 1.2.1 there is a positive real number λ_x with $\lambda x \in D(U)$ whenever $\lambda \in [-\lambda_x, \lambda_x]$, and hence $\lambda(gx) = g(\lambda x) \in D(U)$ for every $\lambda \in [-\lambda_x, \lambda_x]$ and $g \in G$. Without loss of generality, we can assume that $\lambda_x < 1$. For each $x \in X_1$, set

$$\begin{aligned} U^+(x) &= \bigwedge\Big\{\frac{1}{\lambda}U(\lambda x) : 0 < \lambda < 1, \lambda x \in D(U)\Big\}, \\ U^-(x) &= \bigvee\Big\{\frac{1}{\kappa}(-U(-\kappa x)) : 0 < \kappa < 1, -\kappa x \in D(U)\Big\}. \end{aligned} \tag{1.10}$$

Note that $U^-(x)$, $U^+(x) \neq \emptyset$ for each $x \in X_1$, thanks to the existence of $\lambda_x \in (0,1)$. Now we claim that $U^-(x) \leq U^+(x)$ for all $x \in X_1$.

Pick arbitrarily $x \in X_1$. As U is convex, for any λ, $\kappa \in (0,1)$ with λx, $-\kappa x \in D(U)$, it is

$$0 \leq U(0) = U\Big(\frac{1}{2}\lambda\kappa x - \frac{1}{2}\lambda\kappa x\Big) \leq \frac{1}{2}U\Big(\lambda\kappa x\Big) + \frac{1}{2}U\Big(-\lambda\kappa x\Big),$$

and hence $-U(-\lambda\kappa x) \leq U(\lambda\kappa x)$. From this it follows that

$$\frac{1}{\kappa}(-U(-\kappa x)) = \frac{1}{\lambda\kappa}(-U(-\lambda\kappa x)) \leq \frac{1}{\lambda\kappa}U(\lambda\kappa x) = \frac{1}{\lambda}U(\lambda x). \tag{1.11}$$

Taking in (1.11) the infimum with respect to λ and the supremum with respect to κ, we deduce $U^-(x) \leq U^+(x)$, that is the claim.

By [92, Theorem 2.1] applied with $X = X_1, Y = R$, $M = \{0\}$, $T_M = 0$ (the identically zero function) (see, in particular, formulas (2.3)–(2.5) and the final part of the proof, and see also [92, Corollary 2.6]), there is a linear function $L^* \in R^X$ (not necessarily G-invariant or G-equivariant) such that

$$\begin{aligned} &U^-(x) \leq L^*(x) \leq U^+(x) \text{ for every } x \in X_1 \text{ and} \\ &L^*(x) \leq U^+(x) \leq U(x) \text{ for any } x \in D(U). \end{aligned} \tag{1.12}$$

Since G is right amenable, there is a right G-invariant mean $\nu : \mathcal{P}(G) \to [0,1]$. For every $x \in X_1$, put

$$L'(x) = (C)\int_G L^*(gx)\, d\nu(g), \tag{1.13}$$

$$L''(x) = (\rho)\int_G g^{-1}(L^*(gx))\, d\nu(g). \tag{1.14}$$

We show that L' is well-defined on X_1. When U is G-subinvariant, since $D(U)$ is G-invariant, for each $g \in G$ and $x \in X_1$ it is

$$\begin{aligned} U^+(gx) &= \bigwedge\Big\{\frac{1}{\lambda}U(\lambda gx) : 0 < \lambda < 1, \lambda gx \in D(U)\Big\} = \\ &= \bigwedge\Big\{\frac{1}{\lambda}U(g(\lambda x)) : 0 < \lambda < 1, g(\lambda x) \in D(U)\Big\} \leq \\ &\leq \bigwedge\Big\{\frac{1}{\lambda}U(\lambda x) : 0 < \lambda < 1, \lambda x \in D(U)\Big\} = U^+(x), \end{aligned} \tag{1.15}$$

by the definition of infimum and since $g(\lambda x) \in D(U)$ whenever $\lambda x \in D(U)$. Similarly as in (1.15), it is possible to see that

$$U^-(gx) \geq U^-(x) \quad \text{for all } g \in G \text{ and } x \in X_1. \tag{1.16}$$

Taking into account that X_1 is G-invariant, from (1.12) applied to gx, (1.15) and (1.16) we get

$$U^-(x) \leq U^-(gx) \leq L^*(gx) \leq U^+(gx) \leq U^+(x). \tag{1.17}$$

By passing to the (C)-integral, since $\nu(G) = 1$, from (1.17) we obtain

$$\begin{aligned} U^-(x) &= (C)\int_G U^-(x)\,d\nu(g) \leq (C)\int_G U^-(gx)\,d\nu(g) \leq \\ &\leq (C)\int_G L^*(gx)\,d\nu(g) = L'(x) \leq \\ &\leq (C)\int_G U^+(gx)\,d\nu(g) \leq (C)\int_G U^+(x)\,d\nu(g) = U^+(x). \end{aligned} \tag{1.18}$$

Thus, $L'(x) \in R$ for any $x \in X_1$, that is L' is well-defined.

We now prove that L'' is well-defined. Note that, when we deal with equivariance, G is a group, and hence for each $g \in G$ and $x \in X$, we have that $gx \in D(U)$ if and only if $x \in D(U)$, because $D(U)$ is G-invariant. Moreover, by virtue of Proposition 1.2.9, for every $g \in G$ and $x \in X_1$ it is

$$\begin{aligned} g(U^+(x)) &= g\Big(\bigwedge\Big\{\frac{1}{\lambda}U(\lambda x) : 0 < \lambda < 1, \lambda x \in D(U)\Big\}\Big) = \\ &= \bigwedge\Big\{\frac{1}{\lambda}g(U(\lambda x)) : 0 < \lambda < 1, \lambda gx \in D(U)\Big\} = \\ &= \bigwedge\Big\{\frac{1}{\lambda}U(\lambda gx) : 0 < \lambda < 1, \lambda gx \in D(U)\Big\} = U^+(gx). \end{aligned} \tag{1.19}$$

Similarly as in (1.19), it is possible to see that $g(U^-(x)) = U^-(gx)$ for any $g \in G$ and $x \in X_1$. Thus, U^+ and U^- are G-equivariant. From this and (1.12) applied to gx, for each fixed $g \in G$ and $x \in X_1$, since g^{-1} is linear and positive, we get

$$U^-(x) = g^{-1}(U^-(gx)) \leq g^{-1}(L^*(gx)) \leq g^{-1}(U^+(gx)) = U^+(x). \tag{1.20}$$

Taking the (ρ)-integral, since $\nu(G) = 1$, from (1.20) we deduce

$$\begin{aligned} U^-(x) &= (\rho)\int_G U^-(x)\,d\nu(g) = (\rho)\int_G g^{-1}(U^-(gx))\,d\nu(g) \leq \\ &\leq (\rho)\int_G g^{-1}(L^*(gx))\,d\nu(g) = L''(x) \leq \\ &\leq (\rho)\int_G g^{-1}(U^+(gx))\,d\nu(g) = (\rho)\int_G U^+(x)\,d\nu(g) = U^+(x). \end{aligned} \tag{1.21}$$

Therefore, $L''(x) \in R$ for every $x \in X_1$, and hence L'' is well-defined.

We note that $L'(x) \le U(x)$ (resp., $L''(x) \le U(x)$) for each $x \in D(U)$: indeed, this is a consequence of (1.12) and (1.18) (resp., (1.12) and (1.21)).

Now we prove the G-invariance of L' on X_1. By [29, Théorème 2], for any $f \in l_b(G, R)$ and $h \in G$ it is

$$(C) \int_G f(g)\, d\nu(g) = (C) \int_G f_h(g)\, d\nu(g), \tag{1.22}$$

where $f_h : G \to R$ is defined by $f_h(g) = f(gh)$ for any $g \in G$.

Fix arbitrarily $x \in X_1$ and $h \in G$, and set $f_x(g) = L^*(gx)$, $g \in G$. From (1.22) we obtain

$$\begin{aligned} L'(x) &= (C) \int_G L^*(gx)\, d\nu(g) = (C) \int_G f_x(g)\, d\nu(g) = \\ &= (C) \int_G ((f_x)_h)(g)\, d\nu(g) = \\ &= (C) \int_G f_x(gh)\, d\nu(g) = (C) \int_G L^*((gh)x)\, d\nu(g) = \\ &= (C) \int_G L^*(g(hx))\, d\nu(g) = L'(hx). \end{aligned} \tag{1.23}$$

Hence, L' is G-invariant on X_1.

Now we show the G-equivariance of L'' on X_1. Pick arbitrarily $x \in X_1$ and $h \in G$. Thanks to Lemma 1.2.7 applied to h^{-1}, the linearity of h^{-1} and the G-invariance of ν, we obtain

$$\begin{aligned} h^{-1}(L''(hx)) &= h^{-1}\Big((\rho) \int_G g^{-1}(L^*(g(hx)))\, d\nu(g)\Big) = \\ &= (\rho) \int_G h^{-1}(g^{-1}(L^*(g(hx))))\, d\nu(g) = \\ &= (\rho) \int_G (gh)^{-1}(L^*(g(hx)))\, d\nu(gh) = \\ &= (\rho) \int_G g^{-1}(L^*(gx))\, d\nu(g) = L''(x), \end{aligned} \tag{1.24}$$

getting the G-equivariance of L'' on X_1.

Now we prove that L' and L'' are linear on X_1. Choose arbitrarily α_1, $\alpha_2 \in \mathbb{R}$ and $x_1, x_2 \in X_1$. By the linearity of L^*, of the g's and of the (C)- and

(ρ)-integrals, we have

$$\begin{aligned}
L'(\alpha_1 x_1 + \alpha_2 x_2) &= (C)\int_G L^*(g(\alpha_1 x_1 + \alpha_2 x_2))\, d\nu(g) = \\
&= \alpha_1 (C)\int_G L^*(gx_1)\, d\nu(g) + \\
&+ \alpha_2 (C)\int_G L^*(gx_2)\, d\nu(g) = \alpha_1 L'(x_1) + \alpha_2 L'(x_2), \\
L''(\alpha_1 x_1 + \alpha_2 x_2) &= (\rho)\int_G g^{-1}(L^*(g(\alpha_1 x_1 + \alpha_2 x_2)))\, d\nu(g) = \\
&= \alpha_1 (\rho)\int_G g^{-1}(L^*(gx_1))\, d\nu(g) + \\
&+ \alpha_2 (\rho)\int_G g^{-1}(L^*(gx_2))\, d\nu(g) = \\
&= \alpha_1 L''(x_1) + \alpha_2 L''(x_2).
\end{aligned}$$

Let X_2 be an *algebraic complement* of X_1 in X, that is a subspace of X such that every element $x \in X$ can be expressed in a unique way as $x = x_1 + x_2$, where $x_i \in X_i$, $i = 1, 2$. Such a space X_2 does exist (see also [56, Theorem 7.3.3]).

Now we define $T : X \to R$ (resp., $L : X \to R$) by $T(x) = L'(x_1)$ (resp., $L(x) = L''(x_1)$). It is not difficult to check that T (resp., L) is well-defined and linear on X, since L' (resp., L'') is linear on X_1. Note that $U^-(x_1) \le L'(x_1) = T(x) \le U^+(x_1)$ (resp., $U^-(x_1) \le L''(x_1) = L(x) \le U^+(x_1)$) for any $x \in X$, and $T(x) = L'(x) \le U(x)$ (resp., $L(x) = L''(x) \le U(x)$) for each $x \in D(U)$. Furthermore, for every $x \in X$, $x = x_1 + x_2$ and $h \in G$, we get $T(hx) = L'(hx_1) = T(hx)$ (resp., $h(L(x)) = h(L''(x_1)) = L''(hx_1) = L(hx)$), since h is linear, X_1 and X_2 are G-invariant and L' is G-invariant (resp., L'' is G-equivariant) on X_1. Therefore, T is G-invariant (resp., L is G-equivariant) on X. This completes the proof. □

We prove the existence of affine manifolds, separating two nonempty sets of a product space, by extending [54, Theorem 4.1] to invariance and equivariance. For each convex subset $A \subset X \times R$, set

$$P_X(A) = \{x \in X :\ \text{there exists } y \in R \text{ with } (x, y) \in A\}. \tag{1.25}$$

Theorem 1.3.2. *Let A, $B \neq \emptyset$ be two subsets of $X \times R$, such that $A - B$ is convex. Set $\mathcal{D} = P_X(A - B)$, and suppose that $0 \in \mathcal{D}$ and*

$$y_1 \ge y_2 \text{ whenever } (x, y_1) \in A \text{ and } (x, y_2) \in B. \tag{1.26}$$

If

$$(hx, y) \in A - B \text{ for all } (x, y) \in A - B \text{ and } h \in G \tag{1.27}$$

(resp., if A and B are G-invariant), then there exist $T \in \mathcal{L}_{\text{inv}}(X, R)$ (resp., $\mathcal{L}_{\text{equiv}}(X, R)$) and $u_0 \in R$ such that the affine manifold

$$H = \{(x, y) \in X \times R : T(x) - y = u_0\}$$

separates A and B.

Proof. First of all, it is not difficult to check that the convexity and the G-invariance of $A - B$ imply those of $\mathcal{D}$, and the set $C = \bigcup_{\lambda > 0} \lambda (A - B)$ is convex.

Now we claim that, in the context of invariance,

$$\text{for every } h \in G, (hx, y) \in C \text{ whenever } (x, y) \in C. \tag{1.28}$$

Indeed, if $(x, y) \in C$ and $h \in G$, then there are $\lambda^* > 0$, $x^* \in X$, $y^* \in R$, with $x = \lambda^* x^*$, $y = \lambda^* y^*$, and $(x^*, y^*) \in A - B$. From (1.27) and the linearity of h we deduce $(hx^*, y^*) \in A - B$ and $(hx, y) = \lambda^* (hx^*, y^*) \in C$.

Now we claim that, in the setting of equivariance,

$$\text{for every } h \in G, (hx, hy) \in C \text{ if and only if } (x, y) \in C. \tag{1.29}$$

Indeed, if $(x, y) \in C$ and $h \in G$, then there exist $\lambda_0 > 0$, $x' \in X$, $y' \in R$, such that $x = \lambda_0 x'$, $y = \lambda_0 y'$, and $(x', y') \in A - B$. Since A and B are G-invariant and h is linear, it is $(hx', hy') \in A - B$, and hence $(hx, hy) = \lambda_0 (hx', hy') \in C$. This proves the "if part" of (1.29). The "only if" part follows from the "if" part, by exchanging h with h^{-1}.

We construct a convex and G-subinvariant (resp., G-equivariant) function $U : P_X(A - B) \to R$, which will satisfy the hypotheses of Theorem 1.3.1.

For every $x \in \mathcal{D}$, set $E_x = \{y \in R : (x, y) \in C\}$. Note that for each $x \in \mathcal{D}$ there is $y \in R$ with $(x, y) \in A - B \subset C$, and hence

$$E_x \neq \emptyset \text{ for all } x \in \mathcal{D}. \tag{1.30}$$

Now we claim that $E_0 \subset R^+$. Pick arbitrarily $y \in E_0$. Then $(0, y) \in C$, and so there exists a positive real number $\overline{\lambda}$ with $(0, \overline{\lambda} y) \in A - B$. This implies the existence of $\overline{x} \in X$ and y_1, y_2 in R such that $y = y_1 - y_2$, $(\overline{x}, \overline{\lambda} y_1) \in A$ and $(\overline{x}, \overline{\lambda} y_2) \in B$. From (1.26) we deduce $\overline{\lambda} y_1 \geq \overline{\lambda} y_2$, and hence $y \geq 0$, getting the claim.

Now we prove that

(1.3.2.1) the set E_x is lower order bounded for each $x \in \mathcal{D}$.

Fix $x \in \mathcal{D}$, and choose arbitrarily $y \in E_x$. Since $0 \in int(\mathcal{D})$, there is a real number $\lambda_x \in (0, 1)$ with $\lambda x \in \mathcal{D}$ whenever $|\lambda| \leq \lambda_x$. Thus, the set $E_{-\lambda_x x}$ is well-defined and non-empty. Let $y_x \in E_{-\lambda_x x}$. Since $(-\lambda_x, y_x) \in C$, $(x, y) \in C$ and C is convex, we get

$$\left(0, \frac{1}{1 + \lambda_x} y_x + \frac{\lambda_x}{1 + \lambda_x} y\right) = \frac{1}{1 + \lambda_x}(-\lambda_x x, y_x) + \frac{\lambda_x}{1 + \lambda_x}(x, y) \in C,$$

and hence $\dfrac{1}{1+\lambda_x} y_x + \dfrac{\lambda_x}{1+\lambda_x} y \in E_0 \subset R^+$. Therefore, we obtain $y \geq -\dfrac{y_x}{\lambda_x}$, and (1.3.2.1) follows from the arbitrariness of y.

Now, define $U : \mathcal{D} \to R$ by setting $U(x) = \bigwedge E_x$, $x \in \mathcal{D}$. We prove that U satisfies the hypotheses of Theorem 1.3.1.

First of all, observe that U is well-defined, thanks to (1.3.2.1). Now we prove that U is convex. Pick arbitrarily $x_1, x_2 \in \mathcal{D}$ and $\lambda \in [0, 1]$, and choose arbitrarily $y_1 \in E_{x_1}$, $y_2 \in E_{x_2}$. Then, (x_1, y_1) and (x_2, y_2) belong to C. Since C is convex, $(\lambda x_1 + (1-\lambda)x_2, \lambda y_1 + (1-\lambda)y_2) \in C$, and hence $\lambda y_1 + (1-\lambda)y_2 \in E_{\lambda x_1 + (1-\lambda)x_2}$. Therefore,

$$\bigwedge E_{\lambda x_1 + (1-\lambda)x_2} \leq \lambda y_1 + (1-\lambda) y_2. \tag{1.31}$$

By the arbitrariness of $y_1 \in E_{x_1}$, from (1.31) we get

$$\bigwedge E_{\lambda x_1 + (1-\lambda)x_2} \leq \lambda \bigwedge E_{x_1} + (1-\lambda) y_2 \tag{1.32}$$

for each $y_2 \in E_{x_2}$. Since y_2 was chosen arbitrarily, from (1.32) we obtain

$$\begin{aligned} U(\lambda x_1 + (1-\lambda)x_2) &= \bigwedge E_{\lambda x_1 + (1-\lambda)x_2} \leq \\ \leq \; & \lambda \bigwedge E_{x_1} + (1-\lambda) \bigwedge E_{x_2} = \lambda\, U(x_1) + (1-\lambda)\, U(x_2), \end{aligned} \tag{1.33}$$

that is the convexity of U. From (1.33) and the arbitrariness of x_1, x_2 and λ we deduce that U is convex.

Now we claim the G-subinvariance of U in the setting of invariance. From (1.28) and the definition of infimum, for every $x \in \mathcal{D}$ and $h \in G$ it is

$$\begin{aligned} U(hx) &= \bigwedge E_{hx} = \bigwedge \{y \in R : (hx, y) \in C\} \leq \\ &\leq \bigwedge \{y \in R : (x, y) \in C\} = U(x), \end{aligned}$$

getting the claim.

Now we prove the G-equivariance of U in the context of equivariance. Pick arbitrarily $h \in G$. Taking into account (1.29) and Proposition 1.2.9, we deduce

$$\begin{aligned} h(U(x)) &= h\Big(\bigwedge E_x\Big) = \bigwedge h(E_x) = \bigwedge \{hy \in R : (x, y) \in C\} = \\ &= \bigwedge \{y \in R : (x, h^{-1}y) \in C\} = \bigwedge \{y \in R : (hx, y) \in C\} = U(hx) \end{aligned} \tag{1.34}$$

for all $x \in \mathcal{D}$, getting the G-equivariance of U.

We claim that $U(0) = 0$. First note that, since $E_0 \subset R^+$, it is

$$U(0) = \bigwedge E_0 \geq 0.$$

Now, let $\overline{y} \in E_0$. Since $(0, \overline{y}) \in C$, by the definition of C we get $(0, \lambda \overline{y}) \in C$ for each positive real number λ. Thus, we obtain

$$0 \leq U(0) \leq \lambda \overline{y} \text{ for all } \lambda > 0. \tag{1.35}$$

Taking the infimum as $\lambda > 0$ at the right hand of (1.35), we deduce that $U(0) = 0$.

By Theorem 1.3.1 applied with $\mathcal{D}$ and U, there is $T \in \mathcal{L}_{\text{inv}}(X, R)$ (resp., $\mathcal{L}_{\text{equiv}}(X, R)$) with $T(x) \leq U(x)$ for any $x \in \mathcal{D}$. From this we deduce that

$$T(x_1) - T(x_2) = T(x_1 - x_2) \leq U(x_1 - x_2) \leq y_1 - y_2 \tag{1.36}$$

and hence

$$T(x_1) - y_1 \leq T(x_2) - y_2 \tag{1.37}$$

whenever $(x_1, y_1) \in A$ and $(x_2, y_2) \in B$, by the definition of $E_{x_1 - x_2}$ (note that, in this case, $x_1 - x_2 \in \mathcal{D}$). As R is Dedekind complete, from (1.37) it follows that there exists an element $u_0 \in R$ such that

$$\bigvee \{T(x_1) - y_1 : (x_1, y_1) \in A\} \leq u_0 \leq \bigwedge \{T(x_2) - y_2 : (x_2, y_2) \in B\},$$

getting the assertion. □

Now we prove the following sandwich-type theorem, extending [92, Sandwich Theorem 3.1] to invariance and equivariance.

Theorem 1.3.3. *Let $U : D(U) \to R$, $V : D(V) \to R$ be two convex and G-subinvariant (resp., G-equivariant) functions, where*

$$D(U), D(V) \subset X \text{ are convex and } G\text{-invariant}, 0 \in int(D(U) - D(V)), \tag{1.38}$$

and assume that

$$U(x) + V(x) \geq 0 \text{ whenever } x \in D(U) \cap D(V). \tag{1.39}$$

Then there exist $L \in \mathcal{L}_{\text{inv}}(X, R)$ (resp., $\mathcal{L}_{\text{equiv}}(X, R)$) and $u_0 \in R$ with $L(x) - u_0 \leq U(x)$ for any $x \in D(U)$ and $L(x') - u_0 \geq -V(x')$ whenever $x' \in D(V)$.

Proof. Let U and V be as in the hypotheses of Theorem 1.3.3. By (1.39), we get $-V(x) \leq U(x)$ for all $x \in D(U) \cap D(V)$. Set $A = \{(x, y) \colon x \in D(U), y \in R, y \geq U(x)\}$, $B = \{(x, y) \colon x \in D(V), y \in R, y \leq -V(x)\}$. It is not difficult to check that A and B are nonempty, and $A - B$ is convex.

Now we prove that, in the context of invariance,

$$\text{for any } h \in G, (hx, y) \in A \text{ (resp., } B) \text{ whenever } (x, y) \in A \text{ (resp., } B). \tag{1.40}$$

Pick arbitrarily $(x, y) \in A$ and $h \in G$. As U is G-subinvariant, we get $y \geq U(x) \geq U(hx)$, and hence $(hx, y) \in A$. The proof that $(hx, y) \in B$ whenever $(x, y) \in B$ is analogous.

Now we claim that, in the equivariance setting,

$$\text{for all } h \in G, (hx, hy) \in A \text{(resp.,} B) \text{ if and only if } (x, y) \in A \text{ (resp.,} B). \tag{1.41}$$

Choose arbitrarily $(x, y) \in A$ and $h \in G$. Since h is positive and U is G-equivariant (see the beginning of the proof of Theorem 1.3.5), we obtain $hy \geq h(U(x)) = U(hx)$ and hence $(hx, hy) \in A$. The converse implication follows by exchanging h with h^{-1}. The proof that $(hx, hy) \in B$ if and only if $(x, y) \in B$ is analogous.

Now we claim that

$$0 \in int(P_X(A - B)). \tag{1.42}$$

Indeed, we get $P_X(A) = D(U)$, $P_X(B) = D(V)$, $P_X(A - B) = P_X(A) - P_X(B) = D(U) - D(V)$. Since, by hypothesis, $0 \in int(D(U) - D(V))$, then we obtain (1.42) (see, e.g., [6], [40, Satz 1 (6)], [54, Theorem 8]). Now we prove that

$$y_1 \geq y_2 \text{ whenever } (x, y_1) \in A \text{ and } (x, y_2) \in B. \tag{1.43}$$

If $(x, y_1) \in A$ and $(x, y_2) \in B$, then $y_2 \leq -V(x) \leq U(x) \leq y_1$, getting (1.43).

Therefore, A and B satisfy the hypotheses of Theorem 1.3.2. Thus, there are $L_* \in \mathcal{L}_{\text{inv}}(X, R)$ (resp., $\mathcal{L}_{\text{equiv}}(X, R)$) and $u_0 \in R$ with $L_*(x) - y \leq u_0$ whenever $(x, y) \in A$ and $L_*(x') - y' \geq u_0$ for all $(x', y') \in B$. In particular, taking $y = U(x)$ and $y' = -V(x')$, we obtain $L_*(x) - u_0 \leq U(x)$ for any $x \in D(U)$ and $L_*(x') - u_0 \geq -V(x')$ for all $x' \in D(V)$. This ends the proof. □

Now, we prove the following monotone extension theorem for linear invariant or equivariant functionals, which extends [92, Corollary 3.11].

Theorem 1.3.4. *Let $Z \subset X$ be a G-invariant subspace endowed with the order inherited by X, such that $0 \in int(Z - X^+)$, and $L_* \in \mathcal{L}_{+,\text{inv}}(Z, R)$ (resp., $\mathcal{L}_{+,\text{equiv}}(Z, R)$). Then there is $L \in \mathcal{L}_{+,\text{inv}}(X, R)$ (resp., $\mathcal{L}_{+,\text{equiv}}(X, R)$), with $L(z) = L_*(z)$ for all $z \in Z$.*

Proof. Let $U = L_*$ with $D(U) = Z$, and $V = 0$ (the identically zero function) with $D(V) = X^+$. Observe that U and V are convex and G-equivariant maps, $D(U)$ and $D(V)$ are convex and G-invariant sets, and $U(x) + V(x) \geq 0$ for all $x \in D(U) \cap D(V)$, since L_* is positive on Z. So, the hypotheses of Theorem 1.3.3 are satisfied, and hence there exist $L \in \mathcal{L}_{\text{inv}}(X, R)$ (resp., $\mathcal{L}_{\text{equiv}}(X, R)$) and $u_0 \in R$ with

$$L(z) - u_0 \leq L_*(z) \text{ for all } z \in Z, \quad L(x) - u_0 \geq 0 \text{ for all } x \in X^+. \tag{1.44}$$

By taking in (1.44) $z = 0$ and $x = 0$, we get $u_0 = 0$. Thus,

$$L_*(z) - L(z) \geq 0 \text{ for each } z \in Z \tag{1.45}$$

and $L(x) \geq 0$ for any $x \in X^+$, and hence $L \in \mathcal{L}_{+,\text{inv}}(X, R)$ (resp., $\mathcal{L}_{+,\text{equiv}}(X, R)$). Since L and L_* are linear, by changing z with $-z$ in (1.45) we obtain $L(z) = L_*(z)$ for all $z \in Z$. This ends the proof. □

Now, as a consequence of Theorem 1.3.3, we present the next Fenchel-type duality theorem, which extends [90, Theorem 2] to invariance and equivariance.

Theorem 1.3.5. *Let $U : D(U) \to R$, $V : D(V) \to R$ be convex and G-subinvariant (resp., G-equivariant) functions, where $D(U)$ and $D(V)$ satisfy* (1.38). *Let*

$$r = \bigwedge\{U(x) + V(x) : x \in D(U) \cap D(V)\}$$

exist in R, where r is as in Problem (1.2.5.1), *and let U^c and V^c be as in Problem* (1.2.5.2). *Then Problem* (1.2.5.2) *has a solution L_*, such that $-U^c(L_*) - V^c(-L_*) = r$.*

Proof. Set $\widetilde{U}(x) = U(x) - r$. Then, $D(\widetilde{U}) = D(U)$ and $\widetilde{U}$ is convex on its domain. In the invariance case, we observe that the G-subinvariance of U implies that of $\widetilde{U}$: indeed, when U is G-subinvariant, for every $h \in G$ it is

$$\widetilde{U}(hx) = U(hx) - r \le U(x) - r = \widetilde{U}(x),$$

getting the G-subinvariance of $\widetilde{U}$.

Now we claim that, in the equivariance setting, the G-equivariance of U implies that of $\widetilde{U}$. Indeed, pick arbitrarily $h \in G$. From the G-equivariance of U and V, the linearity of h and Proposition 1.2.9, it follows that

$$\begin{aligned} hr &= h\Big(\bigwedge\{U(x) + V(x) : x \in D(U) \cap D(V)\}\Big) = \\ &= \bigwedge\{hU(x) + hV(x) : x \in D(U) \cap D(V)\} = \\ &= \bigwedge\{U(hx) + V(hx) : x \in D(U) \cap D(V)\} = \\ &= \bigwedge\{U(hx) + V(hx) : hx \in D(U) \cap D(V)\} = \\ &= \bigwedge\{U(x) + V(x) : x \in D(U) \cap D(V)\} = r, \end{aligned} \tag{1.46}$$

since $x \in D(U) \cap D(V)$ if and only if $hx \in D(U) \cap D(V)$, by virtue of the G-invariance of $D(U) \cap D(V)$ and the existence of h^{-1}. From (1.46) we obtain $h(\widetilde{U}(x)) = h(U(x) - r) = h(U(x)) - hr = U(hx) - r = \widetilde{U}(hx)$, getting the G-equivariance of $\widetilde{U}$.

Now, observe that $r \le U(x) + V(x)$, and hence $\widetilde{U}(x) + V(x) \ge 0$ for all $x \in D(U) \cap D(V)$. So, $\widetilde{U}$ and V satisfy the hypotheses of Theorem 1.3.3. Therefore, there exist $L_* \in \mathcal{L}_{\text{inv}}(X, R)$ (resp., $\mathcal{L}_{\text{equiv}}(X, R)$) and $u_0 \in R$ such that $L_*(x) - u_0 \le \widetilde{U}(x) = U(x) - r$ for all $x \in D(U)$ and $L_*(x') - u_0 \ge -V(x')$ for all $x' \in D(V)$. From this we obtain

$$L_*(x - x') = L_*(x) - L_*(x') \le u_0 + U(x) - r - u_0 + V(x') \tag{1.47}$$

and hence

$$r + L_*(x) - U(x) \le L_*(x') + V(x') \tag{1.48}$$

for each $x \in D(U)$ and $x' \in D(V)$. Thus,

$$\begin{aligned} r &+ \bigvee\{L_*(x) - U(x) : x \in D(U)\} \leq \bigwedge\{L_*(x) + V(x) : x \in D(V)\} \\ &= -\bigvee\{-L_*(x) - V(x) : x \in D(V)\}, \end{aligned}$$

and hence

$$L_* \in D(U^c) \cap D(V^c) \quad \text{and } r \leq -U^c(L_*) - V^c(-L_*). \tag{1.49}$$

Furthermore, observe that

$$-U^c(L) - V^c(-L) \leq -L(x) + U(x) + L(x) + V(x) = U(x) + V(x) \tag{1.50}$$

for every $x \in D(U) \cap D(V)$ and $L \in D(U^c) \cap D(V^c)$. Taking the infimum as $x \in D(U) \cap D(V)$ and the supremum as $L \in D(U^c) \cap D(V^c)$, from (1.49) and (1.50) we obtain

$$\begin{aligned} r &\leq -U^c(L_*) - V^c(-L_*) \leq s = \\ &= \bigvee\{-U^c(L) - V^c(-L) : L \in D(U^c) \cap D(V^c)\} \leq r. \end{aligned}$$

Thus, $r = -U^c(L_*) - V^c(-L_*) = s$, and the supremum in Problem (1.2.5.2) is a maximum, assumed by L_*. This ends the proof. □

Remark 1.3.6. Observe that, in general, the converse of Theorem 1.3.5 does not hold (see, e.g., [90, §2]).

Now we characterize the solutions of Problem (1.2.5.1) in terms of G-invariant (resp., G-equivariant) subgradients, extending [90, Theorem 3].

Theorem 1.3.7. (Optimality condition) *Let U, V, $D(U)$, $D(V)$ be as in Theorem* 1.3.5, *and $x_0 \in D(U) \cap D(V)$. Then, x_0 is a solution of Problem* (1.2.5.1) *if and only if*

$$\partial_{\text{inv}} U(x_0) \cap (-\partial_{\text{inv}} V(x_0)) \neq \emptyset$$

(resp.,

$$\partial_{\text{equiv}} U(x_0) \cap (-\partial_{\text{equiv}} V(x_0)) \neq \emptyset).$$

Proof. We begin with the "only if" part. Let $x_0 \in D(U) \cap D(V)$ be a solution of Problem (1.2.5.1). By Theorem 1.3.5, Problem (1.2.5.2) has a solution L_*, and hence there is $L_* \in \mathcal{L}_{\text{inv}}(X, R)$ (resp., $\mathcal{L}_{\text{equiv}}(X, R)$) such that

$$U(x_0) + V(x_0) = -U^c(L_*) - V^c(-L_*). \tag{1.51}$$

By definition of the conjugate function, by (1.51) we have

$$U(x_0) + V(x_0) \leq U(x) + V(x') - L_*(x) + L_*(x') \tag{1.52}$$

for every $x \in D(U)$ and $x' \in D(V)$. If we take $x = x_0$ and $x' = x_0$ in (1.52), then we get $-L_* \in \partial_{\text{inv}} V(x_0)$ (resp., $\partial_{\text{equiv}} V(x_0)$) and $L_* \in \partial_{\text{inv}} U(x_0)$ (resp., $\partial_{\text{equiv}} U(x_0)$), respectively. This proves the "only if" part.

Now we turn to the "if" part. Let $x_0 \in D(U) \cap D(V)$ and L be such that $L \in \partial_{\text{inv}} U(x_0) \cap (-\partial_{\text{inv}} V(x_0))$ (resp., $\partial_{\text{equiv}} U(x_0) \cap (-\partial_{\text{equiv}} V(x_0))$. Then for all $x \in D(U) \cap D(V)$ it is

$$U(x) - U(x_0) \geq L(x) - L(x_0) \geq V(x_0) - V(x),$$

and thus $U(x) + V(x) \geq U(x_0) + V(x_0)$. Therefore, x_0 is a solution of Problem (1.2.5.1). □

We now turn to the following version of the Hahn-Banach theorem, extending [92, Theorem 2.1] to invariance and equivariance.

Theorem 1.3.8. *Let $U : D(U) \to R$ be convex and G-subinvariant (resp., G-equivariant), $D(U)$ be G-invariant, $U(0) = 0$, and $Z \subset X$ be a G-invariant subspace, with $0 \in int(D(U) - Z)$. Let $T_0 \in \mathcal{L}_{\text{inv}}(Z, R)$* (resp., $\mathcal{L}_{\text{equiv}}(Z, R)$) *be with $T_0(z) \leq U(z)$ whenever $z \in D(U) \cap Z$. Then T_0 admits an extension $T \in \mathcal{L}_{\text{inv}}(X, R)$* (resp., $\mathcal{L}_{\text{equiv}}(X, R)$), *with $T(x) \leq U(x)$ for all $x \in D(U)$.*

Proof. Let Z, T_0, U be as in the hypotheses of Theorem 1.3.8. Put $V = -T_0$. Then $D(V) = D(T_0) = Z$. Since $T_0(z) \leq U(z)$ for each $z \in D(U) \cap Z$ and $T_0(0) = U(0) = 0$, we have

$$0 = U(0) + V(0) = \min\{U(x) + V(x) : x \in D(U) \cap Z\},$$

that is 0 is a solution of Problem (1.2.5.1). By Theorem 1.3.7 there exists $T \in \partial_{\text{inv}} U(0) \cap (-\partial_{\text{inv}} V(0))$ (resp., $\partial_{\text{equiv}} U(0) \cap (-\partial_{\text{equiv}} V(0))$). We have

$$T(x) = T(x) - T(0) \leq U(x) - U(0) = U(x)$$

for all $x \in D(U)$ and

$$-T(z) = -T(z) + T(0) \leq -T_0(z) + T_0(0) = -T_0(z), \tag{1.53}$$

that is

$$T(z) \geq T_0(z) \tag{1.54}$$

for all $z \in Z$. Taking in (1.53) $-z$ instead of z, we obtain

$$T(z) = -T(-z) \leq -T_0(-z) = T_0(z) \tag{1.55}$$

for any $z \in Z$. From (1.54) and (1.55) we deduce that $T(z) = T_0(z)$ for all $z \in Z$, getting the assertion. □

The next version of the Moreau-Rockafellar formula extends [90, Theorem 4] to invariant and equivariant subdifferentials (see also [41, 58]).

Theorem 1.3.9. (Moreau-Rockafellar formula) *Let* U, V, $D(U)$, $D(V)$ *be as in Theorem* 1.3.5, $x_0 \in D(U) \cap D(V)$, *and assume that* $\partial_{\rm inv} U(x_0)$ *(resp.,* $\partial_{\rm equiv} U(x_0)$*)*$\neq \emptyset$ *and* $\partial_{\rm inv} V(x_0)$ *(resp.,* $\partial_{\rm equiv} V(x_0)$ *)*$\neq \emptyset$*. Then*

$$\begin{aligned} \partial_{\rm inv}(U+V)(x_0) &= \partial_{\rm inv}(U)(x_0) + \partial_{\rm inv}(V)(x_0) \qquad (1.56)\\ (\text{resp., } \partial_{\rm equiv}(U+V)(x_0) &= \partial_{\rm equiv}(U)(x_0) + \partial_{\rm equiv}(V)(x_0)). \end{aligned}$$

Proof. Pick arbitrarily $L' \in \partial_{\rm inv}(U+V)(x_0)$ (resp., $\partial_{\rm equiv}(U+V)(x_0)$), and set $V'(x) = V(x) - L'(x)$, $x \in D(V)$. It is not difficult to check that V' is convex. Moreover, by definition of subdifferential, it is

$$L'(x) - L'(x_0) \leq (U+V)(x) - (U+V)(x_0),$$

and hence

$$U(x_0) + V(x_0) - L'(x_0) \leq U(x) + V(x) - L'(x) = U(x) + V'(x) \qquad (1.57)$$

for every $x \in D(U) \cap D(V)$.

Furthermore, it is

$$(V')^c(-L) = \bigvee\{L'(x) - L(x) - V(x) : x \in D(V)\} = V^c(L' - L)$$

for every $L \in D((V')^c) = D(V^c)$.

Thanks to (1.57) and Theorem 1.3.5, there exists $L_* \in \mathcal{L}_{\rm inv}(X, R)$ (resp., $\mathcal{L}_{\rm equiv}(X, R)$) with

$$U(x_0) + V(x_0) - L'(x_0) = -U^c(L_*) - (V')^c(-L_*) = -U^c(L_*) - V^c(L' - L_*),$$

and hence

$$U^c(L_*) + V^c(L' - L_*) = L_*(x_0) - U(x_0) + (L' - L_*)(x_0) - V(x_0). \qquad (1.58)$$

By the definition of the conjugate function, we have

$$U^c(L_*) \geq L_*(x_0) - U(x_0), \; V^c(L' - L_*) \geq L'(x_0) - L_*(x_0) - V(x_0). \qquad (1.59)$$

From (1.58) and (1.59) it follows that

$$U^c(L_*) = L_*(x_0) - U(x_0), \; V^c(L' - L_*) = L'(x_0) - L_*(x_0) - V(x_0),$$

and hence, again by the definition of the conjugate function,

$$L_*(x) - U(x) \leq L_*(x_0) - U(x_0) \text{ for all } x \in D(U),$$

$$L'(x) - L_*(x) - V(x) \leq L'(x_0) - L_*(x_0) - V(x_0) \text{ for all } x \in D(V).$$

Thus, $L_* \in \partial_{\rm inv}(U)(x_0)$ (resp., $\partial_{\rm equiv}(U)(x_0)$), $L' - L_* \in \partial_{\rm inv}(V)(x_0)$ (resp., $\partial_{\rm equiv}(V)(x_0)$), and hence $L' \in \partial_{\rm inv}(U)(x_0) + \partial_{\rm inv}(V)(x_0)$ (resp., $\partial_{\rm equiv}(U)(x_0) + \partial_{\rm equiv}(V)(x_0)$). By the arbitrariness of L', we deduce

$$\partial_{\rm inv}(U+V)(x_0) \subset \partial_{\rm inv}(U)(x_0) + \partial_{\rm inv}(V)(x_0)$$

(resp., $\partial_{\rm equiv}(U+V)(x_0) \subset \partial_{\rm equiv}(U)(x_0) + \partial_{\rm equiv}(V)(x_0)$).
The proof of the converse inclusion is straightforward. □

As a consequence of Theorem 1.3.9, we give the following

Theorem 1.3.10. *Let A_1, $A_2 \subset X$ be convex G-invariant sets with $0 \in int(A_1 - A_2)$. Then*

$$\begin{aligned} (A_1)^*_{\text{inv}}(x_0) + (A_2)^*_{\text{inv}}(x_0) &= (A_1 \cap A_2)^*_{\text{inv}}(x_0) \\ (\textit{resp.}, (A_1)^*_{\text{equiv}}(x_0) + (A_2)^*_{\text{equiv}}(x_0) &= (A_1 \cap A_2)^*_{\text{equiv}}(x_0)) \end{aligned}$$

for all $x_0 \in A_1 \cap A_2$.

Proof. Set $A_0 = A_1 \cap A_2$, and $\mathcal{N}_{A_j}(x) = 0$ for each $x \in A_j$, $j = 0, 1, 2$. It is readily seen that

$$\begin{aligned} \partial_{\text{inv}} \mathcal{N}_{A_j}(x) &= (A_j)^*_{\text{inv}}(x) && (1.60) \\ (\text{resp.}, \partial_{\text{equiv}} \mathcal{N}_{A_j}(x) &= (A_j)^*_{\text{equiv}}(x)) \text{ for all } x \in A_j,\ j = 0, 1, 2, \end{aligned}$$

and

$$\mathcal{N}_{A_0}(x) = \mathcal{N}_{A_1}(x) + \mathcal{N}_{A_2}(x) \text{ for all } x \in A_0. \quad (1.61)$$

Let now $x_0 \in A_0$ be fixed. From Theorem 1.3.9, (1.60) and (1.61) we deduce

$$\begin{aligned} & (A_1)^*_{\text{inv}}(x_0) + (A_2)^*_{\text{inv}}(x_0) = \partial_{\text{inv}} \mathcal{N}_{A_1}(x_0) + \partial_{\text{inv}} \mathcal{N}_{A_2}(x_0) = \\ = \ & \partial_{\text{inv}}(\mathcal{N}_{A_1} + \mathcal{N}_{A_2})(x_0) = \\ = \ & \partial_{\text{inv}} \mathcal{N}_{A_0}(x_0) = (A_0)^*_{\text{inv}}(x_0) \end{aligned}$$

$$\begin{aligned} (\text{resp.}, \quad & (A_1)^*_{\text{equiv}}(x_0) + (A_2)^*_{\text{equiv}}(x_0) = \\ = \ & \partial_{\text{equiv}} \mathcal{N}_{A_1}(x_0) + \partial_{\text{equiv}} \mathcal{N}_{A_2}(x_0) = \\ = \ & \partial_{\text{equiv}}(\mathcal{N}_{A_1} + \mathcal{N}_{A_2})(x_0) = \\ = \ & \partial_{\text{equiv}} \mathcal{N}_{A_0}(x_0) = (A_0)^*_{\text{equiv}}(x_0)). \end{aligned}$$

Thus the assertion follows. □

Now we present a Farkas-type theorem, related to some minimization problems for vector lattice-valued invariant or equivariant functionals, extending earlier results proved in [41, 58–60, 91].

Assumptions 1.3.11. Let D_0, D_1, $D_2, \ldots, D_q$ be convex and G-invariant subsets of a vector lattice X, assume that every $g \in G$ is a positive linear X-valued functional and that $D = \bigcap_{i=0}^{q} D_i \neq \emptyset$, and let $U_i : D_i = D(U_i) \to X$, $i = 0, 1, 2, \ldots, q$, be convex functions. Assume that U_0 is G-subinvariant (resp., G-equivariant) and U_i is G-equivariant for all $i = 1, 2, \ldots, q$. Let $\mathcal{W} = X^q$ be endowed with the "componentwise" order, given by $w = (w_1, w_2, \ldots, w_q) \geq$

$w' = (w'_1, w'_2, \ldots, w'_q)$ iff $w_i \geq w'_i$ for any $i = 1, 2, \ldots, q$, and set $\mathcal{K} = (X^+)^q$. Let

$$H(x) = (U_1(x), U_2(x), \ldots, U_q(x)), \quad x \in \bigcap_{i=1}^{q} D_i. \tag{1.62}$$

Note that the convexity of H follows from that of the U_i's.

We define the action of G on $\mathcal{W}$. For each $g \in G$, $w \in \mathcal{W}$, $w = (w_1, w_2, \ldots, w_q)$, put $gw = (gw_1, gw_2, \ldots, gw_q)$. We say that H is G-*equivariant* iff $H(gx) = g(H(x))$ for every $g \in G$ and $x \in D$. Note that this property is equivalent to the G-equivariance of the U_i's, $i = 1, 2, \ldots, q$.

We consider the following optimization problem:

(1.3.11.1) Find $x_0 \in D$ such that $U_0(x_0) = \min\{U_0(x) : x \in D, U_i(x) \leq 0, i = 1, 2, \ldots, q\}$.

Now we are ready to give the next Farkas-type theorem.

Theorem 1.3.12. *Assume that, for every $x \in D$, it is*

$$U_0(x) \geq 0 \text{ whenever } U_i(x) \leq 0 \text{ for any } i = 1, 2, \ldots, q, \tag{1.63}$$

and that $0 \in int(H(D) + \mathcal{K})$. Then there are $L_i \in \mathcal{L}_{+,\text{inv}}(X, R)$ (resp., $\mathcal{L}_{+,\text{equiv}}(X, R)$), $i = 1, 2, \ldots, q$, with

$$U_0(x) + \sum_{i=1}^{q} L_i(U_i(x)) \geq 0 \text{ for all } x \in D. \tag{1.64}$$

Proof. Let $\mathcal{Y} = H(D) + \mathcal{K}$. We claim that

(1.3.12.1) the set $\mathcal{Y}$ is convex.

For $i = 1, 2$, let $x^{(i)} \in D$, $y^{(i)} \in R$, be such that $y^{(i)} - H(x^{(i)}) \geq 0$. Pick arbitrarily $\lambda \in [0, 1]$, and let $x_\lambda = \lambda\, x^{(1)} + (1-\lambda)\, x^{(2)}$, $y_\lambda = \lambda\, y^{(1)} + (1-\lambda)\, y^{(2)}$. Then $x_\lambda \in D$ (since D is convex), $y_\lambda \in R$. Moreover, thanks to the convexity of H, it is

$$y_\lambda \geq \lambda\, H(x^{(1)}) + (1 - \lambda)\, H(x^{(2)}) \geq H(x_\lambda),$$

Thus $x_\lambda \in \mathcal{Y}$, getting the claim.

Now we show that

(1.3.12.2) the set $\mathcal{Y}$ is G-invariant.

Choose arbitrarily $w \in \mathcal{Y}$ and $h \in G$. We claim that $hw \in \mathcal{Y}$. Indeed, if $w = H(x) + k$, where x and k are suitable elements of D and $\mathcal{K}$, respectively, then $hw = h(H(x)) + hk = H(hx) + hk$, since H is G-equivariant. As D and $\mathcal{K}$ are G-invariant, then $hx \in D$ and $hk \in \mathcal{K}$ (note that the elements of G are increasing homomorphisms). This proves (1.3.12.2).

Now, set

$$\begin{aligned} A &= \{(w, y) \in \mathcal{Y} \times R : \text{ there is } x_0 \in D \text{ with } w \geq H(x_0) \text{ and } y \geq U_0(x_0)\}; \\ B &= \bigcup_{\lambda > 0} \lambda A. \end{aligned} \tag{1.65}$$

We prove that

(1.3.12.3) The sets A and B defined in (1.65) are convex.

Choose arbitrarily $(w^{(1)}, y^{(1)})$, $(w^{(2)}, y^{(2)}) \in A$ and $\lambda \in [0, 1]$, and let $x^{(i)} \in D$ be such that $w_i \geq H(x^{(i)})$ and $y \geq U_0(x^{(i)})$, $i = 1, 2$. Since H is convex, it is

$$H(\lambda x^{(1)} + (1 - \lambda) x^{(2)}) \leq \lambda H(x^{(1)}) + (1 - \lambda) H(x^{(2)}) \leq \lambda w^{(1)} + (1 - \lambda) w^{(2)},$$

and similarly

$$U_0(\lambda x^{(1)} + (1 - \lambda) x^{(2)}) \leq \lambda y^{(1)} + (1 - \lambda) y^{(2)}.$$

Thus, A is convex. From this it is not difficult to deduce the convexity of B.

For each $w \in \mathcal{Y}$, put $S_w = \{y \in R : (w, y) \in B\}$. We claim that

$$S_w \neq \emptyset \text{ for any } w \in \mathcal{Y}. \tag{1.66}$$

Choose arbitrarily $w \in \mathcal{Y}$. Since $0 \in int(\mathcal{Y})$, we find a positive real number λ_0 such that $\lambda w \in \mathcal{Y}$ for all $\lambda \in [-\lambda_0, \lambda_0]$. So there is $x_0 \in D$ such that

$$0 \leq \lambda_0 w - H(x_0) = \lambda_0 \Big(w - \frac{1}{\lambda_0} H(x_0) \Big).$$

Since $(\lambda_0 w, U_0(x_0)) \in A$, we have

$$\Big(w, \frac{1}{\lambda_0} U_0(x_0) \Big) = \frac{1}{\lambda_0} (\lambda_0 w, U_0(x_0)) \in B,$$

getting (1.66).

Now we show that

$$S_0 \subset R^+. \tag{1.67}$$

Let $y \in S_0$. Then, $(0, y) \in B$, namely there are a positive real number λ_0 and an element $x_0 \in D$ such that $U_i(x_0) \leq 0$ for all $i = 1, 2, \ldots, q$ and $\lambda_0 y \geq U_0(x_0)$. By (1.63), $U_0(x_0) \geq 0$, and hence $\lambda_0 y \geq 0$. Thus, $y \geq 0$, and (1.67) follows from the arbitrariness of y.

Now we claim that, if U_0 is G-subinvariant, then

$$S_w \subset S_{hw} \quad \text{for all } w \in \mathcal{Y} \text{ and } h \in G \tag{1.68}$$

and, if U_0 is G-equivariant, then for every $w \in \mathcal{Y}$, $y \in R$ and $h \in G$ we get

$$y \in S_w \text{ if and only if } hy \in S_{hw}. \tag{1.69}$$

Indeed, if $y \in S_w$, then $(w, y) \in B$, and hence there exist a positive real number λ_0 and an element $x_0 \in D$ with

$$\lambda_0 w \geq H(x_0) \text{ and } \lambda_0 y \geq U_0(x_0). \tag{1.70}$$

Choose arbitrarily $h \in G$. By applying h in (1.70) when U_0 is G-subinvariant, since the elements of G are increasing homomorphisms, D is G-invariant and H is G-equivariant, we obtain

$$\lambda_0 hw \geq H(hx_0) \text{ and } \lambda_0 y \geq U_0(hx_0).$$

Thus, $(hw, y) \in B$, and hence $y \in S_{hw}$, getting (1.68).

When U_0 is G-equivariant, we get

$$\lambda_0 hw \geq H(hx_0) \text{ and } \lambda_0 hy \geq U_0(hx_0),$$

and hence $(hw, hy) \in B$, that is $hy \in S_{hw}$. This proves the "only if" part of (1.69). The "if part" follows by changing h with h^{-1}, getting the claim.

Furthermore, arguing analogously as in (1.3.2.1), it is possible to see that

(1.3.12.4) the set S_w is lower order bounded for each $w \in \mathcal{Y}$.

Thus, it makes sense to define a function $p : \mathcal{Y} \to R$, by setting

$$p(w) = \bigwedge S_w, \quad w \in \mathcal{Y}. \tag{1.71}$$

Arguing analogously as in Theorem 1.3.2, one sees that p is well-defined and convex, and $p(0) = 0$.

In the context of invariance, from (1.68) we deduce that $p(hw) = \bigwedge S_{hw} \leq \bigwedge S_w = p(w)$ for each $w \in \mathcal{Y}$ and $h \in G$, getting the G-subinvariance of p.

In the setting of equivariance, from (1.69) and Proposition 1.2.9 we obtain

$$\begin{aligned} h(p(w)) &= h\Big(\bigwedge S_w\Big) = \bigwedge h(S_w) = \bigwedge\{hy \in R : (w, y) \in B\} = \qquad (1.72) \\ &= \bigwedge\{y \in R : (w, h^{-1}y) \in B\} = \bigwedge\{y \in R : (hw, y) \in B\} = p(hw) \end{aligned}$$

for all $w \in \mathcal{Y}$ and $h \in G$, getting the G-equivariance of p.

Thus, by Theorem 1.3.1, we find a linear and G-invariant (resp., G-equivariant) function $T : \mathcal{W} \to R$ with

$$-T(w) \leq p(w) \text{ for all } w \in \mathcal{Y}. \tag{1.73}$$

Now we prove (1.64). Pick arbitrarily $x \in D$ and $k \in \mathcal{K}$, namely $k = (c_1, c_2, \dots, c_q)$ with $c_i \geq 0$ for each $i = 1, \dots, q$. By definition of A, we get $(H(x) + k, U_0(x)) \in A \subset B$, and hence $U_0(x) \in S_{H(x)+k}$. By definition of $S_{H(x)+k}$, we have $U_0(x) - p(H(x) + k) \geq 0$, and a fortiori

$$U_0(x) + T(H(x) + k) \geq 0, \tag{1.74}$$

thanks to (1.73). For every $i = 1, 2, \dots, q$ and $w = (w_1, w_2, \dots, w_q)$, let $T_i(w_i) = T(0, 0, \dots, w_i, 0, \dots, 0)$, where w_i is at the i-th place. Note that the T_i's are G-invariant (resp., G-equivariant) because T is G-invariant (resp., G-equivariant). Since T is linear, we obtain

$$T(H(x) + k) = \sum_{i=1}^{q} T_i(U_i(x) + c_i) = \sum_{i=1}^{q} (T_i(U_i(x)) + T_i(c_i)). \tag{1.75}$$

From (1.74) and (1.75) we deduce

$$U_0(x) + \sum_{i=1}^{q} (T_i(U_i(x)) + T_i(c_i)) \geq 0 \text{ for every } x \in D. \tag{1.76}$$

From (1.76), taking $c_i = 0$ for all $i = 1, 2, \dots, q$, we get (1.64).

Finally, we prove that

(1.3.12.5) T_i is positive, that is $T_i(a_i) \geq 0$ whenever $a_i \in X$, $a_i \geq 0$, for any $i = 1, 2, \dots, q$.

Choose arbitrarily $\overline{x} \in X$, $i \in \{1, 2, \dots, q\}$ and $n \in \mathbb{N}$. If in (1.76) we take $c_i = n\, a_i$ and $c_j = 0$ whenever $j \neq i$, then we get

$$n(T_i(a_i)) = T_i(n\, a_i) \geq -\sum_{i=1}^{q} T_i(U_i(\overline{x})) - U_0(\overline{x}), \tag{1.77}$$

that is

$$T_i(a_i) \geq \frac{1}{n}\Big(-\sum_{i=1}^{q} T_i(U_i(\overline{x})) - U_0(\overline{x})\Big). \tag{1.78}$$

By the arbitrariness of n, as R is Dedekind complete, from (1.78) we get

$$T_i(a_i) \geq \bigvee_{n \in \mathbb{N}} \Big[\frac{1}{n}\Big(-\sum_{i=1}^{q} T_i(U_i(\overline{x})) - U_0(\overline{x})\Big)\Big] = 0$$

since, thanks to (1.64), it is

$$\sum_{i=1}^{q} T_i(U_i(\overline{x})) + U_0(\overline{x}) \geq 0.$$

Thus, (1.3.12.5) is proved. This ends the proof of the theorem. □

A consequence of Theorem 1.3.12 is the next Karush-Kuhn-Tucker-type condition for the existence of saddle points for Problem (1.3.11.1), which extends [91, Theorem 5] to invariance and equivariance.

Theorem 1.3.13. *Under the same hypotheses as in Theorem* 1.3.12, *if* x_0 *is a solution of Problem* (1.3.11.1), *then there exist* $T_i^0 \in \mathcal{L}_{+,\mathrm{inv}}(X,R)$, (*resp.*, $\mathcal{L}_{+,\mathrm{equiv}}(X,R)$), $i = 1, 2, \dots, q$, *with*

$$U_0(x_0) + \sum_{i=1}^{q} T_i(U_i(x_0)) \le U_0(x_0) + \sum_{i=1}^{q} T_i^0(U_i(x_0)) \le U_0(x) + \sum_{i=1}^{q} T_i^0(U_i(x))$$

for every $x \in D$ *and* $T_i \in \mathcal{L}_{+,\mathrm{equiv}}(X,R)$ (*resp.*, $\mathcal{L}_{+,\mathrm{equiv}}(X,R)$).

Proof. Let x_0 be a solution of Problem (1.3.11.1), and define $Z_0 = \{x \in D$ such that $U_i(x) \le 0$ for all $i = 1, 2, \dots, q\}$. Set $U'(x) = U_0(x) - U_0(x_0)$, $x \in D$. Observe that the G-subinvariance of U_0 implies that of U': indeed, in this case it is

$$U'(hx) = U_0(hx) - U_0(x_0) \le U_0(x) - U_0(x_0) = U'(x)$$

for all $x \in D$ and $h \in G$ (note that D is G-invariant).

Now we claim that

(1.3.13.1) if U_0 is G-equivariant, then U' is G-equivariant.

We first show that

$$\text{for any } h \in G, \text{ it is } x \in Z_0 \text{ if and only if } hx \in Z_0. \tag{1.79}$$

Indeed, if $x \in Z_0$, since D is G-invariant, h is linear and positive, and U_i is G-equivariant for all $i \in \{1, 2, \dots, q\}$, we get that $hx \in X_0$ and $U_i(hx) = h(U_i(x)) \le h(0) = 0$ for each $i \in \{1, 2, \dots, q\}$. The converse implication follows by replacing h with h^{-1}.

Now, we prove that

$$\begin{aligned} &\text{for every } h \in G,\ hx_0 \text{ is a solution of Problem (1.3.11.1)} \\ &\text{if and only if } x_0 \text{ is a solution of Problem (1.3.11.1).} \end{aligned} \tag{1.80}$$

We begin with the "if" part. Let x_0 be such that $U_0(x_0) \le V(x)$ whenever $x \in Z_0$. Note that $h(Z_0) = Z_0$, by virtue of (1.79). Choose arbitrarily $x \in Z_0$. There is $x' \in Z_0$ with $x = hx'$. As U_0 is G-equivariant and h is positive, it is $U_0(hx_0) = h(U_0(x_0)) \le h(U_0(x')) \le U_0(hx') = U_0(x)$, which proves the "if" part. The "only if" part follows from the "if" part, by exchanging h with h^{-1} and taking into account (1.79).

Now we show that

$$U_0(hx_0) = U_0(x_0) \quad \text{for any } h \in G. \tag{1.81}$$

Pick arbitrarily $h \in G$. Taking into account that Z_0 is G-invariant, from (1.80) we get that $U_0(hx_0) \le U_0(x_0)$ and $U_0(x_0) \le U_0(hx_0)$, that is (1.81).

Finally, we are in position to prove (1.3.13.1). Choose arbitrarily $h \in G$. From (1.81) and the G-equivariance of V, we obtain

$$\begin{aligned} h(U'(x)) &= h(U_0(x)) - h(U_0(x_0)) \\ &= U_0(hx) - U_0(h_0) = U_0(hx) - U_0(x_0) = U'(hx). \end{aligned}$$

Now we claim that U' and U_i, $i = 1, 2, \ldots, q$, satisfy (1.63). Indeed, we get

$$U_0(x_0) = \min\Big\{U_0(x) : x \in D, U_i(x) \le 0 \text{ for all } i \in \{1, 2, \ldots, q\}\Big\}.$$

Hence, $U'(x) = U_0(x) - U_0(x_0) \ge 0$ for every $x \in D$ such that $U_i(x) \le 0$ for all $i \in \{1, 2, \ldots, q\}$, getting the claim.

By Theorem 1.3.12, U' and U_i, $i = 1, 2, \ldots, q$, satisfy (1.64). Hence there are $T_1^0, T_2^0, \ldots T_q^0 \in \mathcal{L}_{+,\mathrm{inv}}(X, R)$ (resp., $\mathcal{L}_{+,\mathrm{equiv}}(X, R)$) such that

$$\sum_{i=1}^{q} T_i^0(U_i(x)) + U'(x) \ge 0 \tag{1.82}$$

for all $x \in D$. Taking $x = x_0$ in (1.82), we get

$$\sum_{i=1}^{q} T_i^0(U_i(x_0)) \ge 0. \tag{1.83}$$

Since $U_i(x_0) \le 0$ and T_i^0 is positive for each $i = 1, 2, \ldots, q$, from (1.83) we deduce that

$$\sum_{i=1}^{q} T_i^0(U_i(x_0)) = 0. \tag{1.84}$$

Moreover, for each $T_i \in \mathcal{L}_{+,\mathrm{inv}}(X, R)$ (resp., $\mathcal{L}_{+,\mathrm{equiv}}(X, R)$), $i = 1, 2, \ldots, q$, it is

$$U_0(x_0) + \sum_{i=1}^{q} T_i(U_i(x_0)) \le U_0(x_0) = U_0(x_0) + \sum_{i=1}^{q} T_i^0(U_i(x_0)). \tag{1.85}$$

From (1.82) and (1.84), we obtain

$$0 \le \sum_{i=1}^{q} T_i^0(U_i(x)) - \sum_{i=1}^{q} T_i^0(U_i(x_0)) + U_0(x) - U_0(x_0),$$

namely,

$$U_0(x_0) + \sum_{i=1}^{q} T_i^0(U_i(x_0)) \le U_0(x) + \sum_{i=1}^{q} T_i^0(U_i(x)). \tag{1.86}$$

The assertion follows from (1.85) and (1.86). □

A consequence of Theorem 1.3.13 is the following

Theorem 1.3.14. *Let $Z \subset X$ be a G-invariant subspace, $T_1 \in \mathcal{L}(Z, X)$ be a G-equivariant function, and $T_2 \in \mathcal{L}_{\mathrm{inv}}(Z, R)$ (resp., $\mathcal{L}_{\mathrm{equiv}}(Z, R)$) Suppose that*

$$0 \in int(-T_1(Z) + X^+), \qquad \text{and} \tag{1.87}$$

$$T_2(x) \geq 0 \textit{ whenever } x \textit{ belongs to } Z \textit{ and } T_1(x) \leq 0. \tag{1.88}$$

Then there is $T_0 \in \mathcal{L}_{+,\text{inv}}(X, R)$ *(resp.,* $\mathcal{L}_{+,\text{equiv}}(X, R)$*) such that*

$$T_0(T_1(x)) + T_2(x) = 0 \textit{ for any } x \in Z. \tag{1.89}$$

Proof. By Theorem 1.3.13 applied with $q = 1$, $U_1 = T_1$, $U_0 = T_2$ and $x_0 = 0$, there exists $T_0 \in \mathcal{L}_{\text{inv}}(X, R)$ (resp., $\mathcal{L}_{\text{equiv}}(X, R)$) such that

$$T_0(T_1(x)) + T_2(x) \geq 0 \text{ for every } x \in Z. \tag{1.90}$$

Changing x with $-x$ in (1.90), we obtain

$$T_0(T_1(x)) + T_2(x) \leq 0,$$

and hence

$$T_0(T_1(x)) + T_2(x) = 0 \text{ for all } x \in Z.$$

□

The next step is to prove that, for each fixed Dedekind complete vector lattice (resp., Dedekind complete ρ-space), amenability is a necessary condition in order that the presented theorems are always valid with respect to invariance (resp., equivariance). We begin with the next two theorems.

Theorem 1.3.15. *Let R be any fixed Dedekind complete vector lattice. If G satisfies Theorem* 1.3.4 *with respect to R in the setting of invariance, then G is right amenable.*

Proof. Pick $X = \{f \in \mathbb{R}^G,\ f \text{ is bounded}\}$, $Z = \{f \in \mathbb{R}^G,\ f \text{ is constant}\}$, and let $G \subset X^X$ be a semigroup of positive linear maps $h \in X^X$ defined by setting $(hf)(g) = f(gh)$, $f \in X$, $h \in G$. Choose an element $y \in R^+$, $y \neq 0$, and define $\mu : Z \to R$ by $\mu(\mathbf{r}) = ry$, where $\mathbf{r}$ is the constant function, which assumes the value r. It is not difficult to check that $0 \in int(Z - X^+)$, and that μ is a linear, positive and G-invariant R-functional. By Theorem 1.3.4, μ has a linear, positive and G-invariant R-valued extension $\widetilde{\mu}$, defined on X. For every $A \subset G$, put $\widehat{\mu}(A) = \widetilde{\mu}(\chi_A)$, where

$$\chi_A(g) = \begin{cases} 1, & \text{if } g \in A, \\ 0, & \text{if } g \notin A. \end{cases}$$

It is

$$h(\chi_A(g)) = \chi_A(gh) = \begin{cases} 1, & \text{if } gh \in A, \\ 0, & \text{if } gh \notin A. \end{cases}$$

Since $\widetilde{\mu}$ is G-invariant, we obtain that $\widehat{\mu}$ is right G-invariant. Moreover $\widehat{\mu}$ is finitely additive, thanks to the linearity of $\widetilde{\mu}$.

Now, we prove that G is right amenable. Since R is a Dedekind complete vector lattice, R can be embedded as an order dense ideal of $\mathcal{C}_\infty(S)$, where S is a suitable compact and extremely dense topological space, according to Theorem 1.2.6. Since $\widehat{\mu}(G) \geq 0$, $\widehat{\mu}(G) \neq 0$, by Theorem 1.2.6 there is $\overline{s} \in S$ such that $\widehat{\mu}(G)(\overline{s})$ is a strictly positive real number and $\widehat{\mu}(A)(\overline{s})$ is a non-negative real number for each $A \subset G$.

For every $A \subset G$, set $\nu(A) = \dfrac{\widehat{\mu}(A)(\overline{s})}{\widehat{\mu}(G)(\overline{s})}$. It is not difficult to see that ν takes values in $[0,1]$ and $\nu(G) = 1$. Moreover ν is right G-invariant, because $\widehat{\mu}$ is right G-invariant. Furthermore, if A and B are any two disjoint subsets of G, then

$$\nu(A \cup B) = \frac{\widehat{\mu}(A \cup B)(\overline{s})}{\widehat{\mu}(G)(\overline{s})} = \frac{\widehat{\mu}(A)(\overline{s}) + \widehat{\mu}(B)(\overline{s})}{\widehat{\mu}(G)(\overline{s})} = \nu(A) + \nu(B).$$

Therefore, G is right amenable. □

Theorem 1.3.16. *Let R be any fixed Dedekind complete ρ-space. If G satisfies Theorem* 1.3.4 *with respect to R in the context of equivariance, then G is amenable.*

Proof. Take $X = \{f \in R^G,\ f \text{ is bounded}\}$, $Z = \{f \in R^G,\ f \text{ is constant}\}$, and let $G \subset X^X$ be a group of positive linear maps $h \in X^X$ defined by setting $(hf)(g) = h(f(h(g)))$, $f \in X$, $h \in G$. Choose a fixed element $y \in (R^\rho)^+$, $y \neq 0$ (such an element exists, since R is a ρ-space), and for each $r \in R$, put $\mu(\mathbf{r}) = y(r)$, where $\mathbf{r}$ is the constant function, which assumes the value r. Without loss of generality, taking into account the construction of R^ρ given in Subsection 1.2.2, μ can be viewed as a real-valued functional defined on Z, because y can be considered as a linear real functional defined on a suitable order dense ideal of R, and since y is *fixed.* It is not difficult to check that $0 \in int(Z - X^+)$, and that μ is a linear, positive and G-equivariant real functional, since y is linear and positive. By Theorem 1.3.4, μ admits a linear, positive and G-equivariant extension $\widetilde{\mu}$ on the whole of X. Pick a fixed element $u \in R^+$, $u \neq 0$, and for each $A \subset G$ set

$$\zeta_A(g) = \begin{cases} u, & \text{if } g \in A, \\ 0, & \text{if } g \notin A, \end{cases}$$

and $\widehat{\mu}(A) = \widetilde{\mu}(\zeta_A)$. We have

$$(h\zeta_A)(g) = \begin{cases} h\,u, & \text{if } g \in A, \\ 0, & \text{if } g \notin A; \end{cases} \qquad h(\zeta_A(hg)) = \begin{cases} h\,u, & \text{if } hg \in A, \\ 0, & \text{if } hg \notin A. \end{cases}$$

Since $\widetilde{\mu}$ is linear, positive and G-equivariant, we deduce that $\widehat{\mu}$ is finitely additive and G-invariant, and $\widehat{\mu}(G) > 0$.

Now, let $\nu(A) = \dfrac{\widehat{\mu}(A)}{\widehat{\mu}(G)}$. It is not difficult to check that ν is finitely additive and G-invariant, takes values in $[0, 1]$ and $\nu(G) = 1$. Thus, G is amenable. □

Now, we give the following characterization of amenability of (semi)groups.

Theorem 1.3.17. *Let $G \subset X^X$ be a (semi)group. Then theorems* (1.3.1)–(1.3.5), (1.3.7)–(1.3.10) *and* (1.3.12)–(1.3.14) (*with respect to a fixed Dedekind complete vector lattice, in the context of invariance, or with respect to a fixed Dedekind complete ρ-space, in the context of equivariance*) *are equivalent each other and are equivalent to the* (*right*) *amenability of G.*

Proof. Amenability $\Longrightarrow$ (1.3.1) See Theorem 1.3.1.

(1.3.1) $\Longrightarrow$ (1.3.2) See Theorem 1.3.2.

(1.3.2) $\Longrightarrow$ (1.3.3) See Theorem 1.3.3.

(1.3.3) $\Longrightarrow$ (1.3.4) See Theorem 1.3.4.

(1.3.4) $\Longrightarrow$ Amenability See Theorems 1.3.15 and 1.3.16.

(1.3.3) $\Longrightarrow$ (1.3.5) See Theorem 1.3.5.

(1.3.5) $\Longrightarrow$ (1.3.7) See Theorem 1.3.7.

(1.3.7) $\Longrightarrow$ (1.3.8) See Theorem 1.3.8.

(1.3.8) $\Longrightarrow$ (1.3.2) The proof is analogous to that of Theorem 1.3.2, by applying Theorem 1.3.8 with $Z = \{0\}$ (this is possible, because $U(0) = 0$).

(1.3.5) $\Longrightarrow$ (1.3.9) See Theorem 1.3.9.

(1.3.9) $\Longrightarrow$ (1.3.10) See Theorem 1.3.10.

(1.3.1) $\Longrightarrow$ (1.3.12) See Theorem 1.3.12.

(1.3.12) $\Longrightarrow$ (1.3.13) See Theorem 1.3.13.

(1.3.13) $\Longrightarrow$ (1.3.14) See Theorem 1.3.14.

(1.3.10) $\Longrightarrow$ (1.3.4) Let X^+, Z and L_* be as in the hypotheses of Theorem 1.3.4, and Z' be an algebraic complement of Z, that is a subspace Z' of X such that every element $x \in X$ can be uniquely expressed as $x = z + z'$, where $z \in Z$, $z' \in Z'$.

Define $L_{**} : X \to R$ by $L_{**}(x) = -L_*(z)$, $x \in X$. As $L_* \in \mathcal{L}_{+,\text{inv}}(Z, R)$ (resp., $\mathcal{L}_{+,\text{equiv}}(Z, R)$), then $L_{**} \in (X^+ \cap Z)^*(0)$. By Theorem 1.3.10 applied with $A_1 = X^+$, $A_2 = Z$, $x_0 = 0$, we find L', $L'' \in \mathcal{L}_{+,\text{inv}}(X, R)$ (resp., $\mathcal{L}_{+,\text{equiv}}(X, R)$), with $L_{**}(x) = L'(x) + L''(x)$ for any $x \in X$, $L'(x) \le 0$ for all $x \in X^+$ and $L''(z) \le 0$ for all $z \in Z$. In particular, it is $-L''(z) = L''(-z) \le 0$, and hence $L''(z) = 0$, for all $z \in Z$. If $L = -L'$, then $L \in \mathcal{L}_{+,\text{inv}}(X, R)$ (resp., $\mathcal{L}_{+,\text{equiv}}(X, R)$) and $L(z) = L_*(z)$ for any $z \in Z$. So we get the thesis of Theorem 1.3.4.

(1.3.14) $\Longrightarrow$ (1.3.4) Let X^+, Z and L_* be as in the hypotheses of Theorem 1.3.4. Set $T_1(z) = -z$ and $T_2(z) = L_*(z)$ for all $z \in Z$. Note that L_* is positive, and hence $L_*(z) \geq 0$ for all $z \in Z \cap X^+$. By Theorem 1.3.13, there is $L \in \mathcal{L}_{+,\text{inv}}(X, R)$ (resp., $\mathcal{L}_{+,\text{equiv}}(X, R)$) with $L(-z) + L_*(z) = -L(z) + L_*(z) = 0$ for all $z \in Z$. Thus we get the thesis of Theorem 1.3.4. □

Remark 1.3.18. Observe that the Dedekind completeness of R is a necessary condition for our given results (see, e.g., [41, 83]).

1.4 Applications to Set Functions

Now, we give some applications of the presented results to finitely additive and invariant vector lattice-valued set functions. The classical Vitali example (see [85]) shows that it is not possible to define a σ-additive measure on the family of all subsets of the real line, invariant with respect to the group of all translations, extending the Lebesgue measure. Such a pathology does not exist in the finitely additive case, thanks to the Tarski theorem, whose we give an extension to the setting of partially ordered vector spaces and invariance with respect to any amenable semigroup. The next result is a consequence of Theorem 1.3.3.

Theorem 1.4.1. *Let Ω be any infinite set, let $\mathcal{P}(\Omega)$ be the family of all subsets of Ω, let $\mathcal{A} \subset \mathcal{P}(\Omega)$ be an algebra, let $G \subset \Omega^\Omega$ be a left (resp., right) amenable semigroup such that $g^{-1}(A) = \{\omega \in \Omega\colon g\omega \in A\}$ (resp., $\{\omega \in \Omega\colon \omega g \in A\}$) $\in \mathcal{A}$ for all $A \in \mathcal{A}$ and $g \in G$, and let $\psi : \mathcal{A} \to R^+$ be a G-invariant finitely additive set function (that is, $\psi(g^{-1}(A)) = \psi(A)$ for every $A \in \mathcal{A}$ and $g \in G$). Then ψ has a G-invariant finitely additive extension $\widetilde{\psi} : \mathcal{P}(\Omega) \to R^+$.*

Proof. We prove the theorem only in the setting of left amenability, since the case of right amenability is analogous.

Let X be the space of all bounded real-valued functions defined on Ω and let Z be the linear subspace of X generated by all functions of the type χ_A, as A varies in $\mathcal{A}$ (here, $\chi_A(\omega) = 1$ if $\omega \in A$ and $\chi_A(\omega) = 0$ if $\omega \in \Omega \setminus A$). For each $f \in X$, $g \in G$ and $\omega \in \Omega$, put $(gf)(\omega) = f(g\omega)$. Note that, for each $g \in G$, g is an increasing linear homomorphism. For every $f \in Z$, $f = \sum_{i=1}^{n} r_i \chi_{A_i}$ with $A_i \in \mathcal{A}$, $i = 1, 2, \ldots, n$, set $\phi(f) = \sum_{i=1}^{n} r_i \psi(A_i)$. It is not difficult to check that ϕ is well-defined (that is, independent of the representation of f), linear, increasing and G-invariant. Let us define $\phi^+ : X \to R$, $\phi^- : X \to R$ by

$$\begin{aligned} \phi^+(x) &= \bigwedge\Big\{\phi(z) : x \le z, z \in Z\Big\}, \qquad (1.91)\\ \phi^-(x) &= \bigvee\Big\{\phi(z') : x \ge z', z' \in Z\Big\}. \end{aligned}$$

Pick arbitrarily $x \in X$. Thanks to monotonicity of ϕ, for each z, $z' \in Z$ such that $z' \le x \le z$ it is $\phi(z') \le \phi(z)$. Taking the supremum and the infimum in (1.91), we deduce $\phi^-(x) \le \phi^+(x)$. Moreover, it is not difficult to see that $\phi^+(z) = \phi^-(z) = \phi(z)$ whenever $z \in Z$.

Now we prove that ϕ^+ is convex. Choose arbitrarily x_1, $x_2 \in X$ and $\lambda \in [0, 1]$, and take z_1, $z_2 \in Z$ with $x_i \le z_i$, $i = 1, 2$. Then $\lambda x_1 + (1 - \lambda)x_2 \le \lambda z_1 + (1 - \lambda)z_2$. As ϕ is linear, we obtain

$$\phi^+(\lambda x_1 + (1 - \lambda)x_2) \le \phi(\lambda z_1 + (1 - \lambda)z_2) = \lambda\phi(z_1) + (1 - \lambda)\phi(z_2).$$

By arbitrariness of z_1 and z_2, we get

$$\phi^+(\lambda x_1 + (1 - \lambda)x_2) \le \lambda\phi^+(x_1) + (1 - \lambda)\phi^+(x_2).$$

Thus, ϕ^+ is convex. Analogously, it is possible to prove that $-\phi^-$ is convex.

Now we claim that ϕ^+ is G-subinvariant. Pick arbitrarily $x \in X$ and $g \in G$. For every $z \in Z$ such that $x \le z$, it is $gx \le gz$. Thus, $\phi^+(gx) \le \phi(gz) = \phi(z)$. Taking the infimum, we deduce $\phi^+(gx) \le \phi(x)$, getting the claim. Similarly, it is possible to check that $-\phi^-$ is G-subinvariant.

By virtue of Theorem 1.3.3, there exist $\widetilde{\phi} \in \mathcal{L}_{\text{inv}}(X, R)$ and $u_0 \in R$ such that

$$\phi^-(x) \le \widetilde{\phi}(x) - u_0 \le \phi^+(x) \qquad (1.92)$$

for all $x \in X$. As $\phi^+(0) = \phi^-(0) = 0$, by (1.92) used with $x = 0$ we get $u_0 = \widetilde{\phi}(0) = 0$. Thus, $\widetilde{\phi}(z) = \phi(z)$ for every $z \in Z$. If we set $\widetilde{\psi}(C) = \widetilde{\phi}(\chi_C)$, $C \subset \Omega$, then $\widetilde{\psi}$ is the required extension. □

Now, we apply Theorem 1.4.1 to construct finitely additive and invariant extensions of set functions. In the literature, in certain types of problems and investigations (for instance, in stochastic processes), it would be advisable to deal with some kinds of "probabilities", which associate to each event not necessarily a real number, but a real-valued function. Indeed, it is possible to give different values of the knowledge on an event E, depending, for instance, on the time or on some other events associated to E. More generally, it is advisable to associate to each event an element of a Dedekind complete vector lattice R, because, by virtue of Theorem 1.2.6, R can be viewed as a suitable space of continuous extended real-valued functions.

On the other hand, in order to take into account the time, it is advisable also to associate to an event a time value, for example the "current" instant, in which the involved "probabilities" are formulated, taking into account the state of knowledge at this instant. Thus, we can consider as "events" some sets of the type $\prod_{t \in \mathcal{T}} B_t$, where $\mathcal{T}$ is the "time set" and the B_t's are subsets of a fixed abstract nonempty set $\widetilde{R}$. Now we present the following

Example 1.4.2. *Let R be a Dedekind complete vector lattice, $\widetilde{R}$ and $\mathcal{T}$ be any two nonempty sets, $\mathcal{B}$ be an algebra of subsets of $\widetilde{R}$, $\Omega = \widetilde{R}^{\mathcal{T}}$ and $\mathcal{C} = \{f \in \Omega$: there are $t \in \mathcal{T}$, $B \in \mathcal{B}$: $f(t) \in B\}$. Observe that $\mathcal{C}$ is not an algebra: indeed, if C_1, $C_2 \in \mathcal{C}$, $t_1 \neq t_2 \in \mathcal{T}$, $B_1 \neq \emptyset$ and $B_2 \neq \emptyset$, then $C_1 \cup C_2 = \{f \in G : f(t_1) \in B_1\} \cup \{f \in G : f(t_2) \in B_2\} \notin \mathcal{C}$.*

It is not difficult to check that the algebra $\mathcal{A}(\mathcal{C})$ generated by $\mathcal{C}$ is the family of all finite (disjoint) unions of sets of the type

$$E = \bigcap_{i \in \Lambda} \{f \in \Omega : f(t_i) \in B_i\}, \tag{1.93}$$

where Λ is a finite subset of $\mathcal{T}$.

Let $G \subset \mathcal{T}^{\mathcal{T}}$ be an amenable semigroup of functions (in this setting a concrete example, when $\mathcal{T} = \mathbb{R}^n$, is the group generated by a given isometry). If $f \in \widetilde{R}^{\mathcal{T}}$, then we define the action of G on $\widetilde{R}^{\mathcal{T}}$ by setting $(\tau f)(t) = f(\tau(t))$, $t \in \mathcal{T}$. It is not difficult to check that the families $\mathcal{C}$ and $\mathcal{A}(\mathcal{C})$ are G-invariant.

Let $u \in R^+$ be such that $u \neq 0$, and for each $t \in \mathcal{T}$ let $P_t : \mathcal{B} \to R$ be a finitely additive set function with $P_{\tau(t)}(B) = P_t(B)$ whenever $B \in \mathcal{B}$ and $\tau \in G$ such that $P_t(\emptyset) = 0$ and with $P_t(\widetilde{R}) = u$ for any $t \in \mathcal{T}$. Note that $\{P_t(B) : t \in \mathcal{T}, B \in \mathcal{B}\} \subset V[u] = \{x \in R$: there exists a positive real number c with $-cu \leq x \leq cu\}$. Since $V[u]$ is Dedekind complete, by virtue of the Kakutani representation theorem, there exist a compact Hausdorff topological space Ξ and an isomorphism ι from $V[u]$ into $C(\Xi) = \{f \in \mathbb{R}^{\Xi}$: f is continuous$\}$, which maps u into $\mathbf{1}$, the function which associates to each element of Ξ the constant value 1 (see also [52, 87]). For every set E as in (1.93), where $\Lambda = \{t_1, t_2, \ldots, t_q\}$, define

$$P(E) = \iota^{-1}(\iota(P_{t_1}(B_1)) \cdot \iota(P_{t_2}(B_2)) \cdot \ldots \cdot \iota(P_{t_q}(B_q))).$$

By construction, $0 \leq P(E) \leq u$. Moreover, we get

$$P(\tau^{-1}(E)) = \iota^{-1}(\iota(P_{\tau(t_1)}(B_1)) \cdot \iota(P_{\tau(t_2)}(B_2)) \cdot \ldots \cdot \iota(P_{\tau(t_q)}(B_q))),$$

for every $\tau \in G$. If $\widetilde{E} \in \mathcal{A}(\mathcal{C})$, $\widetilde{E} = \bigcup_{i=1}^{q} B_i$, where B_i is as in (1.93), $i = 1, 2, \ldots, q$, and the B_i's are pairwise disjoint, then set $\widetilde{P}(\widetilde{E}) = \sum_{i=1}^{q} P(B_i)$. It is not difficult to see that $\widetilde{P}$ is a finitely additive G-invariant set function, defined on $\mathcal{A}(\mathcal{C})$ and with values in $V[u]$. By Theorem 1.4.1, $\widetilde{P}$ has a finitely additive G-invariant extension $\widehat{P}$, taking values in $V[u]$ and defined on the *family of all subsets of* $\widetilde{R}^{\mathcal{T}}$*, and in particular on the sets of the type* $\prod_{t\in\mathcal{T}} B_t$*, as* B_t *varies in* $\mathcal{P}(\widetilde{R})$ *for all* $t \in \mathcal{T}$ *and does not belong necessarily to* $\mathcal{B}$. □

Bibliography

[1] Ch. D. Aliprantis, O. Burkinshaw, Locally Solid Riesz Spaces with Applications to Economics, Math. Surveys and Monographs 105, American Mathematical Society, Providence, Rhode Island (2003).

[2] Ch. D. Aliprantis, O. Burkinshaw, Positive Operators, Springer, Dordrecht (2006).

[3] Q. H. Ansari, E. Kőbis, J. -C. Yao, Vector Variational Inequalities and Vector Optimization - Theory and Applications, Springer Int. Publ. AG, Cham (2018).

[4] F. Asdrubali, G. Baldinelli, F. Bianchi, D. Costarelli, A. Rotili, M. Seracini, G. Vinti, Detection of thermal bridges from thermographic images by means of image processing approximation algorithms, Appl. Math. Comput. 317 (2018) 160–171.

[5] G. Aubert, P. Kornprobst, Mathematical Problems in Image Processing – Partial Differential Equations and the Calculus of Variations, Second Edition, Springer, Berlin-Heidelberg-New York (2006).

[6] J. Bair, R. Fourneau, Étude géométrique des espaces vectoriels – Une introduction, Lecture Notes Math. 489, Springer-Verlag, Berlin-Heidelberg-New York (1975).

[7] A. Balbás, R. Balbás, S. Mayoral, Portfolio choice and optimal hedging with general risk functions: a simplex-like algorithm, European J. Oper. Res. 192(2) (2009) 603–620.

[8] D. Bartl, Farkas' Lemma, other theorems of the alternative, and linear programming in infinite-dimensional spaces: a purely linear-algebraic approach, Linear Multilinear Algebra 55 (2007) 327–353.

[9] M. S. Bazaraa, H. D. Sherali, C. M. Shetty, Nonlinear Programming: Theory and algorithms, Wiley-Interscience, John Wiley & Sons, Inc., Hoboken, New Jersey (2006).

[10] M. Bergounioux, Introduction au traitement mathématique des images – méthodes déterministes, Springer, Berlin-Heidelberg-New York (2015).

[11] A. Boccuto, Riesz spaces, integration and sandwich theorems, Tatra Mt. Math. Publ. 3 (1993) 213–230.

[12] A. Boccuto, Hahn-Banach-type theorems and applications to optimization for partially ordered vector space-valued invariant operators, Real Anal. Exchange 44(2) (2019) 333–368.

[13] A. Boccuto, Hahn-Banach and sandwich theorems for equivariant vector lattice-valued operators and applications, Tatra Mt. Math. Publ. 76 (2020) 11–34.

[14] A. Boccuto, D. Candeloro, Sandwich theorems, extension principles and amenability, Atti Sem. Mat. Fis. Univ. Modena 42 (1994) 257–271.

[15] A. Boccuto, D. Candeloro, Integral and Ideals in Riesz Spaces, Inform. Sci. 179 (2009) 2891–2902.

[16] A. Boccuto, X. Dimitriou, Convergence Theorems for Lattice Group-Valued Measures, Bentham Science Publ., Sharjah, United Arab Emirates (2015).

[17] A. Boccuto, I. Gerace, V. Giorgetti, A blind source separation technique for document restoration, SIAM J. Imaging Sci. 12(2) (2019) 1135–1162.

[18] A. Boccuto, I. Gerace, F. Martinelli, Half-quadratic image restoration with a non-parallelism constraint, J. Math. Imaging Vision 59(2) (2017) 270–295.

[19] A. Boccuto, I. Gerace, P. Pucci, Convex approximation technique for interacting line elements deblurring: a new approach, J. Math. Imaging Vision 44(2) (2012) 168–184.

[20] A. Boccuto, A. R. Sambucini, The monotone integral with respect to Riesz space-valued capacities, Rend. Mat. (Roma) 16 (1996) 491-524.

[21] J. M. Borwein, J. D. Vanderwerff, Convex Functions: Constructions, Characterizations and Counterexamples, Cambridge Univ. Press, Cambridge (2010).

[22] J. M. Borwein, Q. J. Zhu, A survey of subdifferential calculus with applications, Nonlinear Anal. 38 (1999) 687-773; Addendum to: “A survey of subdifferential calculus with applications”, 49 (2002) 295–296.

[23] S. Boyd, L. Vandenberghe, Convex Optimization, Cambridge University Press, Cambridge (2004).

[24] G. E. Bredon, Introduction to Compact transformation groups, Academic Press, Inc., New York (1972).

[25] H. Brézis, Functional Analysis, Sobolev Spaces and Partial Differential Equations, Springer, New York-Dordrecht-Heidelberg-London (2011).

[26] G. Buskes, The Hahn-Banach theorem surveyed, Dissertationes Math. 327 (1993) 1–35.

[27] D. Candeloro, R. Mesiar, A. R. Sambucini, A special class of fuzzy measures: Choquet integral and applications, Fuzzy Sets Syst. 355 (2019) 83–99.

[28] Z. Chbani, Z. Mazgouri, H. Riahi, From convergence of dynamical equilibrium systems to bilevel hierarchical Ky Fan minimax inequalities and applications, Minimax Theory Appl. 4(2) (2019) 231–270.

[29] W. Chojnacki, Sur un théorème de Day, un théorème de Mazur-Orlicz et une généralisation de quelques théorèmes de Silverman, Colloq. Math. 50 (1986) 257–262.

[30] F. Cluni, D. Costarelli, A. M. Minotti, G. Vinti, Enhancement of thermographic images as tool for structural analysis in earthquake engineering, NDT & E International 70(4) (2015) 60–72.

[31] F. Cluni, D. Costarelli, A. M. Minotti, G. Vinti, Applications of Sampling Kantorovich Operators to Thermographic Images for Seismic Engineering, J. Comput. Anal. Appl. 19(4) (2015) 602–617.

[32] T. S. Cohen, M. Welling, Group Equivariant Convolutional Networks, In: Proceedings of The 33rd International Conference on Machine Learning (ICML) 48 (2016) 2990–2999.

[33] G. Coletti, R. Scozzafava, Probabilistic Logic in a Coherent Setting, Kluwer Acad. Publ., Dordrecht (2002).

[34] C. Constantinescu, W. Filter, K. Weber, Advanced Integration Theory, Kluwer Acad. Publ., Dordrecht (1998).

[35] D. Costarelli, M. Seracini, G. Vinti, Approximation Problems for Digital Image Processing and Applications. In: O. Gervasi et al. (eds) Computational Science and Its Applications ICCSA 2018. ICCSA 2018. Lecture Notes in Computer Science, vol 10960. Springer, Cham. (2018).

[36] D. Costarelli, M. Seracini, G. Vinti, A segmentation procedure of the previous area of the aorta artery from CT images without contrast medium, Math. Meth. Appl. Sci. 43(1) (2020) 114–133.

[37] M. M. Day, Amenable semigroups, Illinois J. Math. 1(4) (1957) 509–544.

[38] N. Dinh, T. Mo, Generalizations of the Hahn-Banach theorem revisited, Taiwanese J. Math. 19(4) (2015) 1285–1304.

[39] M. L. Eaton, Group invariance applications in statistics, NSF-CBMS Regional Conference Series in Probability and Statistics, 1, Institute of Mathematical Statistics, Hayward, CA (1989).

[40] K. H. Elster, R. Nehse, Konjugierte Operatoren und Subdifferentiale, Math. Operationsforsch. Statist. 6(4) (1975) 641–657.

[41] K. H. Elster, R. Nehse, Necessary and sufficient conditions for the order-completeness of partially ordered vector spaces, Math. Nachr. 81 (1978) 301–311.

[42] J. Farkas, Theorie der einfachen Ungleichungen, J. Reine Angewandte Math. 124 (1902) 1–27.

[43] W. Filter, Representations of archimedean Riesz spaces - a survey, Rocky Mt. J. Math. 24(3) (1994) 771–851.

[44] D. H. Fremlin, Topological Riesz spaces and measure theory, Cambridge University Press, New York (1974).

[45] D. H. Fremlin, Measure theory, Vol. 2: Broad foundations, Torres Fremlin, Colchester (2003).

[46] B. Fuchssteiner, W. Lusky, Convex Cones, North-Holland, Amsterdam (1981).

[47] Z. Gajda, Sandwich theorems and amenable semigroups of transformations, Grazer Math. Ber. 316 (1992) 43–58.

[48] I. Goodfellow, Y. Bengio, A. Courville, Deep Learning, MIT Press, Cambridge, MA (2016).

[49] J. Gwinner, Results of Farkas type, Numer. Funct. Anal. Optim. 9 (1987) 471–520.

[50] J. Jahn, Vector Optimization – Theory, Applications and Extensions, Second Edition, Springer-Verlag, Berlin-Heidelberg (2011).

[51] O. Kallenberg, Probabilistic Symmetries and Invariance Principles, Springer, New York (2005).

[52] S. Kakutani, Concrete Representation of Abstract (M)-Spaces (A characterization of the Space of Continuous Functions), Annals Math. II Ser. 42(4) (1941) 994–1024.

[53] J. Kawabe, The Choquet integral in Riesz space, Fuzzy Sets Syst. 159 (2008) 629–645.

[54] T. Kawasaki, M. Toyoda, T. Watanabe, The Hahn-Banach theorem and the separation theorem in a partially ordered vector space, J. Nonlinear Anal. Optim. 2(1) (2011) 111–117.

[55] R. Kondor, S. Trivedi, On the generalization of equivariance and convolution in neural networks to the action of compact groups. In: Proceedings of the 35 th International Conference on Machine Learning, Stockholm, Sweden 80 (2018) 2747–2755.

[56] G. Köthe, Topological Vector Spaces I, Springer-Verlag, Berlin-Heidelberg-New York (1969).

[57] H. Kuhn, A. W. Tucker, Non-linear programming. In: Proceedings of the Second Berkeley Symposium on Mathematical Statistical Problems, Univ. California Press (1951) 481–492.

[58] A. G. Kusraev, S. S. Kutateladze, Subdifferentials: Theory and Applications, Kluwer Academic Publ., Dordrecht (1995).

[59] S. S. Kutateladze, Boolean models and simultaneous inequalities, Vladikavkaz Math. J. 11(3) (2009) 44–50.

[60] S. S. Kutateladze, The Farkas Lemma revisited, Sibirsk. Mat. Zh. 51(1) (2010) 98-109 (Russian); English translation in: Sib. Math. J. 51(1) (2010) 78–87.

[61] Z. Lipecki, Compactness and extreme points of the set of quasi-measure extensions of a quasi-measure, Dissertationes Math. 493 (2013) 1–59.

[62] W. A. J. Luxemburg, J. J. Masterson, An extension of the concept of the order dual of a Riesz space, Canad. J. Math. 19 (1976) 488–498.

[63] W. A. J. Luxemburg, A. C. Zaanen, Riesz Spaces I, North-Holland Publ. Co., Amsterdam (1971).

[64] F. Maeda, T. Ogasawara, Representation of vector lattices, J. Sci. Hiroshima Univ. Ser. A 12 (1942) 17–35. (Japanese)

[65] O. L. Mangasarian, Nonlinear Programming, Soc. Industr. Appl. Math., New York (1994).

[66] R. Nehse, Some general separation theorems, Math. Nachr. 84 (1978) 319–327.

[67] J. Nocedal, S. J. Wright, Numerical Optimization, Second Edition, Springer, New York (2006).

[68] A. L. T. Paterson, Amenability, Amer. Math. Soc., Providence, Rhode Island (1988).

[69] A. L. Peressini, Ordered Topological Vector Spaces, Harper & Row, New York, Evanston, and London (1967).

[70] G. Ch. Pflug, W. Rőmisch, Modeling, Measuring and Managing Risk, World Scientific Publishing Co. Pte. Ltd., Hackensack, NJ (2007).

[71] J. Rainwater, A note on projective resolutions, Proc. Amer. Math. Soc. 10(5) (1959) 734–735.

[72] B. Ricceri, On a minimax theorem: an improvement, a new proof and an overview of its applications, Minimax Theory Appl. 2(1) (2017) 99–152.

[73] B. Riečan, T. Neubrunn, Integral, Measure, and Ordering, Kluwer Acad. Publ./Ister Science, Dordrecht/Bratislava (1997).

[74] R. T. Rockafellar, R. J. B Wets, Variational Analysis, Springer-Verlag, Berlin-Heidelberg-New York (1998).

[75] A. Ruszczyński, A. Shapiro, Optimality and duality in stochastic programming. In: Handbooks Oper. Res. Management Sci. 10, Elsevier Sci. B. V., Amsterdam (2003), pp 65–139.

[76] A. R. Sambucini, The Choquet integral with respect to fuzzy measures and applications, Math. Slovaca 67(6) (2017) 1427–1450.

[77] L. M. Schmitt, An equivariant version of the Hahn-Banach theorem, Houston J. Math. 18(3) (1992) 429–447.

[78] R. Silverman, Invariant means and cones with vector interiors, Trans. Amer. Math. Soc. 88(1) (1958) 75–79.

[79] R. Silverman, T. Yen, Addendum to: "Invariant means and cones with vector interiors", Trans. Amer. Math. Soc. 88(2) (1958) 327–330.

[80] S. Simons, The Hahn-Banach-Lagrange theorem, Optimization 56(1-2) (2007) 149–169.

[81] S. Simons, From Hahn-Banach to Monotonicity, Second Edition, Springer-Verlag, New York (2008).

[82] M. Théra, Subdifferential calculus for convex operators, J. Math. Anal. Appl. 80 (1981) 78–91.

[83] T.-O. To, The equivalence of the least upper bound property and the Hahn-Banach extension property in ordered linear spaces, Proc. Amer. Math. Soc. 30(2) (1971) 287–295.

[84] L. S. Vandenberghe, S. Boyd, Semidefinite programming, SIAM Review, 38(1) (1996) 49–95.

[85] G. Vitali, On the problem of the measure of the points of a line (Italian), Tipografia Gamberini e Parmeggiani, Bologna (1905).

[86] B. Z. Vulikh, Introduction to the Theory of Partially Ordered Spaces, Wolters-Noordhoff Sci. Publ., Groningen (1967).

[87] J. D. M. Wright, The measure extension problem for vector lattices, Ann. Inst. Fourier (Grenoble) 21 (1971) 65–85.

[88] J. D. M. Wright, Paradoxical decompositions of the cube and injectivity, Bull. London Math. Soc. 22 (1990) 18–24.

[89] A. C. Zaanen, Riesz spaces II, North-Holland Publ. Co., Amsterdam (1983).

[90] J. Zowe, A duality theorem for a convex programming problem in order complete vector lattices, J. Math. Anal. Appl. 50 (1975) 273–287.

[91] J. Zowe, The saddle point theorem of Kuhn and Tucker in ordered vector spaces, J. Math. Anal. Appl. 57 (1977) 41–55.

[92] J. Zowe, Sandwich theorems for convex operators with values in an ordered vector space, J. Math. Anal. Appl. 66 (1978) 282–296.

Chapter 2

Application of Measure of Noncompactness on Infinite System of Functional Integro-differential Equations with Integral Initial Conditions

Anupam Das

Cotton University, math.anupam@gmail.com

2.1 Introduction

The theory of infinite system of ordinary differential equations is a very important branch of the theory of differential equations in Banach spaces. Infinite system of ordinary differential equations describe many real life problems, which can be found in the theory of neural nets, the theory of branching processes and mechanics etc (see [3, 4, 6, 7]).

In the study of infinite dimensional normed spaces, the measure of noncompactness plays a very important role which was introduced by Kuratowski [8]. The idea of measure of noncompactness has been used by many authors in

DOI: 10.1201/9781003178200-2

obtaining the existence of solutions of infinite system of integral equations and differential equations (see [2]). Mursaleen and Mohiuddine [9] proved existence theorems for the infinite system of differential equations in the space ℓ_p. On the other hand, existence theorems for the infinite systems of linear equations in ℓ_1 and ℓ_p were given by Alotaibi et al. [10]. Mursaleen and Alotaibi [11] proved existence theorems for the infinite system of differential equations in some BK-spaces.

2.1.1 Preliminaries

Let M and S be subsets of a metric space (X, d). If $\epsilon > 0$, then the set S is called $\epsilon - net$ of M if for any $x \in M$ there exists $s \in S$, such that $d(x, s) < \epsilon$. If S is finite, then the $\epsilon - net$ S of M is called finite $\epsilon - net$. The set M is said to be totally bounded if it has a finite $\epsilon - net$ for every $\epsilon > 0$. A subset M of a metric space X is said to be compact if every sequence (x_n) in M has a convergent subsequence and the limit of that subsequence is in M. The set M is called relatively compact if the closure $\bar{M}$ of M is a compact set. If a set M is relatively compact, then M is totally bounded. If the metric space (X, d) is complete, then the set M is relatively compact if and only if it is totally bounded.

If $x \in X$ and $r > 0$, then the open ball with center at x and radius r is denoted by $B(x, r)$, where $B(x, r) = \{y \in X : d(x, y) < r\}$. If X is a normed space, then we denote by B_X the closed unit ball in X and by S_X the unit sphere in X. Let $\mathcal{M}_X$ or simply $\mathcal{M}$ be the family of all non-empty and bounded subsets of a metric space (X, d), and let $\mathcal{M}_X^c$ or simply $\mathcal{M}^c$ be the subfamily of $\mathcal{M}_X$ consisting of all closed sets. Further, let $\mathcal{N}_X$ or simply $\mathcal{N}$ be the family of all nonempty and relatively compact subsets of (X, d). Let $d_H : \mathcal{M} \times \mathcal{M} \to \mathbb{R}$ be the function defined by

$$d_H(A, B) = \max\left\{\sup_{x \in A} d(x, B), \sup_{y \in B} d(y, A)\right\},$$

where $A, B \in \mathcal{M}_X$. The function d_H is called the Hausdorff distance, and $d_H(A, B)$ is the Hausdorff distance of sets A, B.

Let X and Y be infinite-dimensional complex Banach spaces and denote the set of bounded linear operators from X into Y by $B(X, Y)$. We put $B(X) = B(X, X)$. For T in $B(X, Y)$, $N(T)$ and $R(T)$ denote the null space and the range space of T, respectively. A linear operator L from X to Y is called compact (or completely continuous) if $D(L) = X$ for the domain of L, and for every sequence $(x_n) \in X$ such that $\| x_n \| \leq C$, the sequence $(L(x_n))$ has a subsequence which converges in Y. A compact operator is bounded. An operator L in $B(X, Y)$ is of finite rank if $dim R(L) < \infty$. An operator of finite rank is clearly compact. Let $F(X, Y), C(X, Y)$ denote the set of all finite rank and compact operators from X to Y, respectively. Set $F(X) = F(X, X)$ and $C(X) = C(X, X)$.

If F is a subset of X, then the intersection of all convex sets that contain F is called convex cover or convex hull of F denoted by $co(F)$.

Let Q be a non-empty and bounded subset of a normed space X. Then the convex closure of Q, denoted by $Co(Q)$, is the smallest convex and closed subset of X that contains Q. Note that $Co(Q) = \bar{co}(Q)$.

2.1.2 Kuratowski measure of noncompactness

Definition 2.1.1. *[1] Let (X, d) be a metric space and Q a bounded subset of X. Then the Kuratowski measure of noncompactness (α-measure or set measure of non-compactness) of Q, denoted by $\alpha(Q)$, is the infimum of the set of all numbers $\epsilon > 0$ such that Q can be covered by a finite number of sets with diameters $\epsilon > 0$, that is,*

$$\alpha(Q) = \inf\left\{\epsilon > 0 : Q \subset \cup_{i=1}^{n} S_i, S_i \subset X, diam(S_i) < \epsilon \ (i = 1, 2, ..., n)\ , n \in \mathbb{N}\right\}$$

The function α is called Kuratowski measure of noncompactness was introduced by Kuratowski [8]. Clearly

$$\alpha(Q) \leq diam(Q) \text{ for each bounded subset } Q \text{ of } X.$$

Lemma 2.1.2. *[1] Let Q, Q_1 and Q_2 be bounded subsets of a complete metric space (X, d). Then*

$$\begin{aligned}
&\alpha(Q) = 0 \; if \text{ and only } if \; \bar{Q} \text{ is compact, } (regularity),\\
&\alpha(Q) = \alpha(\bar{Q}) \text{ (invariance under passage to the closure)},\\
&Q_1 \subset Q_2, \text{ implies } \alpha(Q_1) \leq \alpha(Q_2) \text{ (monotonicity)},\\
&\alpha(Q_1 \cup Q_2) = \max\left\{\alpha(Q_1), \alpha(Q_2)\right\} \text{ (maximum property)},\\
&\alpha(Q_1 \cap Q_2) \leq \min\left\{\alpha(Q_1), \alpha(Q_2)\right\}.
\end{aligned}$$

Lemma 2.1.3. *[1] Let Q, Q_1 and Q_2 be bounded subsets of a normed space X. Then*

$$\begin{aligned}
&\alpha(Q_1 + Q_2) \leq \alpha(Q_1) + \alpha(Q_2),\\
&\alpha(Q + x) = \alpha(Q) \; for \text{ each } x \in X,\\
&\alpha(\lambda Q) = |\lambda|\,\alpha(Q) \; for \text{ each } \lambda \in \mathbb{F},\\
&\alpha(Q) = \alpha(Co(Q)),
\end{aligned}$$

where $\mathbb{F}$ is the field of scalars.

2.1.3 Axiomatic approach to the concept of a measure of noncompactness

Suppose E is a real Banach space with the norm $\| \,.\, \|$. Let $B(x_0, r)$ be a closed ball in E centered at x_0 and with radius r. If X is a non-empty subset of E then by $\bar{X}$ and Conv(X), we denote the closure and convex closure of X.

Moreover, let $\mathcal{M}_E$ denote the family of all non-empty and bounded subsets of E and $\mathcal{N}_E$ its subfamily consisting of all relatively compact sets.

We consider the definition of the concept of a measure of noncompactness defined by Banaś and Lecko [12].

Definition 2.1.4. *A function $\mu : \mathcal{M}_E \to [0,\infty)$ will be called a measure of noncompactness if it satisfies the following conditions:*

(i) the family $ker\ \mu = \{X \in \mathcal{M}_E : \mu(X) = 0\}$ *is non-empty and* $ker\ \mu \subset \mathcal{N}_E$.

(ii) $X \subset Y \implies \mu(X) \leq \mu(Y)$.

(iii) $\mu(\bar{X}) = \mu(X)$.

(iv) $\mu(ConvX) = \mu(X)$.

(v) $\mu(\lambda X + (1-\lambda)Y) \leq \lambda\mu(X) + (1-\lambda)\mu(Y)$ *for* $\lambda \in [0,1]$.

(vi) if $X_n \in \mathcal{M}_E, X_n = \bar{X}_n, X_{n+1} \subset X_n$ *for* $n = 1,2,3,...$ *and* $\lim_{n\to\infty} \mu(X_n) = 0$ *then* $\bigcap_{n=1}^{\infty} X_n \neq \phi$.

The family $ker\mu$ is said to be the *kernel of measure* μ.

A measure μ is said to be the sublinear if it satisfies the following conditions:

(1) $\mu(\lambda X) = |\lambda|\,\mu(X)$ for $\lambda \in \mathbb{R}$.

(2) $\mu(X+Y) \leq \mu(Y) + \mu(Y)$.

A sublinear measure of noncompactness μ satisfying the condition:

$$\mu(X \cup Y) = \max\{\mu(\lambda X), \mu(\lambda Y)\}$$

and such that $ker\mu = \mathcal{N}_E$ is said to be regular.

2.1.4 Hausdorff measure of noncompactness

Definition 2.1.5. *[2] Let (X,d) be a metric space, Q be a bounded subset of X and $B(x,r) = \{y \in X : d(x,y) < r\}$. Then the Hausdorff measure of noncompactness $\chi(Q)$ of Q is defined by*

$$\chi(Q) := \inf\left\{\epsilon > 0 : Q \subset \bigcup_{i=1}^{n} B(x_i, r_i), x_i \in X, r_i < \epsilon\ (i = 1,2,\ldots,n), n \in \mathbb{N}\right\}.$$

The definition of the Hausdorff measure of noncompactness of the set Q it is not supposed that centers of the balls that cover Q belong to Q. Hence it can equivalently be stated as follows:

$$\chi(Q) = \inf\{\epsilon > 0 : Q\ has\ a\ finite\ \epsilon - net\ in\ X\}.$$

Then the following results were obtained in [1, 2].

Lemma 2.1.6. *[2] Let Q, Q_1 and Q_2 be bounded subsets of the complete metric space (X, d). Then*

$$\chi(Q) = 0 \; if \; and \; only \; if \; \bar{Q} \; is \; compact, \tag{2.1}$$

$$\chi(Q) = \chi(\bar{Q}), \tag{2.2}$$

$$Q_1 \subset Q_2, \; implies \; \chi(Q_1) \leq \chi(Q_2), \tag{2.3}$$

$$\chi(Q_1 \cup Q_2) = \max\{\chi(Q_1), \chi(Q_2)\}, \tag{2.4}$$

$$\chi(Q_1 \cap Q_2) \leq \min\{\chi(Q_1), \chi(Q_2)\}. \tag{2.5}$$

Now we point out the well-known result of Goldenštein et al. [5]. Let X be a Banach space with a Schauder basis $\{e_1, e_2, ...\}$. Then each element $x \in X$ has a unique representation $x = \sum_{i=1}^{\infty} \phi_i(x) e_i$ where the functions ϕ_i are the basis functionals. Let $P_n : X \to X$ be the projector onto the linear span of $\{e_1, e_2, ..., e_n\}$, that is, $P_n(x) = \sum_{i=1}^{n} \phi_i(x) e_i$.

Theorem 2.1.7. *[2] Let X be a BK-space with Schauder basis (b_n), $Q \in \mathcal{M}_X$, $P_n : X \to X$ $(n \in \mathbb{N})$ be the projector onto the linear span of $\{e_1, e_2, ..., e_n\}$ and I be the identity operator on X. Then*

$$\frac{1}{a} \limsup_{n\to\infty} \left(\sup_{x\in Q} \| (I - P_n)(x) \| \right) \leq \chi(Q) \leq \limsup_{n\to\infty} \left(\sup_{x\in Q} \| (I - P_n)(x) \| \right),$$

where $a = \limsup_{n\to\infty} \| I - P_n \|$.

We say that a norm $\| . \|$ on a sequence space is monotone if $x, \bar{x} \in X$ with $|x_k| \leq |\bar{x}_k|$ for all k implies $\| x \| \leq \| \bar{x} \|$.

Theorem 2.1.8. *[1] Let X be a BK-space with AK and monotone norm, $Q \in \mathcal{M}_X$ and $P_n : X \to X$ $(n \in \mathbb{N})$ be the operator (projection) defined by $P_n(x_1, x_2, ...) = x^{[n]} = (x_1, x_2, ..., x_n, 0, 0, ...)$ for all $x = (x_1, x_2, ...) \in X$. Then*

$$\chi(Q) = \lim_{n\to\infty} \left(\sup_{x\in Q} \| (I - P_n)(x) \| \right).$$

Consider the following sequence spaces, are Banach spaces with their respective norms,

$$c_0 = \left\{ x \in \omega : \lim_{k\to\infty} x_k = 0, \; \| x \|_\infty = \sup_k |x_k| \right\}$$

the space of all convergent sequences;

$$\ell_p = \left\{ x \in \omega : \sum_{k=0}^{\infty} |x_k|^p < \infty \, (1 \leq p < \infty), \; \| x \|_p = \left(\sum_{k=0}^{\infty} |x_k|^p \right)^{1/p} \right\}$$

the space of all absolutely p-summable series.

In the Banach space $(c_0, \| \cdot \|_{c_0})$, the Hausdorff measure of noncompactness χ can be formulated as follows (see [2]).

$$\chi_{c_0}(B) = \lim_{n \to \infty} \left[\sup_{u \in B} \left\{ \sup_{k \geq n} \mid u_k \mid \right\} \right] \tag{2.6}$$

where $u = (u_i)_{i=1}^{\infty} \in c_0$ and $B \in \mathcal{M}_{c_0}$. The measure χ_{c_0} is regular.

Banaś and Mursaleen [2] defined the Hausdorff measure of noncompactness χ (in Theorem 5.18 (a)) for the Banach space $(\ell_p, \| \cdot \|_{\ell_p}), (1 \leq p < \infty)$ as follows.

$$\chi_{\ell_p}(B) = \lim_{n \to \infty} \left[\sup_{u \in B} \left(\sum_{k=n}^{\infty} \mid u_k \mid^p \right)^{1/p} \right] \tag{2.7}$$

where $u = (u_i)_{i=1}^{\infty} \in \ell_p$ and $B \in \mathcal{M}_{\ell_p}$.

Let us denote by $C(I, c_0)$ the space of all continuous functions on I with values in c_0. Then $C(I, c_0)$ is also a Banach space with norm $\| x(t) \|_{C(I,c_0)} = \sup \{ \| x(t) \|_{c_0} : t \in I \}$ where $x(t) \in C(I, c_0)$.

For any non-empty bounded subset $\hat{E}$ of $C(I, c_0)$ and $t \in I$, let $\hat{E}(t) = \left\{ x(t) : x \in \hat{E} \right\}$. Now, using (2.6), we conclude that the Housdorff measure of noncompactness for $\hat{E} \subset C(I, c_0)$ can be defined by

$$\chi_{C(I,c_0)}(\hat{E}) = \sup \left\{ \chi_{c_0}(\hat{E}(t)) : t \in I \right\}.$$

Similarly we can define $C(I, \ell_1)$ the space of all continuous functions defined on I with values in ℓ_1. Then $C(I, \ell_1)$ is also a Banach space with the norm $\| x(t) \|_{C(I,\ell_1)} = \sup \{ \| x(t) \|_{\ell_1} : t \in I \}$ where $x(t) \in C(I, \ell_1)$.

Now, using (2.7), we conclude that the Housdorff measure of noncompactness for $\hat{E} \subset C(I, \ell_1)$ can be defined by

$$\chi_{C(I,\ell_1)}(\hat{E}) = \sup \left\{ \chi_{\ell_1}(\hat{E}(t)) : t \in I \right\}.$$

2.1.5 Condensing operators, compact operators and related results

Definition 2.1.9. *Let μ_1 and μ_2 be measures of noncompactness on Banach spaces E and F, respectively. Let $L : E \to F$ be an operator. Then*

(a) L is called (μ_1, μ_2)-contractive operator with constant $k > 0$ (or simply k-(μ_1, μ_2)-contractive) if L is continuous and

$$\mu_2(L(Q)) \leq k\mu_1(Q) \ for \ each \ Q \in \mathcal{M}_E.$$

In particular, if $E = F$ and $\mu_1 = \mu_2 = \mu$ then we say that L is a k-μ-contractive operator.

(b) *L is called* (μ_1, μ_2)*-condensing operator with constant* $k > 0$ *(or simply* k-(μ_1, μ_2)*-condensing) if L is continuous and*

$$\mu_2(L(Q)) < k\mu_1(Q) \; for \; each \; non-precompact \; Q \in \mathcal{M}_E.$$

In particular, if $E = F$ *and* $\mu_1 = \mu_2 = \mu$ *then we say that L is a* k-μ*-condensing operator. Moreover, if* $k = 1$*, we say that L is a* μ*-condensing operator.*

If an operator L is (μ_1, μ_2)-contractive, then the number $\| L \|_{\mu_1,\mu_2}$ defined by

$$\| L \|_{\mu_1,\mu_2} = \inf \{k \geq 0 : \mu_2(L(Q)) \leq k\mu_1(Q) \; for \; each \; Q \in \mathcal{M}_E\} \tag{2.8}$$

is called (μ_1, μ_2)-operator norm of L, or (μ_1, μ_2)-measure of noncompactness of L, or simply measures of noncompactness of L. If $\mu_1 = \mu_2 = \mu$, then we write $\| L \|_\mu$ instead of $\| L \|_{\mu,\mu}$ which we call as the μ-norm of L.

In infinite-dimensional spaces E and F, for any arbitrary measure of noncompactness μ, $\| L \|_\mu$ may be expressed by the equivalent formula

$$\| L \|_\mu = \sup \left\{ \frac{\mu(L(Q))}{\mu(Q)} : Q \in \mathcal{M}_E, \; \mu(Q) > 0 \right\}. \tag{2.9}$$

Definition 2.1.10. *Let X and Y be Banach spaces and* χ_1 *and* χ_2 *be Hausdorff measures on X and Y. Then, the operator* $L : X \to Y$ *is called* $(\chi_1, \chi_2) -$ *bounded if* $L(Q)$ *is bounded subset of Y for every* $Q \in \mathcal{M}_X$ *and there exists a positive constant K such that* $\chi_2(L(Q)) \leq K\chi_1(Q)$ *for every* $Q \in \mathcal{M}_X$*. If an operator L is* $(\chi_1, \chi_2)-$*bounded then the number* $\| L \|_{(\chi_1,\chi_2)} = \inf \{K \geq 0 : \chi_2(L(Q)) \leq K\chi_1(Q) \; for \; all \; bounded \; Q \subset X\}$ *is called* (χ_1, χ_2)*-measure of noncompactness of L. In particular, if* $\chi_1 = \chi_2 = \chi$*, then we write*

$$\| L \|_{(\chi,\chi)} = \| L \|_\chi .$$

The idea of compact operators between Banach spaces is closely related to the Hausdorff measure of noncompactness, and it can be given, as follows. Let X and Y be Banach spaces and $L \in B(X, Y)$. Then, the Hausdorff measure of noncompactness $\|L\|_\chi$ of L can be given by $\|L\|_\chi = \chi(L(S_X))$, where $S_X = \{x \in X : \|x\| = 1\}$ and we have L is compact if and only if $\|L\|_\chi = 0$. We also have $\|L\| = \sup_{x \in S_X} \|Lx\|_Y$.

Definition 2.1.11. *[13] Let* E_1 *and* E_2 *be two Banach spaces and let* μ_1 *and* μ_2 *be arbitrary measure of noncompactness on* E_1 *and* E_2*, respectively. An operator f from* E_1 *to* E_2 *is called a* $(\mu_1, \mu_2)-$*condensing operator if it is continuous and* $\mu_2(f(D)) < \mu_1(D)$ *for every set* $D \subset E_1$ *with compact closure.*

Remark 2.1.12. *If* $E_1 = E_2$ *and* $\mu_1 = \mu_2 = \mu$*, then f is called a* $\mu-$*condensing operator.*

Theorem 2.1.13. *[14] Let Ω be a nonempty, closed, bounded and convex subset of a Banach space E and let $f : \Omega \to \Omega$ be a continuous mapping such that there exists a constant $k \in [0,1)$ with the property $\mu_2(f(\Omega)) < k\mu_1(\Omega)$. Then f has a fixed point in Ω.*

Definition 2.1.14. *[15] Let (X,d) be a metric space. Then a mapping T on X is said to be a Meir-Keeler contraction if for any $\epsilon > 0$, there exists $\delta > 0$ such that*

$$\epsilon \leq d(x,y) < \epsilon + \delta \implies d(Tx,Ty) < \epsilon, \forall x,y \in X.$$

Theorem 2.1.15. *[15] Let (X,d) be a complete metric space. If $T : X \to X$ is a Meir-Keeler contraction, then T has a unique fixed point.*

Definition 2.1.16. *[16] Let C be a nonempty subset of a Banach space E and let μ be an arbitrary measure of noncompactness on E. We say that an operator $T : C \to C$ is a Meir-Keeler condensing operator if for any $\epsilon > 0$, there exists $\delta > 0$ such that*

$$\epsilon \leq \mu(X) < \epsilon + \delta \implies \mu(T(X)) < \epsilon$$

for any bounded subset X of C.

Theorem 2.1.17. *[16] Let C be a nonempty, bounded, closed and convex subset of a Banach space E and let μ be an arbitrary measure of noncompactness on E. If $T : C \to C$ is a continuous and Meir-Keeler condensing operator, then T has at least one fixed point and the set of all fixed points of T in C is compact.*

Theorem 2.1.18. *(Shauder [14]) Let $\mathfrak{D}$ be a nonempty, closed and convex subset of a Banach space $\bar{\mathfrak{E}}$. Then every compact, continuous map $\mathfrak{T} : \mathfrak{D} \to \mathfrak{D}$ has at least one fixed point.*

Theorem 2.1.19. *(Darbo [14]) Let $\mathfrak{Z}$ be a nonempty, bounded, closed and convex subset of a Banach space $\bar{\mathfrak{E}}$. Let $\mathfrak{S} : \mathfrak{Z} \to \mathfrak{Z}$ be a continuous mapping. Assume that there is a constant $k \in [0,1)$ such that*

$$\vartheta(\mathfrak{S}\mathfrak{M}) \leq k\vartheta(\mathfrak{M}), \ \mathfrak{M} \subseteq \mathfrak{Z}.$$

Then $\mathfrak{S}$ has a fixed point.

We study the following infinite system of integro-differential equations,

$$y_i^{'}(t) = u_i(t, y(t)) + \int_0^t K_i(t, s, y(s))\, ds, \tag{2.10}$$

where $y_i(0) = y_i^0$, $t \in [0,T] = I$. Also u_i and K_i are continuous functions for all $i \in \mathbb{N}$ and $y(t) = (y_i(t))_{i=1}^{\infty}$.

A function $y_i(t)$ is a solution of (2.10) if and only if $y_i(t)$ is a solution of the following infinite system of integral equation

$$y_i(t) = y_i^0 + \int_0^t u_i(s, y(s))ds + \int_0^t \int_0^s K_i(s, r, y(r))drds, \qquad (2.11)$$

where $i = 1, 2, 3, ...$ and $t \in I$.

The solution of the infinite system in the sequence space ℓ_1 has been studied by Aghajani and Pourhadi [18] and in the sequence spaces c_0 and ℓ_1 has been studied by Mursaleen and Rizvi [17]. In this article, we establish the existence of solution of the infinite system (2.10) for the sequence spaces c_0, ℓ_1.

2.2 Existence of Solution $C(I, c_0)$

Assumptions

(i) The functions $u : I \times C(I, c_0) \to C(I, c_0)$ is continuous where $u(y(t)) = (u_i(t, y(t)))_{i=1}^{\infty}$ and there exists a continuous nonnegative function $a_i(t)$ on I $(i \in \mathbb{N})$ and a constant b_1 such that

$$|u_i(t, y(t))| \leq a_i(t) + b_1 |y_i(t)|$$

for all $y(t) \in C(I, c_0)$, $i \in \mathbb{N}$. Also, $(a_i(t))$ converges uniformly on I to zero function.

(ii) The functions $K : I \times I \times C(I, c_0) \to C(I, c_0)$ is continuous where $K(t, s, y(s)) = (K_i(t, s, y(s)))_{i=1}^{\infty}$ is continuous. Also there exists nonnegative continuous functions $b(t, s)$, $d_i(t, s)$ on $I \times I (i \in \mathbb{N})$ such that

$$|K_i(t, s, y(s))| \leq d_i(t, s) + b(t, s) |y_i(s)|,$$

for all $y(t) \in C(I, c_0)$, $i \in \mathbb{N}$. Also, $(d_i(t, s))$ converges uniformly on $I \times I$ to zero function.

(iii) Define an operator F on $I \times C(I, c_0)$ to $C(I, c_0)$ as follows,

$$(t, y(t)) \to (Fy)(t) = ((F_i y)(t))_{i=1}^{\infty},$$

where

$$(F_i y)(t) = y_i^0 + \int_0^t u_i(s, y(s))ds + \int_0^t \int_0^s K_i(s, r, y(r))drds.$$

The family of functions $((Fy)(t))_{t \in I}$ is equicontinuous at each point od the space c_0.

(iv) Let $y^0 = \sup_{i\in\mathbb{N}} |y_i^0|$, $A = \sup_{i\in\mathbb{N},t\in I} a_i(t)$, $D = \sup_{i\in\mathbb{N};t,s\in I} d_i(t,s)$ and $B_1 = \sup_{t,s\in I} b(t,s)$. Also, $Tb_1 + T^2 B_1 \leq 1$ and $\lim_{i\to\infty} y_i^0 = 0$.

Theorem 2.2.1. *Under the hypothesis (i)–(iv), infinite system (2.10) has at least one solution $y(t) = (y_i(t)) \in C(I, c_0)$ for all $t \in [0,T]$.*

Proof. Define the operator $F : C(I,c_0) \to C(I,c_0)$ as

$$(Fy)(t) = \left\{ y_i^0 + \int_0^t u_i(s, y(s))ds + \int_0^t \int_0^s K_i(s,r,y(r))drds \right\}_{i=1}^{\infty}. \tag{2.12}$$

By assumption (iii), F is well defined and for arbitrary fixed $t \in I$,

$$\begin{aligned}
\| y(t) \|_{c_0} &= \| (Fy)(t) \|_{c_0} \\
&= \sup_{i\geq 1} |(F_i y)(t)| \\
&= \sup_{i\geq 1} \left| y_i^0 + \int_0^t u_i(s,y(s))ds + \int_0^t \int_0^s K_i(s,r,y(r))drds \right| \\
&\leq \sup_{i\geq 1} \left\{ |y_i^0| + \int_0^t |u_i(s,y(s))|\, ds + \int_0^t \int_0^s |K_i(s,r,y(r))|\, drds \right\} \\
&\leq \sup_{i\geq 1} \Big\{ |y_i^0| + \int_0^t [a_i(s) + b_1 |y_i(s)|]\, ds \\
&\quad + \int_0^t \int_0^s [d_i(s,r) + b(s,r) |y_i(r)|]\, drds \Big\} \\
&\leq y^0 + \int_0^T (A + b_1 \| y(s) \|_{c_0})\, ds \\
&\quad + \int_0^T \int_0^T (D + B_1 \| y(r) \|_{c_0})\, drds \\
&= y^0 + \left(A + b_1 \| y \|_{C(I,c_0)}\right) T + \left(D + B_1 \| y(t) \|_{C(I,c_0)}\right) T^2.
\end{aligned}$$

Therefore,

$$\| y \|_{C(I,c_0)} \leq y^0 + \left(A + b_1 \| y \|_{C(I,c_0)}\right) T + \left(D + B_1 \| y \|_{C(I,c_0)}\right) T^2,$$

which gives

$$\| y \|_{C(I,c_0)} \leq \frac{y^0 + TA + T^2 D}{1 - Tb_1 - TB_1^2} = r_0(say).$$

Consider $B = \{ y \in C(I,c_0) : \| y \|_{C(I,c_0)} \leq r_0 \}$ which is an non-empty, bounded, closed and convex subset of $C(I,c_0)$. The operator F is bounded on B. For arbitrary fixed $x \in B$,

$$\begin{aligned} \| (Fy)(t) - (Fx)(t) \|_{c_0} &\leq \sup_{i\geq 1}\left\{ \int_0^t |u_i(s,y(s)) - u_i(s,x(s))|\, ds \right. \\ &\quad \left. + \int_0^t \int_0^s |K_i(s,r,y(s)) - K_i(s,r,x(s))| \right\} \\ &\leq \int_0^T \sup_{i\geq 1} |u_i(s,y(s)) - u_i(s,x(s))|\, ds \\ &\quad + \int_0^T \int_0^T \sup_{i\geq 1} |K_i(s,r,y(s)) - K_i(s,r,x(s))|\, drds. \end{aligned}$$

Using the assumptions, for any $x \in B$ and for $\epsilon > 0$ there exists $\delta > 0$ such that for $\| y - x \|_{C(I,c_0)} < \epsilon$ gives for arbitrary fixed $t \in I$,

$$\| u(y(t)) - u(x(t)) \|_{c_0} < \delta = \frac{\epsilon}{2T}$$

and

$$\| K(Y(t)) - K(x(t)) \|_{c_0} < \delta = \frac{\epsilon}{2T^2},$$

for each $t \in I$ and $y \in B$.

Therefore,

$$\| (Fy)(t) - (Fx(t)) \|_{c_0} < T.\frac{\epsilon}{2T} + T^2.\frac{\epsilon}{2T^2} = \epsilon,$$

which gives

$$\| Fy - Fx \|_{C(I,c_0)} < \epsilon.$$

Therefore, F is continuous on B.

Now, we establish the continuity of Fy in I. Let $t_0 \in I$ and $\epsilon > 0$ be arbitrary then for $|t - t_0| < \epsilon$ we have

$$\begin{aligned} \| (Fy)(t) - (Fx)(t_0) \|_{c_0} &= \sup_{i\geq 1}\left| \int_{t_0}^t u_i(s,y(s))ds + \int_0^t \int_0^s K_i(s,r,y(r))drds \right. \\ &\quad \left. - \int_0^{t_0} \int_0^s K_i(s,r,y(r))drds \right| \\ &\leq \sup_{i\geq 1}\left\{ \int_{t_0}^t |u_i(s,y(s))|\, ds \right. \\ &\quad \left. + \int_{t_0}^t \int_0^s |K_i(s,r,y(r))|\, drds \right\}. \end{aligned}$$

As $\epsilon \to 0$, we have

$$\| (Fy)(t) - (Fy(t_0)) \|_{c_0} \to 0.$$

Therefore, Fy is continuous with respect to t.

Again,

$$\begin{aligned}
\chi_{c_0}(FB) &= \lim_{n\to\infty} \sup_{y\in B} \max_{i\geq n} |(Fy)_i(t)| \\
&\leq \lim_{n\to\infty} \sup_{y\in B} \max_{i\geq n} \left\{ |y_i^0| + \int_0^t [a_i(s) + b_1 |y_i(s)|] \, ds \right. \\
&\quad \left. + \int_0^t \int_0^s [d_i(s,r) + b(s,r) |y_i(r)|] \, drds \right\} \\
&\leq \lim_{n\to\infty} \sup_{y\in B} \max_{i\geq n} \left\{ b_1 \int_0^T |y_i(s)| \, ds + B_1 \int_0^T \int_0^T |y_i(s)| \, drds \right\} \\
&= (b_1 T + B_1 T^2) \chi_{C(I,c_0)}(B).
\end{aligned}$$

Therefore,

$$\sup_{t\in I} \chi_{c_0}(FB) \leq (b_1 T + B_1 T^2) \chi_{C(I,c_0)}(B),$$

gives

$$\chi_{C(I,c_0)}(FB) \leq (b_1 T + B_1 T^2) \chi_{C(I,c_0)}(B).$$

As $(b_1 T + B_1 T^2) < 1$, therefore by Darbo's fixed point theorem, F has at least one fixed point in $B \subset C(I, c_0.)$ □

2.3 Existence of Solution $C(I, \ell_1)$

Assumptions

(i) The functions $u : I \times C(I, \ell_1) \to C(I, \ell_1)$ is continuous where $u(t) = (u_i(t, y(t)))_{i=1}^{\infty}$ and there exists a continuous nonnegative function $\hat{a}_i(t)$ on I $(i \in \mathbb{N})$ and a constant $\hat{b}_1$ such that

$$|u_i(t, y(t))| \leq \hat{a}_i(t) + \hat{b}_1 |y_i(t)|,$$

for all $y(t) \in C(I, \ell_1)$, $i \in \mathbb{N}$. Also, $\sum_{i=1}^{\infty} a_i(t)$ converges uniformly on I.

(ii) The functions $K : I \times I \times C(I, \ell_1) \to C(I, \ell_1)$ is continuous where $K(t, s, y(s)) = (K_i(t, s, y(s)))_{i=1}^{\infty}$. Also there exists nonnegative continuous functions $\hat{b}(t, s)$, $\hat{d}_i(t, s)$ on $I \times I (i \in \mathbb{N})$ such that

$$|K_i(t, s, y(s))| \leq \hat{d}_i(t, s) + \hat{b}(t, s) |y_i(s)|,$$

for all $y(t) \in C(I, \ell_1)$, $i \in \mathbb{N}$. Also, $\sum_{i=1}^{\infty} d_i(t, s)$ converges uniformly on $I \times I$.

(iii) Define an operator F on $I \times C(I, \ell_1)$ to $C(I, \ell_1)$ as follows:

$$(t, y(t)) \to (Fy)(t) = ((F_i y)(t))_{i=1}^{\infty},$$

where

$$(F_i y)(t) = y_i^0 + \int_0^t u_i(s, y(s))ds + \int_0^t \int_0^s K_i(s, r, y(r))drds.$$

The family of functions $((Fy)(t))_{t\in I}$ is equicontinuous at each point of the space ℓ_1.

(iv) Let $\hat{y}_0 = \sum_{i=1}^{\infty} |y_i^0|$, $\hat{A}(t) = \sum_{i=1}^{\infty} \hat{a}_i(t)$, $\hat{D}(t,s) = \sum_{i=1}^{\infty} d_i(t,s)$, $\hat{A} = \sup_{t\in I} \hat{A}(t)$, $\hat{D} = \sup_{t,s\in I} \hat{D}(t,s)$ and $\hat{B}_1 = \sup_{t,s\in I} \hat{b}(t,s)$. Also, $T\hat{b}_1 + T^2\hat{B}_1 \leq 1$.

Theorem 2.3.1. *Under the hypothesis (i)–(iv), infinite system (2.10) has at least one solution $y(t) = (y_i(t)) \in C(I, \ell_1)$ for all $t \in [0, T]$.*

Proof. Define the operator $F : C(I, \ell_1) \to C(I, \ell_1)$ as

$$(Fy)(t) = \left\{ y_i^0 + \int_0^t u_i(s, y(s))ds + \int_0^t \int_0^s K_i(s, r, y(r))dr \right\}_{i=1}^{\infty}. \qquad (2.13)$$

By assumption (iii), F is well defined and for arbitrary fixed $t \in I$,

$$\begin{aligned}
\| y(t) \|_{\ell_1} &= \| (Fy)(t) \|_{\ell_1} \\
&= \sum_{i=1}^{\infty} |(F_i y)(t)| \\
&= \sum_{i=1}^{\infty} \left| y_i^0 + \int_0^t u_i(s, y(s))ds + \int_0^t \int_0^s K_i(s, r, y(r))drds \right| \\
&\leq \sum_{i=1}^{\infty} \left\{ |y_i^0| + \int_0^t |u_i(s, y(s))|\, ds + \int_0^t \int_0^s |K_i(s, r, y(r))|\, drds \right\} \\
&\leq \sum_{i=1}^{\infty} \left\{ |y_i^0| + \int_0^t \left[\hat{a}_i(s) + \hat{b}_1 |y_i(s)| \right] ds \right. \\
&\quad \left. + \int_0^t \int_0^s \left[\hat{d}_i(s, r) + \hat{b}(s, r) |y_i(r)| \right] drds \right\} \\
&\leq y^0 + \int_0^T \left(\hat{A} + \hat{b}_1 \| y(s) \|_{\ell_1} \right) ds \\
&\quad + \int_0^T \int_0^T \left(\hat{D} + \hat{B}_1 \| y(r) \|_{\ell_1} \right) drds \\
&= y^0 + \left(\hat{A} + \hat{b}_1 \| y \|_{C(I,\ell_1)} \right) T + \left(\hat{D} + \hat{B}_1 \| y(t) \|_{C(I,\ell_1)} \right) T^2.
\end{aligned}$$

Therefore,

$$\| y \|_{C(I,\ell_1)} \leq y^0 + \left(A + b_1 \| y \|_{C(I,\ell_1)}\right) T + \left(D + B_1 \| y \|_{C(I,\ell_1)}\right) T^2$$

which gives

$$\| y \|_{C(I,\ell_1)} \leq \frac{y^0 + T\hat{A} + T^2\hat{D}}{1 - T\hat{b}_1 - T\hat{B}_1^2} = \hat{r}_0(say).$$

Consider $\hat{B} = \left\{ y \in C(I,\ell_1) : \| y \|_{C(I,\ell_1)} \leq r_0 \right\}$ which is an non-empty, bounded, closed and convex subset of $C(I,\ell_1)$. The operator F is bounded on $\hat{B}$. For arbitrary fixed $x \in \hat{B}$,

$$\begin{aligned}
\| (Fy)(t) - (Fx)(t) \|_{\ell_1} &\leq \sum_{i=1}^{\infty} \left\{ \int_0^t |u_i(s, y(s)) - u_i(s, x(s))| \, ds \right. \\
&\quad \left. + \int_0^t \int_0^s |K_i(s, r, y(s)) - K_i(s, r, x(s))| \right\} \\
&\leq \int_0^T \sum_{i=1}^{\infty} |u_i(s, y(s)) - u_i(s, x(s))| \, ds \\
&\quad + \int_0^T \int_0^T \sum_{i=1}^{\infty} |K_i(s, r, y(s)) - K_i(s, r, x(s))| \, drds.
\end{aligned}$$

Using the assumptions, for any $x \in \hat{B}$ and for $\epsilon > 0$ there exists $\delta > 0$ such that for $\| y - x \|_{C(I,\ell_1)} < \epsilon$ gives for arbitray fixed t,

$$\| u(y(t)) - u(x(t)) \|_{\ell_1} < \delta = \frac{\epsilon}{2T}$$

and

$$\| K(y(t)) - K(x(t)) \|_{\ell_1} < \delta = \frac{\epsilon}{2T^2},$$

for each $t \in I$ and $y \in \hat{B}$. Therefore, we have

$$\| (Fy)(t) - (Fx(t)) \|_{\ell_1} < T.\frac{\epsilon}{2T} + T^2.\frac{\epsilon}{2T^2} = \epsilon.$$

So,

$$\| Fy - Fx \|_{C(I,\ell_1)} < \epsilon.$$

Thus F is continuous on $\hat{B}$.

Now, we establish the continuity of Fy in I. Let $t_0 \in I$ and $\epsilon > 0$ be arbitrary then for $|t - t_0| < \epsilon$, we have

$$\begin{aligned}
\| (Fy)(t) - (Fx)(t_0) \|_{\ell_1} &= \sum_{i=1}^{\infty} \left| \int_{t_0}^{t} u_i(s, y(s))ds + \int_0^t \int_0^s K_i(s, r, y(r))drds \right. \\
&\quad \left. - \int_0^{t_0} \int_0^s K_i(s, r, y(r))drds \right| \\
&\leq \sum_{i=1}^{\infty} \left\{ \int_{t_0}^{t} |u_i(s, y(s))| \, ds \right. \\
&\quad \left. + \int_{t_0}^{t} \int_0^s |K_i(s, r, y(r))| \, drds \right\}.
\end{aligned}$$

As $\epsilon \to 0$, $\| (Fy)(t) - (Fy(t_0)) \|_{\ell_1} \to 0$. Therefore, Fy is continuous with respect to t. Again,

$$\begin{aligned}
\chi_{\ell_1}(F\hat{B}) &= \lim_{n\to\infty} \sup_{y\in\hat{B}} \sum_{i\geq n} |(Fy)_i(t)| \\
&\leq \lim_{n\to\infty} \sup_{y\in\hat{B}} \sum_{i\geq n} \left\{ |y_i^0| + \int_0^t \left[\hat{a}_i(s) + \hat{b}_1 \, |y_i(s)| \right] ds \right. \\
&\quad \left. + \int_0^t \int_0^s \left[\hat{d}_i(s, r) + \hat{b}(s, r) \, |y_i(r)| \right] drds \right\} \\
&\leq \lim_{n\to\infty} \sup_{y\in B} \sum_{i\geq n} \left\{ \hat{b}_1 \int_0^T |y_i(s)| \, ds + \hat{B}_1 \int_0^T \int_0^T |y_i(s)| \, drds \right\} \\
&= (\hat{b}_1 T + \hat{B}_1 T^2) \chi_{C(I,\ell_1)}(B).
\end{aligned}$$

Therefore,

$$\sup_{t\in I} \chi_{\ell_1}(FB) \leq (\hat{b}_1 T + \hat{B}_1 T^2) \chi_{C(I,\ell_1)}(B)$$

gives

$$\chi_{C(I,\ell_1)}(FB) \leq (\hat{b}_1 T + \hat{B}_1 T^2) \chi_{C(I,\ell_1)}(B).$$

As $\hat{b}_1 T + \hat{B}_1 T^2 < 1$, therefore by Darbo's fixed point theorem, F has at least one fixed point in $B \subset C(I, \ell_1)$. □

2.4 Illustrative Example

In this section, we illustrate our results with the help of examples.

Example 2.4.1. *Let us consider the following system of integro-differential equations*

$$\frac{d}{dt}y_i(t) = \frac{t}{i^2} + \sum_{n\geq i} \frac{y_i(t)}{2n^2 i^2} + \int_0^t \frac{y_i(s)}{20i^2} ds, \qquad (2.14)$$

with $y_i(0) = 0,\ t \in I = [0,1]$. *Here*

$$u_i(t, y(t)) = \frac{t}{i^2} + \sum_{n\geq i} \frac{y_i(t)}{2n^2 i^2} \quad \text{and} \quad K_i(t,s,y(s)) = \frac{y_i(s)}{20i^2}.$$

Now,

$$\begin{aligned} |u_i(t, y(t))| &\leq \frac{t}{i^2} + |y_i(t)| \sum_{n\geq i} \frac{1}{2n^2 i^2} \\ &\leq \frac{t}{i^2} + \frac{\pi^2}{12} |y_i(t)| \end{aligned}$$

Therefore, $a_i(t) = \frac{t}{i^2}$ *is a continuous nonnegative function on* I $(i \in \mathbb{N})$ *and* $(a_i(t))$ *is uniformly convergent to zero on* I. *Also, as* $i \to \infty$, *we have* $u_i \to 0$.

Let $\epsilon > 0$ *and arbitrary* $z(t) = (z_i(t))_{i=1}^{\infty} \in C(I, c_0)$ *for arbitrary fixed* $t \in I$ *with*

$$\| y - z \|_{C(I,c_0)} < \delta = \frac{12\epsilon}{\pi^2},$$

we get

$$\begin{aligned} |u_i(t, y(t)) - u_i(t, z(t))| &= \left| \sum_{n\geq i} \frac{y_i(t)}{2n^2 i^2} - \sum_{n\geq i} \frac{z_i(t)}{2n^2 i^2} \right| \\ &\leq \sum_{n\geq i} \frac{1}{2n^2 i^2} |y_i(t) - z_i(t)| \\ &< \delta \sum_{n\geq i} \frac{1}{2n^2 i^2} \\ &< \frac{\delta \pi^2}{12} = \epsilon. \end{aligned}$$

Therefore, $\| u(t, y(t)) - u(t, z(t)) \|_{c_0} < \epsilon$ *for* $\| y(t) - z(t) \|_{c_0} < \delta$ *for all* $t \in I$. *Hence,* $u : I \times C(I, c_0) \to C(I, c_0)$ *is continuous.*

Also,

$$|K_i(t,s,y(s))| = \frac{|y_i(s)|}{20i^2} \leq \frac{|y_i(s)|}{20},$$

where

$$d_i(t,s) = 0 \quad \text{and} \quad b(t,s) = \frac{1}{20}.$$

It is obvious that $K : I \times I \times C(I, c_0) \to C(I, c_0)$ *is continuous.*

Here,

$$y^0 = 0,\ A = 1,\ b_1 = \frac{\pi^2}{12},\ D = 0 \ \ \textit{and} \ \ B_1 = \frac{1}{20}.$$

Thus $Tb_1 + B_1T^2 = \frac{\pi^2}{12} + \frac{1}{20} < 1$. *Hence, by Theorem 11.2.4, the system (2.14) has a solution in* $C(I, c_0)$.

2.5 Conclusion

The main contribution of this article is that applying measure of noncompactness, solvability of infinite system of integro-differential equations has been established with integral initial conditions in Banach spaces will be proved and also illustrated the findings with the help of examples. This approach can be applied to infinite system of fractional integro-differential equations for future work.

Bibliography

[1] J. Banaś, K. Goebel, Measure of noncompactness in Banach spaces, Lecture Notes in Pure and Applied Mathematics, Vol. 60, Marcel Dekker, New York (1980).

[2] J. Banaś, M. Mursaleen, Sequence Spaces and Measures of Noncompactness with Applications to Differential and Integral Equations, New Delhi, Springer (2014).

[3] R. Bellman, Methods of Nonlinear Analysis II, Academic Press, New York (1973).

[4] K. Deimling, Ordinary differential equations in Banach spaces, Lecture Notes in Mathematics, Vol. 596, Springer, Berlin (1977).

[5] L. S. Goldenštein, I. T. Gohberg, A. S. Markus, Investigations of some properties of bounded linear operators with their q-norms, Učen. Zap. Kishinevsk. Univ. 29 (1957) 29–36.

[6] M. Mursaleen, Application of measure of noncompactness to infinite system of differential equations, Canad. Math. Bull. 56 (2013) 388–394.

[7] M. N. O. Poreli, On the neural equations of Cowan and Stein, Utilitas Math. 2 (1972) 305–315.

[8] K. Kuratowski, Sur les espaces complets, Fund. Math. 15 (1930) 301–309.

[9] M. Mursaleen, S. A. Mohiuddine, Applications of measures of noncompactness to the infinite system of differential equations in ℓ_p spaces, Nonlinear Anal. 75 (2012) 2111–2115.

[10] A. Alotaibi, M. Mursaleen, S. A. Mohiuddine, Application of measure of noncompactness to infinite system of linear equations in sequence spaces, Bull. Iranian Math. Soc. 41 (2015) 519–527.

[11] M. Mursaleen and Abdullah Alotaibi, Infinite system of differential equations in some BK-spaces, Abst. Appl. Anal. Vol. 2012, Article ID 863483, 20 pages.

[12] J. Banaś, M. Lecko, Solvability of infinite systems of differential equations in Banach sequence spaces, J. Comput. Appl. Math. 137 (2001) 363–375.

[13] R. R. Akhmerov, M. I. Kamenskii, A. S. Potapov, A. E. Rodkina, B. N. Sadovskii, Measure of noncompactness and condensing operators, Operator Theory: Advances and Applications, vol. 55, Birkhäuser Verlag, Basel, 1992. Translated from the 1986 Russian original by A. Iacob.

[14] G. Darbo, Punti uniti in trasformazioni a codominio non compatto (Italian), Rend. Sem. Mat. Univ. Padova 24 (1955) 84–92.

[15] A. Meir, E. Keeler, A theorem on contraction mappings, J. Math. Anal. Appl. 28 (1969) 326–329.

[16] A. Aghajani, M. Mursaleen, A. S. Haghighi, Fixed point theorems for Meir-Keeler condensing operators via measure of noncompactness, Acta. Math. Sci. 35(3) (2015) 552–566.

[17] M. Mursaleen, S. M. H. Rizvi, Solvability of infinite systems of second order differential equations in c_0 and ℓ_1 by Meir-Keeler condensing operators, Proc. Amer. Math. Soc. 144(10) (2016) 4279–4289.

[18] A. Aghajani, E. Pourhadi, Application of measure of noncompactness to ℓ_1-solvability of infinite systems of second order differential equations, Bull. Belg. Math. Soc. Simon Stevin 22 (2015) 105–118.

Chapter 3

λ-Statistical Convergence of Interval Numbers of Order α

Ayhan Esi
Malatya Turgut Ozal University, aesi23@hotmail.com

Ayten Esi
Adiyaman University, aytenesi@yahoo.com

3.1 Introduction

The idea of statistical convergence for ordinary sequences was introduced by Fast [2] in 1951. Schoenberg [3] studied statistical convergence as a summability method and listed some of elementary properties of statistical convergence. Both of these authors noted that if bounded sequence is statistically convergent, then it is Cesàro summable. Existing work on statistical convergence appears to have been restricted to real or complex sequence, but several authors extended the idea to apply to sequences of fuzzy numbers and also introduced and discussed the concept of statistically convergent sequences of fuzzy numbers.

Interval arithmetic was first suggested by Dwyer [7] in 1951. Development of interval arithmetic as a formal system and evidence of its value as a computational device was provided by Moore [10] in 1959, and Moore and Yang [11] 1962. Furthermore, Moore the authors of [7–9,12] have developed applications to differential equations.

Chiao in [4] introduced sequence of interval numbers and defined usual convergence of sequences of interval number. Şengönül and Eryilmaz in [6] introduced and studied bounded and convergent sequence spaces of interval numbers and showed that these spaces are complete metric space. Recently, Esi in [6] and [7] defined and studied λ-statistical and lacunary statistical convergence of interval numbers, respectively and different types of notions of interval sequence spaces were introduced and investigated by Debnath et al. [18–22]

DOI: 10.1201/9781003178200-3

and Esi et al. [23–25] and many others. A set consisting of a closed interval of real numbers x such that $a \leq x \leq b$ is called an interval number. A real interval can also be considered as a set. Thus we can investigate some properties of interval numbers, for instance arithmetic properties or analytical properties. We denote the set of all real valued closed intervals by IR. Any elements of IR is a closed interval and denoted by $\overline{A}$. That is, $\overline{A} = \{x \in \mathbb{R} : a \leq x \leq b\}$. An interval number $\overline{A}$ is a closed subset of real numbers [4]. Let x_l and x_r be first and last points of $\overline{A}$ interval number, respectively. For $\overline{A}, \overline{B} \in$ IR, we have $\overline{A} = \overline{B} \Leftrightarrow x_{1_l}{=}x_{2_l}$, $x_{1_r}{=}x_{2_r}$. $\overline{A} + \overline{B} = \{x \in \mathbb{R} : x_{1_l} + x_{2_l} \leq x \leq x_{1_r} + x_{2_r}\}$, and if $\alpha \geq 0$, then $\alpha\overline{A} = \{x \in \mathbb{R} : \alpha x_{1_l} \leq x \leq \alpha x_{1_r}\}$ and if $\alpha < 0$, then $\alpha\overline{A} = \{x \in \mathbb{R} : \alpha x_{1_r} \leq x \leq \alpha x_{1_l}\}$,

$$\overline{A}.\overline{B} = \{x \in \mathbb{R} : \min\{x_{1_l}.x_{2_l}, x_{1_l}.x_{2_r}, x_{1_r}.x_{2_l}, x_{1_r}.x_{2_r}\} \leq x \leq \max\{x_{1_l}.x_{2_l}, x_{1_l}.x_{2_r}, x_{1_r}.x_{2_l}, x_{1_r}.x_{2_r}\}\}.$$

The set of all interval numbers IR is a complete metric space defined by

$$d\left(\overline{A}, \overline{B}\right) = \max\{|x_{1_l} - x_{2_l}|, |x_{1_r} - x_{2_r}|\} \text{ [10].}$$

In the special case $\overline{A} = [a, a]$ and $\overline{B} = [b, b]$, we obtain usual metric of R.

Let us define transformation $f : \mathbb{N} \to$IR by $k \to f(k) = \overline{A}$, $\overline{A} = \left(\overline{A}_k\right)$. Then $\overline{A} = \left(\overline{A}_k\right)$ is called sequence of interval numbers. The $\overline{A}_k$ is called k^{th} term of sequence $\overline{A} = \left(\overline{A}_k\right)$. w^i denotes the set of all interval numbers with real terms and the algebraic properties of w^i can be found in [13].

Now, we give the definition of convergence of interval numbers:

Definition 3.1.1. *([4]) A sequence $\overline{A} = \left(\overline{A}_k\right)$ of interval numbers is said to be convergent to the interval number $\overline{A}_o$ if for each $\varepsilon > 0$ there exists a positive integer k_o such that $d\left(\overline{A}_k, \overline{A}_o\right) < \varepsilon$ for all $k \geq k_o$ and we denote it by* $\lim_k \overline{A}_k = \overline{A}_o$.

Thus, $\lim_k \overline{A}_k = \overline{A}_o \Leftrightarrow \lim_k x_{k_l} = x_{o_l}$ and $\lim_k x_{k_r} = x_{o_r}$.

3.2 Main Results

Definition 3.2.1. *Let $\lambda = (\lambda_n)$ be a non-decreasing sequence of positive numbers such that $\lambda_{n+1} \leq \lambda_n + 1$, $\lambda_1 = 1$, $\lambda_n \to \infty$ as $n \to \infty$ and $I_n = [n - \lambda_n + 1, n]$. Let α be any real number such that $0 < \alpha \leq 1$. The sequence $\overline{A} = \left(\overline{A}_k\right)$ of interval numbers is said to be strongly λ-summable of order α if there is an interval number $\overline{A}_o$ such that*

$$\lim_{n\to\infty} \frac{1}{\lambda_n^\alpha} \sum_{k\in I_n} d\left(\overline{A}_k, \overline{A}_o\right) = 0.$$

In this case, we say that the sequence $\overline{A} = (\overline{A}_k)$ of interval numbers is strongly λ-summable to interval numbers $\overline{A}_o$ of order α. The set of all strongly λ-summable to interval sequences of order α will be denoted by $[\overline{w}^\alpha]_\lambda$.

Remark 3.2.2. *If $\lambda_n = n$, then strongly λ-summable of order α reduces to strongly Cesàro summable of order α defined as follows:*

$$\lim_{n\to\infty} \frac{1}{n^\alpha} \sum_{k\in I_n} d\left(\overline{A}_k, \overline{A}_o\right) = 0.$$

The set of all strongly Cesàro summable to interval sequences of order α will be denoted by $[\overline{w}^\alpha]$. If $\lambda_n = n$ and $\alpha = 1$, then strongly λ-summable reduces to strongly Cesàro summable of interval numbers which was defined by Esi [5].

Definition 3.2.3. *Let the sequence $\lambda = (\lambda_n)$ of real numbers be defined as in Definition 3.2.1 and $\alpha \in (0,1]$ be given. A sequence $\overline{A} = (\overline{A}_k)$ of interval numbers is said to be statistically λ-convergent to interval number $\overline{A}_o$ of order α if for every $\varepsilon > 0$*

$$\lim_{n\to\infty} \frac{1}{\lambda_n^\alpha} \left|\left\{k \in I_n : d\left(\overline{A}_k, \overline{A}_o\right) \geq \varepsilon \right\}\right| = 0.$$

In this case, we write $\overline{s}_\lambda^\alpha - \lim \overline{A}_k = \overline{A}_o$. The set of all statistically λ-convergent interval sequences of order α will be denoted by $\overline{s}_\lambda^\alpha$. We will write $\overline{s}_{\lambda,\overline{0}}^\alpha$ to denote the set of statistically λ-null sequences of interval numbers of order α.

Remark 3.2.4. *If $\lambda_n = n$, then statistically λ-convergence of order α reduces to statistically convergence of order α as follows:*

$$\lim_{n\to\infty} \frac{1}{n^\alpha} \left|\left\{k \in I_n : d\left(\overline{A}_k, \overline{A}_o\right) \geq \varepsilon \right\}\right| = 0.$$

In this case, we write $\overline{s}^\alpha - \lim \overline{A}_k = \overline{A}_o$. The set of all statistically convergent interval sequences of order α will be denoted by $\overline{s}^\alpha$. If $\alpha = 1$, then statistically λ-convergence of order α reduces to statistically λ-convergence which was defined by Esi [5].

It is evident that $\overline{s}_{\lambda,\overline{0}}^\alpha \subset \overline{s}_\lambda^\alpha$ for each $\alpha \in (0,1]$. The statistical λ-convergence of interval numbers of order $\alpha = 1$ is same with the statistical λ-convergence, that is $\overline{s}_{\lambda,\overline{0}}^\alpha = \overline{s}_\lambda^\alpha$ for $\alpha = 1$. The statistical λ-convergence of interval numbers of order α is well-defined for $0 < \alpha \leq 1$. But it is not well-defined for $\alpha > 1$ in general.

Example 3.2.5. *Let $\overline{A} = (\overline{A}_k)$ interval sequence be defined as follows*

$$\overline{A}_k = \begin{cases} \overline{1}, & k = 2n,\ n = 1, 2, \ldots \\ \overline{0}, & \text{otherwise} \end{cases}.$$

Then both

$$\lim_{n\to\infty}\frac{1}{\lambda_n^{\alpha}}\left|\left\{k\in I_n:\ d\left(\overline{A}_k,\overline{1}\right)\geq\varepsilon\right\}\right|\leq\lim_{n\to\infty}\frac{[\lambda_n]+1}{2\lambda_n^{\alpha}}=0$$

and

$$\lim_{n\to\infty}\frac{1}{\lambda_n^{\alpha}}\left|\left\{k\in I_n:\ d\left(\overline{A}_k,\overline{0}\right)\geq\varepsilon\right\}\right|\leq\lim_{n\to\infty}\frac{[\lambda_n]+1}{2\lambda_n^{\alpha}}=0$$

for $\alpha>1$, such that $\overline{A}=\left(\overline{A}_k\right)$ statistically λ-convergent of order α both to interval number $\overline{1}$ and $\overline{0}$, but this is impossible.

Theorem 3.2.6. *Let $0<\alpha\leq 1$ and $\overline{A}=\left(\overline{A}_k\right)$ and $\overline{B}=\left(\overline{B}_k\right)$ be sequences of interval numbers.*

(i) If $\overline{s}_{\lambda}^{\alpha}-\lim\overline{A}_k=\overline{A}_o$ and $b\in\mathbb{R}$, then $\overline{s}_{\lambda}^{\alpha}-\lim b\overline{A}_k=b\overline{A}_o$.

(ii) If $\overline{s}_{\lambda}^{\alpha}-\lim\overline{A}_k=A_o$ and $\overline{s}_{\lambda}^{\alpha}-\lim\overline{B}_k=\overline{B}_o$, then $\overline{s}_{\lambda}^{\alpha}-\lim\left(\overline{A}_k+\overline{B}_k\right)=\overline{A}_o+\overline{B}_o$.

Proof. (i) Let $b\in\mathbb{R}$. We have $d\left(b\overline{A}_k,b\overline{A}_o\right)=|b|\,d\left(\overline{A}_k,\overline{A}_o\right)$. For a given $\varepsilon>0$

$$\frac{1}{\lambda_n^{\alpha}}\left|\left\{k\in I_n:\ d\left(b\overline{A}_k,b\overline{A}_o\right)\geq\varepsilon\right\}\right|\leq\frac{1}{\lambda_n^{\alpha}}\left|\left\{k\in I_n:\ d\left(\overline{A}_k,\overline{A}_o\right)\geq\frac{\varepsilon}{|b|}\right\}\right|.$$

Hence $\overline{s}_{\lambda}^{\alpha}-\lim b\overline{A}_k=b\overline{A}_o$.

(ii) Suppose that $\overline{s}_{\lambda}^{\alpha}-\lim\overline{A}_k=\overline{A}_o$ and $\overline{s}_{\lambda}^{\alpha}-\lim\overline{B}_k=\overline{B}_o$. We have

$$d\left(\overline{A}_k+\overline{B}_k,\overline{A}_o+\overline{B}_o\right)\leq d\left(\overline{A}_k,\overline{A}_o\right)+d\left(\overline{B}_k,\overline{B}_o\right).$$

Therefore given $\varepsilon>0$, we have

$$\begin{aligned}&\frac{1}{\lambda_n^{\alpha}}\left|\left\{k\in I_n:d\left(\overline{A}_k+\overline{B}_k,\overline{A}_o+\overline{B}_o\right)\geq\varepsilon\right\}\right|\\ &\leq\frac{1}{\lambda_n^{\alpha}}\left|\left\{k\in I_n:d\left(\overline{A}_k,\overline{A}_o\right)+d\left(\overline{B}_k,\overline{B}_o\right)\geq\varepsilon\right\}\right|\\ &\leq\frac{1}{\lambda_n^{\alpha}}\left|\left\{k\in I_n:\ d\left(\overline{A}_k,\overline{A}_o\right)\geq\frac{\varepsilon}{2}\right\}\right|+\frac{1}{\lambda_n^{\alpha}}\left|\left\{k\in I_n:\ d\left(\overline{B}_k,\overline{B}_o\right)\geq\frac{\varepsilon}{2}\right\}\right|.\end{aligned}$$

Thus, $\overline{s}_{\lambda}^{\alpha}-\lim\left(\overline{A}_k+\overline{B}_k\right)=\overline{A}_o+\overline{B}_o$. □

Note that $\overline{s}_{\lambda}^{\alpha}$ is different from $\overline{s}_{\lambda}$ defined in Esi [5] in general. If we take $\lambda_n=n^{\alpha}$ for $0<\alpha<1$ then $\overline{s}^{\alpha}\subset\overline{s}_{\lambda}$. If $\lambda_n=n$ and $\alpha=1$ then $\overline{s}^{\alpha}=\overline{s}_{\lambda}=\overline{s}$ (see Esi [5]), that is the statistical convergence of interval numbers of order α, statistical convergence and statistical λ-convergence of interval numbers coincide when $\lambda_n=n$ and $\alpha=1$.

It is easy to see that every convergent sequence of interval numbers is statistically convergent of interval numbers of order α, that is $c^{i}\subset\overline{s}^{\alpha}$ for each $0<\alpha\leq 1$, where the space c^{i} was defined in Şengönül and Eryılmaz [6]. But it follows from the following example that the converse does not hold.

Example 3.2.7. *Let the interval number sequence $\overline{A} = \left(\overline{A}_k\right)$ defined by*

$$\overline{A}_k = \begin{cases} \overline{1}, & k = n^3,\, n = 1, 2, \dots \\ \overline{0}, & \text{otherwise} \end{cases}.$$

Then $\overline{A} = \left(\overline{A}_k\right)$ is statistically convergent of order $\alpha > \frac{1}{3}$, but it is not convergent.

Theorem 3.2.8. $0 < \alpha \leq \beta \leq 1$. *Then $\overline{s}_\lambda^\alpha \subset \overline{s}_\lambda^\beta$ and the inclusion is strict for some α and β such that $\alpha \leq \beta$.*

Proof. If $0 < \alpha \leq \beta \leq 1$, then

$$\frac{1}{\lambda_n^\beta}\left|\left\{k \in I_n : d\left(\overline{A}_k, \overline{A}_o\right) \geq \varepsilon\right\}\right| \leq \frac{1}{\lambda_n^\alpha}\left|\left\{k \in I_n : d\left(\overline{A}_k, \overline{A}_o\right) \geq \varepsilon\right\}\right|$$

for every $\varepsilon > 0$, which gives that $\overline{s}_\lambda^\alpha \subset \overline{s}_\lambda^\beta$. It follows from the following example that the inclusion is strict. □

Example 3.2.9. *Consider the interval sequence $\overline{A} = \left(\overline{A}_k\right)$ defined by*

$$\overline{A}_k = \begin{cases} \overline{k}, & k = m^2,\, m = 1, 2, \dots \\ \overline{0}, & \text{otherwise} \end{cases}.$$

Let $\lambda_n = n$ and then for every ε $(0 < \varepsilon \leq 1)$

$$\frac{1}{n^\beta}\left|\left\{k \in I_n : d\left(\overline{A}_k, \overline{0}\right) \geq \varepsilon\right\}\right| \leq \frac{[\sqrt{n}]}{n^\beta} \to 0 \text{ as } n \to \infty \text{ for } \frac{1}{2} < \beta \leq 1$$

i.e., $\overline{A} = \left(\overline{A}_k\right) \in \overline{s}_\lambda^\beta$ but $\overline{A} = \left(\overline{A}_k\right) \notin \overline{s}_\lambda^\alpha$ for $0 < \alpha \leq \frac{1}{2}$.

If we take $\beta = 1$ in Theorem 3.2.8 then we obtain the following result.

Corollary 3.2.10. *If an interval sequence $\overline{A} = \left(\overline{A}_k\right)$ is statistically λ-convergent to interval number $\overline{A}_o$ of order α, then it is statistically λ-convergent to $\overline{A}_o$, that is $\overline{s}_\lambda^\alpha \subset \overline{s}_\lambda$ for each α $(0 < \alpha \leq 1)$ and the inclusion is strict.*

The following result is a consequence of Theorem 3.2.8.

Corollary 3.2.11. *(i) $\overline{s}_\lambda^\beta = \overline{s}_\lambda^\alpha$ if and only if $\alpha = \beta$,*
(ii) $\overline{s}_\lambda^\beta = \overline{s}_\lambda \Leftrightarrow \alpha = 1$.

Theorem 3.2.12. *$\overline{s}_\lambda^\alpha \subset \overline{s}$ for all λ for each $\alpha \in (0, 1]$.*

Proof. It is easy to see that $\overline{s}_\lambda \subset \overline{s}$ for all λ, since $\frac{\lambda_n}{n}$ is bounded by 1 (Mursaleen [15]). From Corollary 3.2.10, we have $\overline{s}_\lambda \subset \overline{s}$. □

Theorem 3.2.13. *$\overline{s} \subset \overline{s}_\lambda^\alpha$ if and only if*

$$\lim_{n\to\infty}\inf \frac{\lambda_n^\alpha}{n} > 0. \tag{3.1}$$

Proof. For a given $\varepsilon > 0$, we have

$$\left\{k \le n : d\left(\overline{A}_k, \overline{A}_o\right) \ge \varepsilon\right\} \supset \left\{k \in I_n : d\left(\overline{A}_k, \overline{A}_o\right) \ge \varepsilon\right\}.$$

Therefore

$$\begin{aligned}\frac{1}{n}\left|\left\{k \le n : d\left(\overline{A}_k, \overline{A}_o\right) \ge \varepsilon\right\}\right| &\ge \frac{1}{n}\left|\left\{k \in I_n : d\left(\overline{A}_k, \overline{A}_o\right) \ge \varepsilon\right\}\right| \\ &= \frac{\lambda_n^\alpha}{n}\frac{1}{\lambda_n^\alpha}\left|\left\{k \in I_n : d\left(\overline{A}_k, \overline{A}_o\right) \ge \varepsilon\right\}\right|.\end{aligned}$$

Taking limit as $n \to \infty$ and using (3.1), we get $\overline{s}-\lim \overline{A}_k = \overline{A}_o \Rightarrow \overline{s}_\lambda^\alpha-\lim \overline{A}_k = \overline{A}_o$.

Conversely, suppose that $\liminf_{n\to\infty} \frac{\lambda_n^\alpha}{n} > 0$. We can choose a subsequence $(n(j))_{j=1}^\infty$ such that

$$\frac{\lambda_{n(j)}^\alpha}{n(j)} < \frac{1}{j}.$$

Define an interval number sequence $\overline{A} = \left(\overline{A}_k\right)$ by

$$\overline{A}_i = \begin{cases} \overline{1}, & i \in I_{n(j)};\ j = 1, 2, 3, \ldots \\ \overline{0}, & \text{otherwise} \end{cases}.$$

Then $\overline{A} = \left(\overline{A}_k\right) \in \overline{s}$ but on the other hand $\overline{A} = \left(\overline{A}_k\right) \notin \overline{s}_\lambda$. From Corollary 3.2.10, since $\overline{s}_\lambda^\alpha \subset \overline{s}_\lambda$, we have $\overline{A} = \left(\overline{A}_k\right) \notin \overline{s}_\lambda^\alpha$. Hence (3.1) is necessary. □

Theorem 3.2.14. *Let $0 < \alpha \le \beta \le 1$. Then $[\overline{w}^\alpha]_\lambda \subset \left[\overline{w}^\beta\right]_\lambda$ and the inclusion is strict for some α and β such that $\alpha < \beta$.*

Proof. Let $\overline{A} = \left(\overline{A}_k\right) \in [\overline{w}^\alpha]_\lambda$. Then given α and β such that $0 < \alpha \le \beta \le 1$, we may write

$$\frac{1}{\lambda_n^\beta}\sum_{k\in I_n} d\left(\overline{A}_k, \overline{A}_o\right) \le \frac{1}{\lambda_n^\alpha}\sum_{k\in I_n} d\left(\overline{A}_k, \overline{A}_o\right)$$

which gives $[\overline{w}^\alpha]_\lambda \subset \left[\overline{w}^\beta\right]_\lambda$. To show that the inclusion is strict consider the following example. □

Example 3.2.15. *Consider the interval sequence $\overline{A} = \left(\overline{A}_k\right)$ defined by*

$$\overline{A}_k = \begin{cases} \overline{1}, & k = m^2, m = 1, 2, \ldots \\ \overline{0}, & \textit{otherwise} \end{cases}.$$

Let $\lambda_n = n$ for all $n \in \mathbb{N}$. Then it is easy to see that

$$\frac{1}{n^\beta}\sum_{k\in I_n} d\left(\overline{A}_k, \overline{0}\right) \le \frac{\sqrt{n}}{n^\beta} = \frac{1}{n^{\beta-\frac{1}{2}}}.$$

Since $\frac{1}{n^{\beta-\frac{1}{2}}} \to 0$ *as* $n \to \infty$, *then* $[\overline{w}^{\beta}] - \lim \overline{A}_k = \overline{A}_o = \overline{0}$, *i.e.*, $\overline{A} = (\overline{A}_k) \in [\overline{w}^{\beta}]$ *for* $\frac{1}{2} < \beta \leq 1$. *But since*

$$\frac{\sqrt{n}-1}{n^{\alpha}} \leq \frac{1}{n^{\alpha}} \sum_{k \in I_n} d\left(\overline{A}_k, \overline{0}\right)$$

and $\frac{\sqrt{n}-1}{n^{\alpha}} \to \infty$ *as* $n \to \infty$, *then* $\overline{A} = (\overline{A}_k) \notin [\overline{w}^{\alpha}]$ *for* $0 < \alpha < \frac{1}{2}$. *This completes the proof.*

The following result is a consequence of Theorem 3.2.14.

Corollary 3.2.16. *Let* $0 < \alpha \leq \beta \leq 1$. *Then*

(i) $[\overline{w}^{\alpha}]_{\lambda} = [\overline{w}^{\beta}]_{\lambda}$ *if and only if* $\alpha = \beta$,

(ii) $[\overline{w}^{\alpha}]_{\lambda} \subset [\overline{w}]_{\lambda}$ *for each* $\alpha \in (0,1]$.

Theorem 3.2.17. *Let* $0 < \alpha \leq \beta \leq 1$. *Then* $[\overline{w}^{\alpha}]_{\lambda} \subset \overline{s}_{\lambda}^{\beta}$.

Proof. Let $\overline{A} = (\overline{A}_k) \in [\overline{w}^{\alpha}]_{\lambda}$ and $\varepsilon > 0$, we have

$$\begin{aligned}
\frac{1}{\lambda_n^{\alpha}} \sum_{k \in I_n} d\left(\overline{A}_k, \overline{A}_o\right) &= \frac{1}{\lambda_n^{\alpha}} \sum_{\substack{k \in I_n \\ d(\overline{A}_k, \overline{A}_o) \geq \varepsilon}} d\left(\overline{A}_k, \overline{A}_o\right) + \frac{1}{\lambda_n^{\alpha}} \sum_{\substack{k \in I_n \\ d(\overline{A}_k, \overline{A}_o) < \varepsilon}} d\left(\overline{A}_k, \overline{A}_o\right) \\
&\geq \frac{1}{\lambda_n^{\alpha}} \sum_{\substack{k \in I_n \\ d(\overline{A}_k, \overline{A}_o) \geq \varepsilon}} d\left(\overline{A}_k, \overline{A}_o\right) \\
&= \frac{1}{\lambda_n^{\alpha}} \left|\left\{k \in I_n : \ d\left(\overline{A}_k, \overline{A}_o\right) \geq \varepsilon\right\}\right| \varepsilon \\
&\geq \frac{1}{\lambda_n^{\beta}} \left|\left\{k \in I_n : \ d\left(\overline{A}_k, \overline{A}_o\right) \geq \varepsilon\right\}\right| \varepsilon.
\end{aligned}$$

So, it follows that $\overline{A} = (\overline{A}_k) \in \overline{s}_{\lambda}^{\beta}$. This completes the proof. □

If we take $\alpha = \beta$ in Theorem 3.2.17, we obtain the following result.

Corollary 3.2.18. *Let* $\alpha \in (0,1]$. *Then* $[\overline{w}^{\alpha}]_{\lambda} \subset \overline{s}_{\lambda}^{\alpha}$.

Corollary 3.2.19. *Let* $\alpha \in (0,1]$. *Then* $[\overline{w}^{\alpha}]_{\lambda} \subset \overline{s}_{\lambda}$ *and the inclusion is strict for* $\alpha \in (0,1)$.

Proof. It is consequence of Corollary 3.2.10 and Corollary 3.2.18. □

Now, we give some definitions for interval sequences as follows: An interval valued sequence space $\overline{E}$ is said to be solid if $\overline{B} = (\overline{B}_k) \in \overline{E}$ whenever $|\overline{B}_k| \leq |\overline{A}_k|$ for all $k \in \mathbb{N}$ and $\overline{A} = (\overline{A}_k) \in \overline{E}$.

An interval valued sequence space $\overline{E}$ is said to be monotone if $\overline{E}$ contains the canonical pre-image of all its step spaces.

An interval valued sequence space $\overline{E}$ is said to be convergence free if $\overline{B} = (\overline{B}_k) \in \overline{E}$ whenever $\overline{A} = (\overline{A}_k) \in \overline{E}$ and $\overline{A}_k = \overline{0}$ implies $\overline{B} = (\overline{B}_k) = \overline{0}$.

Throughout the study, $p = (p_k)$ is a sequence of bounded strictly positive numbers. We define the following interval valued sequence space:

$$\overline{l}\,(p) = \left\{ \overline{A} = (\overline{A}_k) : \sum_{k=1}^{\infty} \left[d\left(\overline{A}_k, \overline{0}\right) \right]^{p_k} < \infty \right\}$$

and if $p_k = 1$ for all $k \in \mathbb{N}$, then we have

$$\overline{l} = \left\{ \overline{A} = (\overline{A}_k) : \sum_{k=1}^{\infty} d\left(\overline{A}_k, \overline{0}\right) < \infty \right\}.$$

Lemma 3.2.20. *If a sequence space E is solid, then it is monotone.*

Theorem 3.2.21. *The sequence space $\overline{l}\,(p)$ is a complete metric space with respect to the metric defined by*

$$\overline{d}\left(\overline{A}, \overline{B}\right) = \left[\sum_{k=1}^{\infty} \left[d\left(\overline{A}_k, \overline{B}_k\right) \right]^{p_k} \right]^{\frac{1}{M}}$$

where $M = \max(1, \sup_k p_k)$.

Proof. Let $(\overline{A}^i)$ be a Cauchy sequence in $\overline{l}\,(p)$. Then for a given $\varepsilon > 0$, there exists $n_o \in \mathbb{N}$, such that $\overline{d}\left(\overline{A}_k{}^i, \overline{A}_k{}^j\right) < \varepsilon$ for all $i, j \geq n_o$. Then

$$\left[\sum_{k=1}^{\infty} \left[d\left(\overline{A}_k{}^i, \overline{A}_k{}^j\right) \right]^{p_k} \right]^{\frac{1}{M}} < \varepsilon \text{ for all } i, j \geq n_o \tag{3.2}$$

$$\Rightarrow \sum_{k=1}^{\infty} \left[d\left(\overline{A}_k{}^i, \overline{A}_k{}^j\right) \right]^{p_k} < \varepsilon^M \text{ for all } i, j \geq n_o$$

$$\Rightarrow \left[d\left(\overline{A}_k{}^i, \overline{A}_k{}^j\right) \right] < \varepsilon \text{ for all } i, j \geq n_o \text{ and all } k \in \mathbb{N}.$$

This means that $\left(\overline{A}_k{}^i\right)$ is a Cauchy sequence in $\mathbb{R}$. Since $\mathbb{R}$ is a Banach space, $\left(\overline{A}_k^i\right)$ is convergent. Now, let $\lim \overline{A}_k{}^i = \overline{A}_k$ for each $k \in \mathbb{N}$ and $\overline{A} = (\overline{A}_k)$. Taking limit as $j \to \infty$ in (3.2), we have $\overline{d}\left(\overline{A}^i, \overline{A}\right) < \varepsilon$ for all $i \geq n_o$.

Now for all $i \geq n_o$

$$\overline{d}\left(\overline{A}, \overline{0}\right) \leq \overline{d}\left(\overline{A}, \overline{A}^i\right) + \overline{d}\left(\overline{A}^i, \overline{0}\right) < \infty.$$

Thus $\overline{A} = (\overline{A}_k) \in \overline{l}\,(p)$ and so $\overline{l}\,(p)$ is complete. This completes the proof. ◻

Theorem 3.2.22. *The sequence space $l(p)$ is solid as well as monotone.*

Proof. Let $\overline{A} = (\overline{A}_k) \in \overline{l}\,(p)$ and $\overline{B} = (\overline{B}_k)$ be two interval valued sequences such that $|\overline{B}_k| \leq |\overline{A}_k|$ for all $k \in \mathbb{N}$. Then

$$\sum_{k=1}^{\infty} \left[\overline{d}\left(\overline{A}, \overline{0}\right)\right]^{p_k} < \infty$$

and

$$\sum_{k=1}^{\infty} \left[\overline{d}\left(\overline{B}_k, \overline{0}\right)\right]^{p_k} \leq \sum_{k=1}^{\infty} \left[\overline{d}\left(\overline{A}_k, \overline{0}\right)\right]^{p_k} < \infty$$

Thus $\overline{B} = (\overline{B}_k) \in \overline{l}\,(p)$ and so $\overline{l}\,(p)$ is solid. Also by Lemma 3.2.20, it follows that the space $\overline{l}\,(p)$ is monotone. This completes the proof. □

Theorem 3.2.23. *The sequence space $\overline{l}\,(p)$ is not convergence free in general.*

Proof. The result follows from the following example. □

Example 3.2.24. *Let $p_k = 2$ for all $k \in \mathbb{N}$. We consider the interval sequence $\overline{A} = (\overline{A}_k)$ as follows:*
$\overline{A}_k = \left[\frac{-1}{k^2}, 0\right]$, for all $k \in \mathbb{N}$. Then

$$\sum_{k=1}^{\infty} \left[\overline{d}\left(\overline{A}_k, \overline{0}\right)\right]^{p_k} \leq \sum_{k=1}^{\infty} \left(\frac{1}{k^2}\right)^2 < \infty$$

so $\overline{A} = (\overline{A}_k) \in \overline{l}\,(p)$. Now let us consider the interval sequence $\overline{B} = (\overline{B}_k)$ defined as follows:
$\overline{B}_k = [-k^2, k^2]$, for all $k \in \mathbb{N}$. Then

$$\sum_{k=1}^{\infty} \left[\overline{d}\left(\overline{B}_k, \overline{0}\right)\right]^{p_k} \leq \sum_{k=1}^{\infty} \left(\frac{1}{k^2}\right)^2 = \infty$$

so $\overline{B} = (\overline{B}_k) \notin \overline{l}\,(p)$. Hence $\overline{l}\,(p)$ is not convergence free. This completes the proof.

Theorem 3.2.25.

(i) If $0 < q_k \leq p_k < 1$ for all $k \in \mathbb{N}$, then $\overline{l}\,(p) \subset \overline{l}\,(q)$ and the inclusion is proper.

(ii) If $0 < \inf_k p_k \leq p_k$ for all $k \in \mathbb{N}$, then $\overline{l}\,(p) \subset \overline{l}$ and if $1 < p_k \leq \sup_k p_k < \infty$ for all $k \in \mathbb{N}$, then $\overline{l} \subset \overline{l}\,(p)$.

Proof. (i) Let $\overline{A} = (\overline{A}_k) \in \overline{l}\,(p)$. Then

$$\sum_{k=1}^{\infty} \left[\overline{d}\left(\overline{A}_k, \overline{0}\right)\right]^{p_k} < \infty.$$

So there exists $n_0 \in \mathbb{N}$ such that $\left[d\left(\overline{A}_k, \overline{0}\right)\right]^{p_k} < 1$, for all $k \geq n_0$. Thus $\left[d\left(\overline{A}_k, \overline{0}\right)\right]^{q_k} < \left[d\left(\overline{A}_k, \overline{0}\right)\right]^{p_k}$ for all $k \geq n_0$ and so,

$$\sum_{k=1}^{\infty} \left[\overline{d}\left(\overline{A}_k, \overline{0}\right)\right]^{q_k} < \infty.$$

Thus $\overline{A} = \left(\overline{A}_k\right) \in \overline{l}\,(q)$.

(ii) The first part of the result follows from the inequality

$$\sum_{k=1}^{\infty} \overline{d}(\overline{A}_k, \overline{0}) \leq \sum_{k=1}^{\infty} \left[\overline{d}\left(\overline{A}_k, \overline{0}\right)\right]^{p_k}$$

and the second part of the result follows from following inequality

$$\sum_{k=1}^{\infty} \left[d\left(\overline{A}_k, \overline{0}\right)\right]^{p_k} \leq \sum_{k=1}^{\infty} d\left(\overline{A}_k, \overline{0}\right).$$

□

The inclusion of Theorem 3.2.25(i) is strict and it follows from the following example.

Example 3.2.26. *We consider the interval sequence $\overline{A} = \left(\overline{A}_k\right)$ as follows:*

$\overline{A}_k = \left[\frac{-1}{\sqrt{3k}}, 0\right]$, *for all $k \in \mathbb{N}$*

and

$q_k = 2 + \frac{1}{k}$ *and* $p_k = 1$, *for all $k \in \mathbb{N}$.*

Then

$$\sum_{k=1}^{\infty} \left[\overline{d}\left(\overline{A}_k, \overline{0}\right)\right]^{q_k} = \sum_{k=1}^{\infty} \left(\frac{1}{\sqrt{3k}}\right)^{2+\frac{1}{k}} < \infty,$$

and so $\overline{A} = \left(\overline{A}_k\right) \in \overline{l}\,(q)$, but

$$\sum_{k=1}^{\infty} \left[\overline{d}\left(\overline{A}_k, \overline{0}\right)\right]^{p_k} = \sum_{k=1}^{\infty} \frac{1}{\sqrt{3k}} = \infty,$$

so $\overline{A} = \left(\overline{A}_k\right) \notin \overline{l}\,(p)$.

Bibliography

[1] K.-P. Chiao, Fundamental properties of interval vector max-norm, Tamsui Oxf. J. Math. Sci. 18(2) (2002) 219–233.

[2] R. Çolak, C. A. Bektaş, λ-statistical convergence of order α, Acta Math. Sci. Ser. B Engl. Ed. 31(3) (2011) 953–959.

[3] P. S. Dwyer, Linear Computation, New York, Wiley, 1951.

[4] P. S. Dwyer, Errors of matrix computation, simultaneous equations and eigenvalues, National Bureu of Standards Applied Mathematics Series 29 (1953) 49–58.

[5] A. Esi, λ-sequence spaces of interval numbers, Appl. Math. Inf. Sci. 8(3) (2014) 1099–1102.

[6] A. Esi, Lacunary sequence spaces of interval numbers, Thai J. Math. 10(2) (2012) 445–451.

[7] A. Esi, Strongly almost λ-convergence and statistically almost λ-convergence of interval numbers, Sci. Magna 7(2) (2011) 117–122.

[8] H. Fast, Sur la convergence statistique, Colloq. Math 2 (1951) 241–244.

[9] P. S. Fischer, Automatic propagated and round-off error analysis, paper presented at the 13th National Meeting of the Association of Computing Machinery, June 1958.

[10] A. R. Freedman, J. J. Sember and M. Raphael, Some Cesàro type summability, Proc. London Math. Soc. 37(3) (1978) 508–520.

[11] S. Markov, Quasilinear spaces and their relation to vector spaces, Eletron. J. Math. Comput. 2(1) (2005) 1–21.

[12] R. E. Moore, Automatic Error Analysis in Digital Computation, LSMD-48421, Lockheed Missiles and Space Company (1959).

[13] R. E. Moore, C. T. Yang, Theory of an interval algebra and its application to numeric analysis, RAAG Memories II, Gaukutsu Bunken Fukeyu-kai, Tokyo (1958).

[14] R. E. Moore, C. T. Yang, Interval Analysis I, LMSD-285875, Lockheed Missiles and Space Company (1962).

[15] Mursaleen, λ-statistical convergence, Math. Slovaca 50(1) (2000) 111–115.

[16] I. J. Schoenberg, The integrability of certain functions and related summability methods, Amer. Math. Monthly 66 (1959) 361–375.

[17] M. Şengönül, A. Eryılmaz, On the sequence spaces of interval numbers, Thai J. Math. 8(3) (2010) 503–510.

[18] S. Debnath, B. Sarma, S. Saha, On some sequence spaces of interval vectors, Afr. Mat. 26(5) (2015) 673–678.

[19] S. Debnath, A. J. Datta, S. Saha, Regular matrix of interval numbers based on Fibonacci numbers, Afr. Mat. 26(7) (2015) 1379–1385.

[20] S. Debnath, S. Saha, Matrix transformation on statistically convergent sequence spaces of interval number sequences, Proecciones 35(1) (2016) 187–195.

[21] S. Debnath and S. Saha, On statistically convergent sequence spaces of two dimensional interval vectors, J. Indian Acad. Math. 38(1) (2016) 29–37.

[22] S. Debnath, S. Saha, Some statistically convergent difference sequence spaces of interval numbers, Romai J. 12(1) (2016) 53–59.

[23] A. Esi, S. Debnath, S. Saha, Asymptotically double lacunary statistically equivalent sequences of interval numbers, Proyecciones 35(4) (2016) 469–479.

[24] A. Esi, S. Debnath, S. Saha, Some new sequence spaces of interval number based on Zweier sequences and Fibonacci numbers, Divulgaciones Matematicas 19(2) (2018) 36–42.

[25] A. Esi, S. Debnath, S. Saha, Asymptotically double λ_2-statistically equivalent sequences of interval numbers, Mathematica 62(1) (2020) 39–46.

Chapter 4

Necessary and Sufficient Tauberian Conditions under which Convergence follows from $(A^{r,\delta}, p, q; 1, 1)$, $(A^{r,*}, p, *; 1, 0)$ and $(A^{*,\delta}, *, q; 0, 1)$ Summability Methods of Double Sequences

Çağla Kambak

Ege University, caglakambak@gmail.com

Ibrahim Canak

Ege University, ibrahim.canak@ege.edu.tr

4.1 Introduction

Let $p = (p_m)$ and $q = (q_n)$ be a sequence of nonnegative numbers such that $p_0, q_0 > 0$ and

$$P_m := \sum_{j=0}^{m} p_j \to \infty, m \to \infty, \tag{4.1}$$

DOI: 10.1201/9781003178200-4

$$Q_n := \sum_{k=0}^{n} q_k \to \infty, n \to \infty. \tag{4.2}$$

Given a sequence $x = (x_{mn})$ of real or complex numbers. The $(A^{r,\delta}p, q; 1, 1)$ transform of a sequence x is defined by

$$(A^{r,\delta,p,q}x)_{mn} = \sigma_{mn}^{r,\delta,p,q}(x) = \frac{1}{P_m Q_n} \sum_{j=0}^{m} \sum_{k=0}^{n} (1 + r^j)(1 + \delta^k) p_j q_k x_{jk},\ 0 < r, \delta < 1.$$

A double sequence (x_{mn}) is said to be convergent in Pringsheim's sense (or P-convergent) to l if for every $\epsilon > 0$ there exists a positive integer $n_0 = n_0(\epsilon)$ such that $|x_{mn} - l| < \epsilon$ whenever $m, n \geq n_0$ (see [11]). The number l is called the Pringsheim limit of x and we denote by $P - \lim_{m,n\to\infty} x_{mn} = l$. A double sequence (x_{mn}) is said to be bounded if there exists a positive number M such that $|x_{mn}| < M$ for all m and n. If

$$\lim_{m,n\to\infty} \sigma_{m,n}^{r,\delta,p,q}(x) = l, \tag{4.3}$$

then we say that (x_{mn}) is summable to l by $(A^{r,\delta}, p, q; 1, 1)$ summability method. The $(A^{r,*}, p, *; 1, 0)$ and $(A^{*,\delta}, *, q; 0, 1)$ transformations are defined respectively by

$$(A^{r,*,p,*}x)_{mn} = \sigma_{mn}^{r,*,p,*}(x) = \frac{1}{P_m} \sum_{j=0}^{m} (1 + r^j) p_j x_{jn},\ \ 0 < r < 1$$

and

$$(A^{*,\delta,*,q}x)_{mn} = \sigma_{mn}^{*,\delta,*,q}(x) = \frac{1}{Q_n} \sum_{k=0}^{n} (1 + \delta^k) q_k x_{mk},\ \ 0 < \delta < 1.$$

If

$$\lim_{m,n\to\infty} \sigma_{mn}^{r,*,p,*}(x) = l \tag{4.4}$$

or

$$\lim_{m,n\to\infty} \sigma_{mn}^{*,\delta,*,q}(x) = l, \tag{4.5}$$

then we say that (x_{mn}) is summable to l by $(A^{r,*}, p, *; 1, 0)$ or $(A^{*,\delta}, *, q; 0, 1)$ summability method, respectively. It is plain that (4.1) and (4.2) are a necessary and sufficient condition that every convergent sequence (x_{mn}) be $(A^{r,\delta}, p, q; 1, 1)$ summable to the same limit, respectively.

It is easy to check that if the limit

$$\lim_{m,n\to\infty} x_{mn} = l \tag{4.6}$$

exists and (x_{mn}) is bounded, then we also have (4.3). However, the converse implication is not true in general. If we define the sequence (x_{mn}) by $x_{mn} = (-1)^{mn}((1 + r^m)(1 + \delta^n)p_m q_n)^{-1}$, provided that (4.1) and (4.2), then we have $\sigma_{mn}^{r,\delta,p,q}(x) \to 0$ as $m, n \to \infty$. This shows that (x_{mn}) is $(A^{r,\delta}, p, q; 1, 1)$

summable to zero, but it doesn't converge. We notice that (4.3) implies (4.6) under a certain condition, which is called a Tauberian condition, imposed on (x_{mn}). Any theorem which states that convergence of sequences follows from its $(A^{r,\delta}, p, q; 1, 1)$ summability and some Tauberian condition(s) is said to be a Tauberian theorem for $(A^{r,\delta}, p, q; 1, 1)$ summability method. We have similar situations for $(A^{r,*}, p, *; 1, 0)$ and $(A^{*,\delta}, *, q; 0, 1)$ summability methods as in $(A^{r,\delta}, p, q; 1, 1)$ summability method.

In [13], Talo and Başar have given necessary and sufficent Tauberian conditions for the A^r method which has been introduced by Başar [6]. (We refer the reader to [1–5] for some results related to sequence spaces defined by the domain of the A^r matrices). In this chapter, we extend the results in [13] to $(A^{r,\delta}, p, q; \alpha, \beta)$ summability methods, where $(\alpha, \beta) = (1, 1), (1, 0), (0, 1)$ and obtain necessary and sufficient Tauberian conditions for $(A^{r,\delta}, p, q; \alpha, \beta)$ summability methods under which the existence of the limit (4.6) follows from that of (4.3), (4.4) or (6.40), respectively.

4.2 Auxiliary Results

We need the following lemmas to prove our theorems. We denote the integer part of the product λ and n by $\lambda_n := [\lambda n]$.

Lemma 4.2.1. *([7,10]) If $p = (p_m)$ and $q = (q_n)$ are sequences of nonnegative numbers, then conditions*

$$\liminf_{m\to\infty} \frac{P_{\lambda_m}}{P_m} > 1 \quad \textit{for every} \quad \lambda > 1 \tag{4.7}$$

and

$$\liminf_{m\to\infty} \frac{P_m}{P_{\lambda_m}} > 1 \quad \textit{for every} \quad 0 < \lambda < 1 \tag{4.8}$$

are equivalent. Similarly, conditions

$$\liminf_{n\to\infty} \frac{Q_{\lambda_n}}{Q_n} > 1 \quad \textit{for every} \quad \lambda > 1 \tag{4.9}$$

and

$$\liminf_{n\to\infty} \frac{Q_n}{Q_{\lambda_n}} > 1 \quad \textit{for every} \quad 0 < \lambda < 1 \tag{4.10}$$

are equivalent.

Lemma 4.2.2. *Let (4.7) and (4.9) be satisfied. If a sequence (x_{mn}) is $(A^{r,\delta}, p, q; 1, 1)$ summable to a finite number l, then*

$$\lim_{m,n\to\infty} \frac{1}{(P_{\lambda_m} - P_m)(Q_{\lambda_n} - Q_n)} \sum_{j=m+1}^{\lambda_m} \sum_{k=n+1}^{\lambda_n} (1 + r^j)(1 + \delta^k) p_j q_k x_{jk} = l \tag{4.11}$$

for every $\lambda > 1$ *and*

$$\lim_{m,n\to\infty} \frac{1}{(P_m - P_{\lambda_m})(Q_n - Q_{\lambda_n})} \sum_{j=\lambda_m+1}^{m} \sum_{k=\lambda_n+1}^{n} (1+r^j)(1+\delta^k)p_j q_k x_{jk} = l \tag{4.12}$$

for every $0 < \lambda < 1$.

Proof. Case $\lambda > 1$. Let $\lambda_m > m$, $\lambda_n > n$. By definition of $\sigma_{mn}^{r,\delta,p,q}$,

$$\begin{aligned}
\tau_{mn}^{r,\delta,p,q} &:= \frac{1}{(P_{\lambda_m} - P_m)(Q_{\lambda_n} - Q_n)} \sum_{j=m+1}^{\lambda_m} \sum_{k=n+1}^{\lambda_n} (1+r^j)(1+\delta^k)p_j q_k x_{jk} \\
&= \sigma_{\lambda_m,\lambda_n}^{r,\delta,p,q} + \frac{P_m Q_n}{(P_{\lambda_m} - P_m)(Q_{\lambda_n} - Q_n)} (\sigma_{\lambda_m,\lambda_n}^{r,\delta,p,q} - \sigma_{\lambda_m,n}^{r,\delta,p,q} - \sigma_{m,\lambda_n}^{r,\delta,p,q} + \sigma_{mn}^{r,\delta,p,q}) \\
&\quad + \frac{P_m}{P_{\lambda_m} - P_m}(\sigma_{\lambda_m,\lambda_n}^{r,\delta,p,q} - \sigma_{m,\lambda_n}^{r,\delta,p,q}) + \frac{Q_n}{Q_{\lambda_n} - Q_n}(\sigma_{\lambda_m,\lambda_n}^{r,\delta,p,q} - \sigma_{\lambda_m,n}^{r,\delta,p,q}) \\
&= \sigma_{\lambda_m,\lambda_n}^{r,\delta,p,q} + \frac{1}{\left(\frac{P_{\lambda_m}}{P_m} - 1\right)\left(\frac{Q_{\lambda_n}}{Q_n} - 1\right)} (\sigma_{\lambda_m,\lambda_n}^{r,\delta,p,q} - \sigma_{\lambda_m,n}^{r,\delta,p,q} - \sigma_{m,\lambda_n}^{r,\delta,p,q} + \sigma_{mn}^{r,\delta,p,q}) \\
&\quad + \frac{1}{\left(\frac{P_{\lambda_m}}{P_m} - 1\right)}(\sigma_{\lambda_m,\lambda_n}^{r,\delta,p,q} - \sigma_{m,\lambda_n}^{r,\delta,p,q}) + \frac{1}{\left(\frac{Q_{\lambda_n}}{Q_n} - 1\right)}(\sigma_{\lambda_m,\lambda_n}^{r,\delta,p,q} - \sigma_{\lambda_m,n}^{r,\delta,p,q}).
\end{aligned}$$

Arranging the terms on the last identity gives

$$\begin{aligned}
\tau_{\lambda_m,\lambda_n}^{r,\delta,p,q} &= \sigma_{\lambda_m,\lambda_n}^{r,\delta,p,q} + \frac{1}{\left(\frac{P_{\lambda_m}}{P_m} - 1\right)\left(\frac{Q_{\lambda_n}}{Q_n} - 1\right)} (\sigma_{\lambda_m,\lambda_n}^{r,\delta,p,q} - \sigma_{\lambda_m,n}^{r,\delta,p,q} - \sigma_{m,\lambda_n}^{r,\delta,p,q} + \sigma_{mn}^{r,\delta,p,q}) \\
&\quad + \frac{1}{\left(\frac{P_{\lambda_m}}{P_m} - 1\right)}(\sigma_{\lambda_m,\lambda_n}^{r,\delta,p,q} - \sigma_{m,\lambda_n}^{r,\delta,p,q}) + \frac{1}{\left(\frac{Q_{\lambda_n}}{Q_n} - 1\right)}(\sigma_{\lambda_m,\lambda_n}^{r,\delta,p,q} - \sigma_{\lambda_m,n}^{r,\delta,p,q}).
\end{aligned} \tag{4.13}$$

It follows from Lemma 4.2.1 that

$$0 < \limsup_{m,n\to\infty} \frac{1}{\left(\frac{P_{\lambda_m}}{P_m} - 1\right)} = \frac{1}{\liminf_{m,n\to\infty}\left(\frac{P_{\lambda_m}}{P_m} - 1\right)} < \infty$$

and

$$0 < \limsup_{m,n\to\infty} \frac{1}{\left(\frac{Q_{\lambda_m}}{Q_m} - 1\right)} = \frac{1}{\liminf_{m,n\to\infty}\left(\frac{Q_{\lambda_m}}{Q_m} - 1\right)} < \infty.$$

Since (x_{mn}) is $(A^{r,\delta}, p, q; 1, 1)$ summable to l, we have

$$\limsup_{m,n\to\infty} \frac{1}{\left(\frac{P_{\lambda_m}}{P_m} - 1\right)\left(\frac{Q_{\lambda_n}}{Q_n} - 1\right)} (\sigma_{\lambda_m,\lambda_n}^{r,\delta,p,q} - \sigma_{\lambda_m,n}^{r,\delta,p,q} - \sigma_{m,\lambda_n}^{r,\delta,p,q} + \sigma_{mn}^{r,\delta,p,q}) = 0.$$

The same situation similarly applies to the third and fourth terms in the right hand-side of (4.13). If we take the limit of both sides of (4.13) as $m, n \to \infty$, we have

$$\lim_{m,n\to\infty} \tau_{m,n}^{r,\delta,p,q} = l.$$

Case $0 < \lambda < 1$. By definition of $\sigma_{mn}^{r,\delta,p,q}$ as in the case $\lambda > 1$, we similarly have

$$\gamma_{m,n}^{r,\delta,p,q} = \sigma_{\lambda_m,\lambda_n}^{r,\delta,p,q} + \frac{1}{\left(\frac{P_m}{P_{\lambda_m}} - 1\right)\left(\frac{Q_n}{Q_{\lambda_n}} - 1\right)}(\sigma_{\lambda_m,\lambda_n}^{r,\delta,p,q} - \sigma_{\lambda_m,n}^{r,\delta,p,q} - \sigma_{m,\lambda_n}^{r,\delta,p,q} + \sigma_{mn}^{r,\delta,p,q})$$

$$+ \frac{1}{\left(\frac{P_m}{P_{\lambda_m}} - 1\right)}(\sigma_{\lambda_m,\lambda_n}^{r,\delta,p,q} - \sigma_{m,\lambda_n}^{r,\delta,p,q}) + \frac{1}{\left(\frac{Q_n}{Q_{\lambda_n}} - 1\right)}(\sigma_{\lambda_m,\lambda_n}^{r,\delta,p,q} - \sigma_{\lambda_m,n}^{r,\delta,p,q}), \quad (4.14)$$

where

$$\gamma_{m,n}^{r,\delta,p,q} := \frac{1}{(P_m - P_{\lambda_m})(Q_n - Q_{\lambda_n})} \sum_{j=\lambda_m+1}^{m} \sum_{k=\lambda_n+1}^{n} (1 + r^j)(1 + \delta^k) p_j q_k x_{jk}.$$

It follows from Lemma 4.2.1 and (4.3) that

$$\lim_{m,n\to\infty} \gamma_{m,n}^{r,\delta,p,q} = l.$$

This completes the proof of lemma. □

4.3 Tauberian Theorems for the $(A^{r,\delta}, p, q; 1, 1)$ Summability Method

First, we consider sequences of real numbers and prove the following one-sided Tauberian theorem.

Theorem 4.3.1. *Let (4.7) and (4.9) be satisfied. If (x_{mn}) is a sequence of real numbers which is $(A^{r,\delta}, p, q; 1, 1)$ summable to a finite limit l, then (4.6) holds if and only if the following two conditions are satisfied:*

$$\sup_{\lambda>1} \liminf_{m,n\to\infty} \frac{1}{(P_{\lambda_m} - P_m)(Q_{\lambda_n} - Q_n)} \sum_{j=m+1}^{\lambda_m} \sum_{k=n+1}^{\lambda_n} p_j q_k \times \left((1 + r^j)(1 + \delta^k) x_{jk} - x_{mn}\right) \geq 0 \quad (4.15)$$

and

$$\sup_{0<\lambda<1} \liminf_{m,n\to\infty} \frac{1}{(P_m - P_{\lambda_m})(Q_n - Q_{\lambda_n})} \sum_{j=\lambda_m+1}^{m} \sum_{k=\lambda_n+1}^{n} p_j q_k \times \left(x_{mn} - (1+r^j)(1+\delta^k)x_{jk}\right) \geq 0. \quad (4.16)$$

A sequence (x_{mn}) of real numbers is said to be slowly decreasing in sense $(1,1)$ if

$$\lim_{\lambda\to 1^+} \liminf_{m,n\to\infty} \min_{\substack{m<j\leq\lambda_m \\ n<k\leq\lambda_n}} (x_{jk} - x_{mn}) \geq 0. \quad (4.17)$$

Note that condition (4.17) can be equivalently reformulated as follows:

$$\lim_{\lambda\to 1^-} \liminf_{m,n\to\infty} \min_{\substack{\lambda_m<j\leq m \\ \lambda_n<k\leq n}} (x_{mn} - x_{jk}) \geq 0. \quad (4.18)$$

The right-hand limit in (4.17) exists and can be equivalently replaced by $\sup_{\lambda>1}$. The concept of slow decreasing for sequences of real numbers was introduced by Schmidt [12]. We recall that for sequences (x_m) and (y_m) of real or complex numbers, we write $x_m = O(y_m)$ if there exists some positive number M such that $|x_m| \leq M|y_m|$ for all sufficiently large integers m. We have the following corollary for Theorem 4.3.1.

Corollary 4.3.2. *Let (4.7), (4.9) and* $P_m = O(mp_m)$, $Q_n = O(nq_n)$ *be satisfied. If a sequence* (x_{mn}) *of real numbers is slowly decreasing in sense* $(1,1)$, *then (4.3), (4.4) and (6.40) imply (4.6).*

Remark 4.3.3. *If conditions* (4.3) *and* (4.6) *or equivalently, conditions* (4.3), (4.15) *and* (4.16) *are satisfied, then we necessarily have*

$$\lim_{m,n\to\infty} \frac{1}{(P_{\lambda_m} - P_m)(Q_{\lambda_n} - Q_n)} \sum_{j=m+1}^{\lambda_m} \sum_{k=n+1}^{\lambda_n} p_j q_k \times \left((1+r^j)(1+\delta^k)x_{jk} - x_{mn}\right) = 0 \quad (4.19)$$

for every $\lambda > 1$ *and*

$$\lim_{m,n\to\infty} \frac{1}{(P_m - P_{\lambda_m})(Q_n - Q_{\lambda_n})} \sum_{j=\lambda_m+1}^{m} \sum_{k=\lambda_n+1}^{n} p_j q_k \times \left(x_{mn} - (1+r^j)(1+\delta^k)x_{jk}\right) = 0 \quad (4.20)$$

for every $0 < \lambda < 1$.

Remark 4.3.4. *Theorem 4.3.1 remains true if conditions (4.15) and (4.16) are replaced by their symmetric counterparts:*

$$\inf_{\lambda>1} \limsup_{m,n\to\infty} \frac{1}{(P_{\lambda_m}-P_m)(Q_{\lambda_n}-Q_n)} \sum_{j=m+1}^{\lambda_m} \sum_{k=n+1}^{\lambda_n} p_j q_k \times \left((1+r^j)(1+\delta^k)x_{jk}-x_{mn}\right) \leq 0 \quad (4.21)$$

and

$$\inf_{0<\lambda<1} \limsup_{m,n\to\infty} \frac{1}{(P_m-P_{\lambda_m})(Q_n-Q_{\lambda_n})} \sum_{j=\lambda_m+1}^{m} \sum_{k=\lambda_n+1}^{n} p_j q_k \times \left(x_{mn}-(1+r^j)(1+\delta^k)x_{jk}\right) \leq 0, \quad (4.22)$$

respectively.

Second, we consider sequences of complex numbers and prove the following two-sided Tauberian theorem.

Theorem 4.3.5. *Let (4.7) and (4.9) be satisfied. If (x_{mn}) is a sequence of complex numbers which is $(A^{r,\delta}, p, q; 1, 1)$ summable to a finite limit l, then (x_{mn}) converges to the same limit if and only if one of the following two conditions is satisfied:*

$$\inf_{\lambda>1} \limsup_{m,n\to\infty} \left| \frac{1}{(P_{\lambda_m}-P_m)(Q_{\lambda_n}-Q_n)} \sum_{j=m+1}^{\lambda_m} \sum_{k=n+1}^{\lambda_n} p_j q_k \times \left((1+r^j)(1+\delta^k)x_{jk}-x_{mn}\right) \right| = 0 \quad (4.23)$$

or

$$\inf_{0<\lambda<1} \limsup_{m,n\to\infty} \left| \frac{1}{(P_m-P_{\lambda_m})(Q_n-Q_{\lambda_n})} \sum_{j=\lambda_m+1}^{m} \sum_{k=\lambda_n+1}^{n} p_j q_k \times \left(x_{mn}-(1+r^j)(1+\delta^k)x_{jk}\right) \right| = 0. \quad (4.24)$$

We recall that a sequence (x_{mn}) of complex numbers is said to be slowly oscillating in sense $(1, 1)$ if

$$\lim_{\lambda\to 1^+} \limsup_{m,n\to\infty} \max_{\substack{m<j\leq\lambda_m \\ n<k\leq\lambda_n}} |x_{jk}-x_{mn}| = 0. \quad (4.25)$$

The concept of slow oscillation for sequences of complex numbers was introduced by Hardy [8]. An equivalent reformulation of (4.25) can be given as follows:

$$\lim_{\lambda\to 1^-} \limsup_{m,n\to\infty} \max_{\substack{\lambda_m<j\leq m \\ \lambda_n<k\leq n}} |x_{mn}-x_{jk}| = 0. \quad (4.26)$$

The right-hand limit in (4.25) can be equivalently replaced by $\inf_{\lambda>1}$. We have the following corollary to Theorem 4.3.5.

Corollary 4.3.6. *Let (4.7), (4.9) and $P_m = O(mp_m)$, $Q_n = O(nq_n)$ be satisfied. If a sequence (x_{mn}) of complex numbers is slowly oscillating in sense $(1,1)$, then (4.3), (4.4) and (6.40) imply (4.6).*

4.3.1 Proofs

Proof of Theorem 4.3.1. Necessity. Assume that (4.3), (4.6) and (4.7), (4.9) are satisfied. Then Lemma 4.2.2 yields (4.15) in case $\lambda > 1$ and (4.16) in case $0 < \lambda < 1$.

Sufficiency. Assume that (4.3), (4.7), (4.9) and (4.15), (4.16) are satisfied. First, we consider the case $\lambda > 1$. Let $\epsilon > 0$ be given. By (4.15), there exists some $\lambda > 1$ such that

$$\liminf_{m,n\to\infty} \frac{1}{(P_{\lambda_m} - P_m)(Q_{\lambda_n} - Q_n)} \sum_{j=m+1}^{\lambda_m} \sum_{k=n+1}^{\lambda_n} p_j q_k \times \left((1+r^j)(1+\delta^k)x_{jk} - x_{mn}\right) \geq -\epsilon. \quad (4.27)$$

By (4.13), we have

$$x_{mn} - \sigma^{r,\delta,p,q}_{\lambda_m,\lambda_n} = \frac{1}{\left(\frac{P_{\lambda_m}}{P_m} - 1\right)\left(\frac{Q_{\lambda_n}}{Q_n} - 1\right)} (\sigma^{r,\delta,p,q}_{\lambda_m,\lambda_n} - \sigma^{r,\delta,p,q}_{\lambda_m,n} - \sigma^{r,\delta,p,q}_{m,\lambda_n} + \sigma^{r,\delta,p,q}_{mn})$$
$$+ \frac{1}{\left(\frac{P_{\lambda_m}}{P_m} - 1\right)} (\sigma^{r,\delta,p,q}_{\lambda_m,\lambda_n} - \sigma^{r,\delta,p,q}_{m,\lambda_n}) + \frac{1}{\left(\frac{Q_{\lambda_n}}{Q_n} - 1\right)} (\sigma^{r,\delta,p,q}_{\lambda_m,\lambda_n} - \sigma^{r,\delta,p,q}_{\lambda_m,n})$$
$$- \frac{1}{(P_{\lambda_m} - P_m)(Q_{\lambda_n} - Q_n)} \sum_{j=m+1}^{\lambda_m} \sum_{k=n+1}^{\lambda_n} p_j q_k \left((1+r^j)(1+\delta^k)x_{jk} - x_{mn}\right). \quad (4.28)$$

By (4.3), (4.7) and (4.9), the first three terms on the right hand-side of (4.28) vanish as $m, n \to \infty$.

It follows from (4.28) that

$$\limsup_{m,n\to\infty} (x_{mn} - \sigma^{r,\delta,p,q}_{\lambda_m,\lambda_n}) \leq -\liminf_{m,n\to\infty} \frac{1}{(P_{\lambda_m} - P_m)(Q_{\lambda_n} - Q_n)} \times \sum_{j=m+1}^{\lambda_m} \sum_{k=n+1}^{\lambda_n} p_j q_k \left((1+r^j)(1+\delta^k)x_{jk} - x_{mn}\right) \leq \epsilon.$$

Consequently, we have

$$\limsup_{m,n\to\infty} x_{mn} \leq l + \epsilon. \quad (4.29)$$

Second, we consider the case $0 < \lambda < 1$. Let $\epsilon > 0$ be given. By (4.16), there exists some $0 < \lambda < 1$ such that

$$\liminf_{m,n\to\infty} \frac{1}{(P_m - P_{\lambda_m})(Q_n - Q_{\lambda_n})} \sum_{j=\lambda_m+1}^{m} \sum_{k=\lambda_n+1}^{n} p_j q_k \times \left(x_{mn} - (1+r^j)(1+\delta^k)x_{jk}\right) \geq -\epsilon. \tag{4.30}$$

By (4.14), we have

$$x_{mn} - \sigma^{r,\delta,p,q}_{\lambda_m,\lambda_n} = \frac{1}{\left(\frac{P_m}{P_{\lambda_m}} - 1\right)\left(\frac{Q_n}{Q_{\lambda_n}} - 1\right)}(\sigma^{r,\delta,p,q}_{\lambda_m,\lambda_n} - \sigma^{r,\delta,p,q}_{\lambda_m,n} - \sigma^{r,\delta,p,q}_{m,\lambda_n} + \sigma^{r,\delta,p,q}_{mn})$$
$$+ \frac{1}{\left(\frac{P_m}{P_{\lambda_m}} - 1\right)}(\sigma^{r,\delta,p,q}_{\lambda_m,\lambda_n} - \sigma^{r,\delta,p,q}_{m,\lambda_n}) + \frac{1}{\left(\frac{Q_n}{Q_{\lambda_n}} - 1\right)}(\sigma^{r,\delta,p,q}_{\lambda_m,\lambda_n} - \sigma^{r,\delta,p,q}_{\lambda_m,n})$$
$$+ \frac{1}{(P_{\lambda_m} - P_m)(Q_{\lambda_n} - Q_n)} \sum_{j=m+1}^{\lambda_m} \sum_{k=n+1}^{\lambda_n} p_j q_k \left(x_{mn} - (1+r^j)(1+\delta^k)x_{jk}\right). \tag{4.31}$$

By (4.3), (4.7) and (4.9), the first three terms on the right hand-side of (4.31) vanish as $m, n \to \infty$.

It follows from (4.31) that

$$\liminf_{m,n\to\infty} (x_{mn} - \sigma^{r,\delta,p,q}_{\lambda_m,\lambda_n}) \geq \liminf_{m,n\to\infty} \frac{1}{(P_{\lambda_m} - P_m)(Q_{\lambda_n} - Q_n)} \times \sum_{j=m+1}^{\lambda_m} \sum_{k=n+1}^{\lambda_n} p_j q_k \left((1+r^j)(1+\delta^k)x_{jk} - x_{mn}\right) \geq -\epsilon.$$

Consequently, we have

$$\liminf_{m,n\to\infty} x_{mn} \geq l - \epsilon. \tag{4.32}$$

Combining (4.29) and (4.32) yields

$$l - \epsilon \leq \liminf_{m,n\to\infty} x_{mn} \leq \limsup_{m,n\to\infty} x_{mn} \leq l + \epsilon.$$

Being ϵ arbitrary small, hence (4.6) follows. □

Proof of Corollary 4.3.2. For $\lambda > 1$, we have the following inequality:

$$\frac{1}{(P_{\lambda_m} - P_m)(Q_{\lambda_n} - Q_n)} \sum_{j=m+1}^{\lambda_m} \sum_{k=n+1}^{\lambda_n} p_j q_k \left((1+r^j)(1+\delta^k)x_{jk} - x_{mn}\right)$$
$$\geq \min_{\substack{m<j\leq\lambda_m \\ n<k\leq\lambda_n}} \left((1+r^j)(1+\delta^k)x_{jk} - x_{mn}\right)$$
$$\geq \min_{\substack{m<j\leq\lambda_m \\ n<k\leq\lambda_n}} (x_{jk} - x_{mn}) + \min_{\substack{m<j\leq\lambda_m \\ n<k\leq\lambda_n}} (r^j x_{jk} + \delta^k x_{jk} + r^j\delta^k x_{jk}).$$

We have, by definition of $\sigma_{mn}^{r,\delta,p,q}$,

$$x_{mn} = \frac{P_m Q_n \sigma_{mn}^{r,\delta,p,q} - P_{m-1} Q_n \sigma_{m-1,n}^{r,\delta,p,q} - P_m Q_{n-1} \sigma_{m,n-1}^{r,\delta,p,q} + P_{m-1} Q_{n-1} \sigma_{m-1,n-1}^{r,\delta,p,q}}{p_m q_n (1+r^m)(1+\delta^n)}$$

and

$$\begin{aligned} \frac{x_{mn}}{mn} &= \frac{P_m Q_n (\sigma_{mn}^{r,\delta,p,q} - \sigma_{m-1,n}^{r,\delta,p,q})}{mnp_m q_n (1+r^m)(1+\delta^n)} + \frac{Q_n \sigma_{m-1,n}^{r,\delta,p,q}}{mnq_n (1+r^m)(1+\delta^n)} \\ &- \frac{P_m Q_{n-1} (\sigma_{m,n-1}^{r,\delta,p,q} - \sigma_{m-1,n-1}^{r,\delta,p,q})}{mnp_m q_n (1+r^m)(1+\delta^n)} - \frac{Q_{n-1} \sigma_{m-1,n-1}^{r,\delta,p,q}}{mnq_n (1+r^m)(1+\delta^n)}. \end{aligned}$$

Since (x_{mn}) is $(A^{r,\delta}, p, q; 1, 1)$ summable to l and $P_m = O(mp_m)$, $Q_n = O(nq_n)$, then we have $x_{mn}/mn \to 0$, as $m, n \to \infty$. Therefore, $r^m \delta^n x_{mn} \to 0$, as $m, n \to \infty$. By definitions of $\sigma_{mn}^{r,*,p,*}$ and $\sigma_{mn}^{*,\delta,*,q}$, we have

$$x_{mn} = \frac{P_m \sigma_{mn}^{r,*,p,*} - P_{m-1} \sigma_{m-1,n}^{r,*,p,*}}{mp_m (1+r^m)},$$

$$\frac{x_{mn}}{m} = \frac{P_m (\sigma_{mn}^{r,*,p,*} - \sigma_{m-1,n}^{r,*,p,*})}{mp_m (1+r^m)} + \frac{\sigma_{m-1,n}^{r,*,p,*}}{m(1+r^m)}$$

and

$$x_{mn} = \frac{Q_n \sigma_{mn}^{*,\delta,*,q} - Q_{n-1} \sigma_{m,n-1}^{*,\delta,*,q}}{q_n (1+\delta^n)},$$

$$\frac{x_{mn}}{n} = \frac{Q_n (\sigma_{mn}^{*,\delta,*,q} - \sigma_{m,n-1}^{*,\delta,*,q})}{nq_n (1+\delta^n)} + \frac{\sigma_{m,n-1}^{*,\delta,*,q}}{n(1+\delta^n)}.$$

respectively. Since (x_{mn}) is $(A^{r,*}, p, *; 1, 0)$ and $(A^{*,\delta}, *, q; 0, 1)$ summable to l and $P_m = O(mp_m)$ and $Q_n = O(nq_n)$ respectively, then we have $x_{mn}/m \to 0$ and $x_{mn}/n \to 0$ as $m, n \to \infty$, respectively. Hence, $r^m x_{mn} \to 0$ and $\delta^n x_{mn} \to 0$ as $m, n \to \infty$. Thus, condition (4.17) clearly implies condition (4.15). Similarly, (9.7) implies (4.16). By Theorem 4.3.1, we have (4.6). □

Proof of Theorem 4.3.5. Necessity. The proof is similar to the proof of the necessity part of Theorem 4.3.1.

Sufficiency. Assume that (4.3), (4.7), (4.9) and one of conditions (4.23) and (4.24) is satisfied. Let any $\epsilon > 0$ be given. By (4.23), there exists some $\lambda > 1$ such that

$$\begin{aligned} \limsup_{m,n\to\infty} \Bigg| \frac{1}{(P_{\lambda_m} - P_m)(Q_{\lambda_n} - Q_n)} \sum_{j=m+1}^{\lambda_m} \sum_{k=n+1}^{\lambda_n} p_j q_k \\ \times ((1+r^j)(1+\delta^k) x_{jk} - x_{mn}) \Bigg| < \epsilon. \quad (4.33) \end{aligned}$$

By (4.3) and (4.7), (4.9), we have

$$\limsup_{m,n\to\infty} |x_{mn} - \sigma^{r,\delta,p,q}_{\lambda_m,\lambda_n}| \le \limsup_{m,n\to\infty} \left| \frac{1}{(P_{\lambda_m} - P_m)(Q_{\lambda_n} - Q_n)} \sum_{j=m+1}^{\lambda_m} \sum_{k=n+1}^{\lambda_n} p_j q_k \right.$$
$$\left. \times \left((1+r^j)(1+\delta^k)x_{jk} - x_{mn}\right) \right| \le \epsilon. \quad (4.34)$$

Let any $\epsilon > 0$ be given. By (4.24), there exists some $0 < \lambda < 1$ such that

$$\limsup_{m,n\to\infty} \left| \frac{1}{(P_m - P_{\lambda_m})(Q_n - Q_{\lambda_n})} \sum_{j=\lambda_m+1}^{m} \sum_{k=\lambda_n+1}^{n} p_j q_k \right.$$
$$\left. \times \left(x_{mn} - (1+r^j)(1+\delta^k)x_{jk}\right) \right| < \epsilon. \quad (4.35)$$

By (4.3), (4.7) and (4.9), we have

$$\limsup_{m,n\to\infty} |x_{mn} - \sigma^{r,\delta,p,q}_{\lambda_m,\lambda_n}| \le \limsup_{m,n\to\infty} \left| \frac{1}{(P_m - P_{\lambda_m})(Q_n - Q_{\lambda_n})} \right.$$
$$\left. \times \sum_{j=\lambda_m+1}^{m} \sum_{k=\lambda_n+1}^{n} p_j q_k \left(x_{mn} - (1+r^j)(1+\delta^k)x_{jk}\right) \right| \le \epsilon. \quad (4.36)$$

By (4.34) or (4.36), in either case we obtain

$$\limsup_{m,n\to\infty} |x_{mn} - \sigma^{r,\delta,p,q}_{\lambda_m,\lambda_n}| = 0 \quad (4.37)$$

whence it follows that

$$\lim_{m,n\to\infty} |x_{mn} - \sigma^{r,\delta,p,q}_{\lambda_m,\lambda_n}| = 0. \quad (4.38)$$

Now we conclude (4.6) from (4.3) and (4.38). □

Proof of Corollary 4.3.6. For $\lambda > 1$, we have the following inequality:

$$\left| \frac{1}{(P_{\lambda_m} - P_m)(Q_{\lambda_n} - Q_n)} \sum_{j=m+1}^{\lambda_m} \sum_{k=n+1}^{\lambda_n} p_j q_k \left((1+r^j)(1+\delta^k)x_{jk} - x_{mn}\right) \right|$$
$$\le \max_{\substack{m<j\le\lambda_m \\ n<k\le\lambda_n}} |(1+r^j)(1+\delta^k)x_{jk} - x_{mn}|$$
$$\le \max_{\substack{m<j\le\lambda_m \\ n<k\le\lambda_n}} |x_{jk} - x_{mn}| + \max_{\substack{m<j\le\lambda_m \\ n<k\le\lambda_n}} |r^j x_{jk} + \delta^k x_{jk} + r^j \delta^k x_{jk}|.$$

As in the proof of Corollary 4.3.2, we have $x_{mn}/mn \to 0$ as $m, n \to \infty$, $x_{mn}/m \to 0$ as $m, n \to \infty$ and $x_{mn}/n \to 0$ as $m, n \to \infty$. Hence, $r^m \delta^n x_{mn} \to 0$, $r^m x_{mn} \to 0$ and $\delta^n x_{mn} \to 0$ as $m, n \to \infty$. Thus, condition (4.25) clearly implies condition (4.23). Similarly, (4.26) implies (4.24).By Theorem 4.3.5, we have (4.6).

4.4 Tauberian Theorems for the $(A^{r,*}, p, *; 1, 0)$ Summability Method

First, we consider sequences of real numbers and prove the following one-sided Tauberian theorem. We need the following lemma to prove our main theorems in this section.

Lemma 4.4.1. *Let (4.7) be satisfied. If a sequence* (x_{mn}) *is* $(A^{r,*}, p, *; 1, 0)$ *summable to a finite number* l*, then*

$$\lim_{m,n\to\infty} \frac{1}{P_{\lambda_m} - P_m} \sum_{j=m+1}^{\lambda_m} (1 + r^j) p_j x_{jn} = l \tag{4.39}$$

for every $\lambda > 1$ *and*

$$\lim_{m,n\to\infty} \frac{1}{P_m - P_{\lambda_m}} \sum_{j=\lambda_m+1}^{m} (1 + r^j) p_j x_{jn} = l \tag{4.40}$$

for every $0 < \lambda < 1$.

Proof. Case $\lambda > 1$. By definition of $\sigma_{mn}^{r,*,p,*}$, we have

$$\frac{1}{P_{\lambda_m} - P_m} \sum_{j=m+1}^{\lambda_m} (1 + r^j) p_j x_{jn} = \frac{1}{P_{\lambda_m} - P_m} \left(\sum_{j=0}^{\lambda_m} - \sum_{j=0}^{m} \right) (1 + r^j) p_j x_{jn}$$

$$= \frac{1}{P_{\lambda_m} - P_m} \sum_{j=0}^{\lambda_m} (1 + r^j) p_j x_{jn} - \frac{1}{P_{\lambda_m} - P_m} \sum_{j=0}^{m} (1 + r^j) p_j x_{jn}$$

$$= \frac{P_{\lambda_m}}{P_{\lambda_m}(P_{\lambda_m} - P_m)} \sum_{j=0}^{\lambda_m} (1 + r^j) p_j x_{jn} - \frac{P_m}{P_m(P_{\lambda_m} - P_m)} \sum_{j=0}^{m} (1 + r^j) p_j x_{jn}$$

$$= \left(\frac{1}{P_{\lambda_m}} + \frac{P_m}{P_{\lambda_m}(P_{\lambda_m} - P_m)} \right) \sum_{j=0}^{\lambda_m} (1 + r^j) p_j x_{jn} - \frac{P_m}{P_{\lambda_m} - P_m} \sigma_{mn}^{r,*,p,*}$$

$$= \frac{1}{P_{\lambda_m}} \sum_{j=0}^{\lambda_m} (1+r^j) p_j x_{jn} + \frac{P_m}{P_{\lambda_m}(P_{\lambda_m} - P_m)} \sum_{j=0}^{\lambda_m} (1+r^j) p_j x_{jn} - \frac{P_m}{P_{\lambda_m} - P_m} \sigma_{mn}^{r,*,p,*}$$

$$= \sigma_{\lambda_m,n}^{r,*,p,*} + \frac{P_m}{P_{\lambda_m} - P_m} \sigma_{\lambda_m,n}^{r,*,p,*} - \frac{P_m}{P_{\lambda_m} - P_m} \sigma_{mn}^{r,*,p,*}$$

$$= \sigma_{\lambda_m,n}^{r,*,p,*} + \frac{P_m}{P_{\lambda_m} - P_m} (\sigma_{\lambda_m,n}^{r,*,p,*} - \sigma_{mn}^{r,*,p,*})$$

$$= \sigma_{\lambda_m,n}^{r,*,p,*} + \frac{1}{\frac{P_{\lambda_m}}{P_m} - 1} (\sigma_{\lambda_m,n}^{r,*,p,*} - \sigma_{mn}^{r,*,p,*}). \tag{4.41}$$

It follows from (6.23) that

$$x_{mn} - \sigma^{r,*,p,*}_{\lambda_m,n} = \frac{1}{\frac{P_{\lambda_m}}{P_m} - 1}\left(\sigma^{r,*,p,*}_{\lambda_m,n} - \sigma^{r,*,p,*}_{mn}\right) - \frac{1}{P_{\lambda_m} - P_m}\sum_{j=m+1}^{\lambda_m} p_j\left((1+r^j)x_{jn} - x_{mn}\right). \quad (4.42)$$

By Lemma 4.2.1, we have

$$\limsup_{m,n\to\infty} \frac{1}{\left(\frac{P_{\lambda_m}}{P_m} - 1\right)} = \frac{1}{\liminf_{m,n\to\infty}\left(\frac{P_{\lambda_m}}{P_m} - 1\right)} < \infty. \quad (4.43)$$

Since (x_{mn}) is $(A^{r,*}, p, *; 1, 0)$ summable to l, we have

$$\limsup_{m,n\to\infty} \frac{1}{\frac{P_{\lambda_m}}{P_m} - 1}\left(\sigma^{r,*,p,*}_{\lambda_m,n} - \sigma^{r,*,p,*}_{mn}\right) = 0.$$

If we take limit of both of side of equation (6.23) as $m, n \to \infty$, then we have

$$\lim_{m,n\to\infty} \frac{1}{P_{\lambda_m} - P_m}\sum_{j=m+1}^{\lambda_m} (1+r^j)p_j x_{jn} = l.$$

Case $0 < \lambda < 1$. By definition of $\sigma^{r,*,p,*}_{mn}$, we have

$$x_{mn} - \sigma^{r,*,p,*}_{\lambda_m,n} = \frac{1}{\frac{P_m}{P_{\lambda_m}} - 1}\left(\sigma^{r,*,p,*}_{mn} - \sigma^{r,*,p,*}_{\lambda_m,n}\right) + \frac{1}{P_m - P_{\lambda_m}}\sum_{j=\lambda_m+1}^{m} p_j\left(x_{mn} - (1+r^j)x_{jn}\right). \quad (4.44)$$

□

Theorem 4.4.2. *Let (4.7) be satisfied. If (x_{mn}) is a sequence of real numbers which is $(A^{r,*}, p, *; 1, 0)$ summable to a finite limit l, then (4.6) holds if and only if the following two conditions are satisfied:*

$$\sup_{\lambda>1}\liminf_{m,n\to\infty} \frac{1}{P_{\lambda_m} - P_m}\sum_{j=m+1}^{\lambda_m} p_j\left((1+r^j)x_{jn} - x_{mn}\right) \geq 0 \quad (4.45)$$

and

$$\sup_{0<\lambda<1}\liminf_{m,n\to\infty} \frac{1}{P_m - P_{\lambda_m}}\sum_{j=\lambda_m+1}^{m} p_j\left(x_{mn} - (1+r^j)x_{jn}\right) \geq 0, \quad (4.46)$$

in which case, we necessarily have

$$\lim_{m,n\to\infty} \frac{1}{P_{\lambda_m} - P_m} \sum_{j=m+1}^{\lambda_m} p_j \left((1+r^j)x_{jn} - x_{mn}\right) = 0$$

for all $\lambda > 1$ *and*

$$\lim_{m,n\to\infty} \frac{1}{P_m - P_{\lambda_m}} \sum_{j=\lambda_m+1}^{m} p_j \left(x_{mn} - (1+r^j)x_{jn}\right) = 0$$

for all $0 < \lambda < 1$.

A sequence (x_{mn}) of real numbers is said to be slowly decreasing in sense $(1,0)$ if

$$\lim_{\lambda\to 1^+} \liminf_{m,n\to\infty} \min_{m<j\le\lambda_m} (x_{jn} - x_{mn}) \ge 0. \tag{4.47}$$

Note that condition (4.47) can be equivalently reformulated as follows:

$$\lim_{\lambda\to 1^-} \liminf_{m,n\to\infty} \min_{\lambda_m<j\le m} (x_{mn} - x_{jn}) \ge 0. \tag{4.48}$$

Corollary 4.4.3. *Let (4.7) and* $P_m = O(mp_m)$ *be satisfied. If a sequence* (x_{mn}) *of real numbers is slowly decreasing in sense* $(1,0)$*, then (4.3) implies (4.6).*

Second, we consider sequences of complex numbers and prove the following two-sided Tauberian theorem.

Theorem 4.4.4. *Let (4.7) be satisfied. If* (x_{mn}) *is a sequence of complex numbers which is* $(A^{r,*}, p, *; 1, 0)$ *summable to a finite limit* l*, then* (x_{mn}) *converges to the same limit if and only if one of the following two conditions is satisfied:*

$$\inf_{\lambda>1} \limsup_{m,n\to\infty} \left| \frac{1}{P_{\lambda_m} - P_m} \sum_{j=m+1}^{\lambda_m} p_j \left((1+r^j)x_{jn} - x_{mn}\right) \right| = 0 \tag{4.49}$$

or

$$\inf_{0<\lambda<1} \limsup_{m,n\to\infty} \left| \frac{1}{P_m - P_{\lambda_m}} \sum_{j=\lambda_m+1}^{m} p_j \left(x_{mn} - (1+r^j)x_{jn}\right) \right| = 0. \tag{4.50}$$

We recall that a sequence (x_{mn}) of complex numbers is said to be slowly oscillating in sense $(1,0)$ if

$$\lim_{\lambda\to 1^+} \limsup_{m,n\to\infty} \max_{m<j\le\lambda_m} |x_{jn} - x_{mn}| = 0. \tag{4.51}$$

An equivalent reformulation of (4.51) can be given as follows:

$$\lim_{\lambda\to 1^-} \limsup_{m,n\to\infty} \max_{\lambda_m<j\le m} |x_{mn} - x_{jn}| = 0. \tag{4.52}$$

Corollary 4.4.5. *Let (4.7) and $P_m = O(mp_m)$ be satisfied. If a sequence (x_{mn}) of complex numbers is slowly oscillating in sense* (1,0)*, then (4.3) implies (4.6).*

4.4.1 Proofs

Proof of Theorem 4.4.2. Necessity. Assume that (4.4) and (4.7) are satisfied. Then Lemma 4.4.1 yields (4.45) in case $\lambda > 1$ and (4.46) in case $0 < \lambda < 1$.

Sufficiency. Assume that (4.4), (4.45) and (4.46) are satisfied. First, we consider the case $\lambda > 1$. Let $\epsilon > 0$ be given. By (4.45), there exists some $\lambda > 1$ such that

$$\liminf_{m,n\to\infty} \frac{1}{P_{\lambda_m} - P_m} \sum_{j=m+1}^{\lambda_m} p_j \left((1+r^j)x_{jn} - x_{mn}\right) \geq -\epsilon. \tag{4.53}$$

It follows from (4.42) that

$$x_{mn} - \sigma^{r,*,p,*}_{\lambda_m,n} = \frac{1}{\frac{P_{\lambda_m}}{P_m} - 1}\left(\sigma^{r,*,p,*}_{\lambda_m,n} - \sigma^{r,*,p,*}_{mn}\right) - \frac{1}{P_{\lambda_m} - P_m} \sum_{j=m+1}^{\lambda_m} p_j \left((1+r^j)x_{jn} - x_{mn}\right). \tag{4.54}$$

By (4.4) and (4.7), the first term on the right hand-side of (4.54) vanishes as $m, n \to \infty$. It follows from (4.54) that

$$\limsup_{m,n\to\infty}(x_{mn} - \sigma^{r,*,p,*}_{\lambda_m,n}) \leq -\liminf_{m,n\to\infty} \frac{1}{P_{\lambda_m} - P_m} \sum_{j=m+1}^{\lambda_m} p_j \times \left((1+r^j)x_{jn} - x_{mn}\right) \leq \epsilon.$$

Consequently, we have

$$\limsup_{m,n\to\infty} x_{mn} \leq l + \epsilon. \tag{4.55}$$

Second, we consider the case $0 < \lambda < 1$. Let $\epsilon > 0$ be given. By (4.46), there exists some $0 < \lambda < 1$ such that

$$\liminf_{m,n\to\infty} \frac{1}{P_m - P_{\lambda_m}} \sum_{j=\lambda_m+1}^{m} p_j \left(x_{mn} - (1+r^j)x_{jn}\right) \geq -\epsilon. \tag{4.56}$$

It follows from (4.44) that

$$x_{mn} - \sigma^{r,*,p,*}_{\lambda_m,n} = \frac{1}{\frac{P_m}{P_{\lambda_m}} - 1}\left(\sigma^{r,*,p,*}_{mn} - \sigma^{r,*,p,*}_{\lambda_m,n}\right) + \frac{1}{P_m - P_{\lambda_m}} \sum_{j=\lambda_m+1}^{m} p_j\left(x_{mn} - (1+r^j)x_{jn}\right). \quad (4.57)$$

Using a similar argument as above, we conclude by (4.57) that

$$\begin{aligned}\liminf_{m,n\to\infty}\left(x_{mn} - \sigma^{r,*,p,*}_{\lambda_m,n}\right) &\geq \liminf_{m,n\to\infty} \frac{1}{P_m - P_{\lambda_m}} \sum_{j=\lambda_m+1}^{m} p_j\left(x_{mn} - (1+r^j)x_{jn}\right) \\ &\geq -\epsilon.\end{aligned}$$

Consequently, we have

$$\liminf_{m,n\to\infty} x_{mn} \geq l - \epsilon. \quad (4.58)$$

Combining (4.55) and (4.58) yields

$$l - \epsilon \leq \liminf_{m,n\to\infty} x_{mn} \leq \limsup_{m,n\to\infty} x_{mn} \leq l + \epsilon.$$

Being ϵ arbitrary small, hence (4.6) follows. □

Proof of Corollary 4.4.3. For $\lambda > 1$, we have the following inequality:

$$\begin{aligned}\frac{1}{P_{\lambda_m} - P_m} \sum_{j=m+1}^{\lambda_m} p_j\left((1+r^j)x_{jn} - x_{mn}\right) &\geq \min_{m<j\leq\lambda_m} \left((1+r^j)x_{jn} - x_{mn}\right) \\ &\geq \min_{m<j\leq\lambda_m} (x_{jn} - x_{mn}) + \min_{m<j\leq\lambda_m} (r^j x_{jk}).\end{aligned}$$

We have

$$x_{mn} = \frac{P_m\sigma^{r,*,p,*}_{mn} - P_{m-1}\sigma^{r,*,p,*}_{m-1,n}}{mp_m(1+r^m)}$$

and

$$\frac{x_{mn}}{m} = \frac{P_m(\sigma^{r,*,p,*}_{mn} - \sigma^{r,*,p,*}_{m-1,n})}{mp_m(1+r^m)} + \frac{\sigma^{r,*,p,*}_{m-1,n}}{m(1+r^m)}.$$

Since (x_{mn}) is $(A^{r,*}, p, *; 1, 0)$ summable to l and $P_m = O(mp_m)$, then we have $x_{mn}/m \to 0$ as $m, n \to \infty$. Therefore, $r^m x_{mn} \to 0$ as $m, n \to \infty$. Hence, condition (4.51) clearly implies (4.49). Similarly, (4.52) implies (4.50). By Theorem 4.4.4, we have (4.6). □

Proof of Theorem 4.4.4. Necessity. The proof of similar to the proof of the necessity part of Theorem 4.3.5.

Sufficiency. Assume that (4.4) and one of conditions (4.49) and (4.50) is satisfied. Let $\epsilon > 0$ be given. By (4.49), there exists some $\lambda > 1$ such that

$$\limsup_{m,n\to\infty} \left| \frac{1}{P_{\lambda_m} - P_m} \sum_{j=m+1}^{\lambda_m} p_j \left((1+r^j)x_{jn} - x_{mn}\right) \right| < \epsilon. \tag{4.59}$$

By (4.4), we have

$$\begin{aligned} \limsup_{m,n\to\infty} |x_{mn} - \sigma^{r,*,p,*}_{\lambda_m,n}| &\le \limsup_{m,n\to\infty} \left| \frac{1}{P_{\lambda_m} - P_m} \sum_{j=m+1}^{\lambda_m} p_j \left((1+r^j)x_{jn} - x_{mn}\right) \right| \\ &\le \epsilon. \end{aligned} \tag{4.60}$$

Let any $\epsilon > 0$ be given. By (4.50), there exists some $0 < \lambda < 1$ such that

$$\limsup_{m,n\to\infty} \left| \frac{1}{P_m - P_{\lambda_m}} \sum_{j=\lambda_m+1}^{m} p_j \left(x_{mn} - (1+r^j)x_{jn}\right) \right| < \epsilon. \tag{4.61}$$

By (4.6), we have

$$\limsup_{m,n\to\infty} |x_{mn} - \sigma^{r,*,p,*}_{\lambda_m,n}| \le \limsup_{m,n\to\infty} \left| \frac{1}{P_m - P_{\lambda_m}} \sum_{j=\lambda_m+1}^{m} p_j \left(x_{mn} - (1+r^j)x_{jn}\right) \right| \le \epsilon. \tag{4.62}$$

By (4.60) or (4.62), in either case we obtain

$$\limsup_{m,n\to\infty} |x_{mn} - \sigma^{r,*,p,*}_{\lambda_m,n}| = 0 \tag{4.63}$$

whence it follows that

$$\lim_{m,n\to\infty} |x_{mn} - \sigma^{r,*,p,*}_{\lambda_m,n}| = 0. \tag{4.64}$$

Now, we conclude (4.6) from (4.4) and (4.64). □

Proof of Corollary 4.4.3. For $\lambda > 1$, we have the following inequality:

$$\begin{aligned} \left| \frac{1}{P_{\lambda_m} - P_m} \sum_{j=m+1}^{\lambda_m} p_j \left((1+r^j)x_{jn} - x_{mn}\right) \right| &\le \max_{m<j\le\lambda_m} |(1+r^j)x_{jn} - x_{mn}| \\ &\le \max_{m<j\le\lambda_m} |x_{jn} - x_{mn}| + \max_{m<j\le\lambda_m} |r^j x_{jk}|. \end{aligned}$$

As in the proof of Corollary 4.4.3, we have $x_{mn}/m \to 0$ as $m, n \to \infty$. Therefore, $r^m x_{mn} \to 0$ as $m, n \to \infty$. Hence, condition (4.51) clearly implies (4.49). Similarly, (4.52) implies (4.50). By Theorem 4.4.4, we have (4.6). □

4.5 Tauberian Theorems for the $(A^{*,\delta}, *, q; 0, 1)$ Summability Method

For $(A^{*,\delta}, *, q; 0, 1)$ summability method, similar results as obtained in Section 4.4 can be derived. We only list these results and omit the proofs of them.

Lemma 4.5.1. *Let (4.9) be satisfied. If a sequence (x_{mn}) is $(A^{*,\delta}, *, q; 0, 1)$ summable to a finite number l, then*

$$\lim_{m,n\to\infty} \frac{1}{Q_{\lambda_n} - Q_n} \sum_{k=n+1}^{\lambda_n} (1+\delta^k) q_k x_{mk} = l \tag{4.65}$$

for every $\lambda > 1$ and

$$\lim_{m,n\to\infty} \frac{1}{Q_n - Q_{\lambda_n}} \sum_{k=\lambda_n+1}^{n} (1+\delta^k) q_k x_{mk} = l \tag{4.66}$$

for every $0 < \lambda < 1$.

Theorem 4.5.2. *Let (4.9) be satisfied. If (x_{mn}) is a sequence of real numbers which is $(A^{*,\delta}, *, q; 0, 1)$ summable to a finite limit l, then (4.6) holds if and only if the following two conditions are satisfied:*

$$\sup_{\lambda>1} \liminf_{m,n\to\infty} \frac{1}{Q_{\lambda_n} - Q_n} \sum_{k=n+1}^{\lambda_n} q_k \left((1+\delta^k) x_{mk} - x_{mn}\right) \geq 0 \tag{4.67}$$

and

$$\sup_{0<\lambda<1} \liminf_{m,n\to\infty} \frac{1}{Q_n - Q_{\lambda_n}} \sum_{k=\lambda_n+1}^{n} q_k \left(x_{mn} - (1+\delta^k) x_{mk}\right) \geq 0, \tag{4.68}$$

in which case we necessarily have

$$\lim_{m,n\to\infty} \frac{1}{Q_{\lambda_n} - Q_n} \sum_{k=n+1}^{\lambda_n} q_k \left((1+\delta^k) x_{mk} - x_{mn}\right) = 0$$

for all $\lambda > 1$ and

$$\lim_{m,n\to\infty} \frac{1}{Q_n - Q_{\lambda_n}} \sum_{k=\lambda_n+1}^{n} q_k \left(x_{mn} - (1+\delta^k) x_{mk}\right) = 0$$

for all $0 < \lambda < 1$.

A sequence (x_{mn}) of real numbers is said to be slowly decreasing in sense $(0,1)$ if

$$\lim_{\lambda\to 1^+} \liminf_{m,n\to\infty} \min_{n<k\le\lambda_n} (x_{mk} - x_{mn}) \ge 0. \tag{4.69}$$

Note that condition (6.22) can be equivalently reformulated as follows:

$$\lim_{\lambda\to 1^-} \liminf_{m,n\to\infty} \min_{\lambda_n<k\le n} (x_{mn} - x_{mk}) \ge 0. \tag{4.70}$$

Corollary 4.5.3. *Let (4.9) and $Q_n = O(nq_n)$ be satisfied. If a sequence (x_{mn}) of real numbers is slowly decreasing in sense $(0,1)$, then (6.40) implies (4.6).*

Theorem 4.5.4. *Let (4.9) be satisfied. If (x_{mn}) is a sequence of complex numbers which is $(A^{*,\delta}, *, q; 0, 1)$ summable, then (x_{mn}) converges to the same limit if and only if one of the following two conditions is satisfied:*

$$\inf_{\lambda>1} \limsup_{m,n\to\infty} \left| \frac{1}{Q_{\lambda_n} - Q_n} \sum_{k=n+1}^{\lambda_n} q_k \left((1+\delta^k)x_{mk} - x_{mn}\right)\right| = 0 \tag{4.71}$$

or

$$\inf_{0<\lambda<1} \limsup_{m,n\to\infty} \left| \frac{1}{Q_n - Q_{\lambda_n}} \sum_{k=\lambda_n+1}^{n} q_k \left(x_{mn} - (1+\delta^k)x_{mk}\right)\right| = 0. \tag{4.72}$$

We recall that a sequence (x_{mn}) of complex numbers is said to be slowly oscillating in sense $(0,1)$ if

$$\lim_{\lambda\to 1^+} \limsup_{m,n\to\infty} \max_{n<k\le\lambda_n} |x_{mk} - x_{mn}| = 0. \tag{4.73}$$

An equivalent reformulation of (4.73) can be given as follows:

$$\lim_{\lambda\to 1^-} \limsup_{m,n\to\infty} \max_{\lambda_n<k\le n} |x_{mn} - x_{mk}| = 0. \tag{4.74}$$

Corollary 4.5.5. *Let (4.9) and $Q_n = O(nq_n)$ be satisfied. If a sequence (x_{mn}) of complex numbers is slowly oscillating in sense $(0,1)$, then (6.40) implies (4.6).*

Bibliography

[1] C. Aydın, F. Başar, On the new sequence spaces which include the spaces c_0 and c, Hokkaido Math. J. 33 (2004) 383–398.

[2] C. Aydın, F. Başar, Some new paranormed sequence spaces, Inform Sci. 160 (2004) 27–40.

[3] C. Aydın, F. Başar, Some new difference sequence spaces, Appl. Math. Comput. 157 (2004) 677–693.

[4] C. Aydın, F. Başar, Some new sequence spaces which include the spaces ℓ_p and ℓ_∞, Demonstr. Math. 38 (2005) 641–656.

[5] C. Aydın, F. Başar, Some generalizations of the sequence space a_p^r, Iran. J. Sci. Technol. Trans. A. Sci. 30 (A2) (2006) 175–190.

[6] F. Başar, A note on the triangle limitation methods, Fırat Univ. Fen. & Müh. Bil. Dergisi. 5 (1993) 113–117.

[7] C. P. Chen, J. M. Hsu, Tauberian theorems for weighted meansof double sequences, Analysis Mathematica. 26 (2000) 243–262.

[8] G. H. Hardy, Divergent Series, Oxford, Clarendon Press (1949).

[9] Ç. Kambak, İ. Çanak, An alternative proof of a Tauberian theorem for the weighted mean method of summability, Natl. Acad. Sci. Lett. 42 (2019) 355–357.

[10] F. Moricz, B. E. Rhoades, Necessary and sufficient Tauberian conditions for certain weighted mean methods of summability. II, Acta Math. Hungar. 102 (2004) 279–285.

[11] A. Pringsheim, Zur Theorie der zweifach unendlichen Zahlenfolgen, Math. Ann. 53 (1900) 289–321.

[12] R. Schmidt, Über divergente Folgen und lineare Mittelbildungen, Math. Z. 22 (1925) 89–152.

[13] Ö. Talo, F. Başar, Necessary and sufficient Tauberian conditions for the A^r method of summability, Math. J. Okayama Univ. 60 (2018) 209–219.

Chapter 5

On New Sequence Spaces Related to Domain of the Jordan Totient Matrix

Emrah Evren Kara

Düzce University, karaeevren@gmail.com

Necip Şimşek

İstanbul Commerce University, nsimsek@ticaret.edu.tr

Merve Ilkhan Kara

Düzce University, merveilkhan@gmail.com

5.1 Introduction and Background

A sequence space X is a vector subspace of the space of all sequences ω. The sets

$$c_0 = \{x = (x_n) \in \omega : \lim_n x_n = 0\},$$

$$c = \{x = (x_n) \in \omega : \lim_n x_n = L \text{ for some } L \in \mathbb{R}\},$$

$$\ell_\infty = \{x = (x_n) \in \omega : \sup_n |x_n| < \infty\},$$

$$\ell_p = \{x = (x_n) \in \omega : \sum_n |x_n|^p < \infty\},$$

$$cs = \{x = (x_n) \in \omega : \lim_n \sum_{k=1}^{n} x_k = L \text{ for some } L \in \mathbb{R}\},$$

DOI: 10.1201/9781003178200-5

$$bs = \{x = (x_n) \in \omega : \sup_n \left| \sum_{k=1}^{n} x_k \right| < \infty\}$$

are some of the classical sequence spaces.

The theory of sequence spaces is a fundamental subject in summability which has several applications essentially in functional analysis. The classical theory concerns with the generalization of the concept of convergence for series and sequences. The purpose is to assign a limit for non-convergent series and sequences by using a transformation which is defined by infinite special matrices. Since the most general linear mapping from a sequence space into another sequence space is given by an infinite matrix, we use matrices instead of a general linear mapping.

For an infinite matrix $A = (a_{nk})$, A-transform of a sequence $x = (x_n) \in \omega$ is the sequence $Ax = ((Ax)_n)$ obtained by the usual matrix product with entries

$$(Ax)_n = \sum_k a_{nk} x_k$$

provided that the series is convergent for each $n \in \mathbb{N}$. If the sequence Ax exists and $Ax \in Y$ for all $x \in X$, then A is called a matrix mapping from the sequence space X into the sequence space Y. (X, Y) denotes the class of all infinite matrices from X into Y.

The matrix domain of A in the space X is defined as

$$X_A = \{x \in \omega : Ax \in X\}$$

which is also a sequence space. By using the domains of some special summability matrices, several new spaces have been introduced. Some examples for the domains of some special matrices in classical sequence spaces can be seen from references [1–14].

Since an infinite matrix can be considered as a linear operator between two sequence spaces in some cases, the study of the theory of matrix transformations in sequence spaces has aroused interest over years and it has linked with particular consequences in summability theory. The theory of matrix transformations provides the necessary and sufficient conditions for a matrix to map a sequence space into another sequence space. For matrix transformations and some applications in summmability theory, we refer to [15–24].

For more about summability theory and its applications, we refer to [25].

The Jordan's function $J_r : \mathbb{N} \to \mathbb{N}$ of order r is an arithmetic function which generalizes the well-known Euler totient function. Here and what follows r is a positive integer and $\mathbb{N}$ is the set of all natural numbers $1, 2, 3, \ldots$. The value $J_r(n)$ equals to the number of r-tuples of positive integers all less than or equal to n that form a co-prime $(r+1)$-tuples together with n. Note that J_1 is Euler's totient function.

The Jordan function J_r is multiplicative, i.e. for $n_1, n_2 \in \mathbb{N}$ with $\gcd(n_1, n_2) = 1$, the relation $J_r(n_1 n_2) = J_r(n_1) J_r(n_2)$ holds.

Let $p_1^{a_1} p_2^{a_2} \ldots p_i^{a_i}$ be the unique prime decomposition of $n \in \mathbb{N}$, then

$$J_r(n) = n^r \left(1 - \frac{1}{p_1^r}\right)\left(1 - \frac{1}{p_2^r}\right) \ldots \left(1 - \frac{1}{p_i^r}\right).$$

Also, the following equations hold:

$$\sum_{k|n} J_r(k) = n^r.$$

$$\sum_{k|n} \frac{\mu(k)}{k^r} = \frac{J_r(n)}{n^r}.$$

Let φ and μ denote the Euler function and the Möbius function, respectively. For every $n \in \mathbb{N}$ with $n > 1$, $\varphi(n)$ is the number of positive integers less than n which are co-prime with n and $\varphi(1) = 1$. If $p_1^{a_1} p_2^{a_2} ... p_i^{a_i}$ is the prime factorization of a natural number $n > 1$, then

$$\varphi(n) = n(1 - \frac{1}{p_1})(1 - \frac{1}{p_2})...(1 - \frac{1}{p_i}).$$

For every $n \in \mathbb{N}$ with $n > 1$, μ is defined as

$$\mu(n) = \begin{cases} (-1)^r & \text{if } n = p_1 p_2 ... p_i, \text{ where } p_1, p_2, ..., p_i \text{ are} \\ & \text{non-equivalent prime numbers} \\ 0 & \text{if } p^2 \mid n \text{ for some prime number } p \end{cases}$$

and $\mu(1) = 1$. If $p_1^{a_1} p_2^{a_2} ... p_i^{a_i}$ is the prime factorization of a natural number $n > 1$, then

$$\sum_{k|n} k\mu(k) = (1 - p_1)(1 - p_2)...(1 - p_i).$$

Also, the equality

$$\sum_{k|n} \mu(k) = 0 \tag{5.1}$$

holds except for $n = 1$.

Firstly, using Euler totient function, a regular infinite matrix $\Phi = (\phi_{nk})$ is defined by İlkhan and Kara [26] as

$$\phi_{nk} = \begin{cases} \frac{\varphi(k)}{n}, & \text{if } k \mid n \\ 0, & \text{if } k \nmid n. \end{cases}$$

Recently, İlkhan [27] introduced the matrix domains of the matrix Φ in the spaces c and c_0.

More recently, using Jordan totient function, İlkhan et all. [28] defined a new regular matrix $\Upsilon^r = (\upsilon_{nk}^r)$as

$$\upsilon_{nk}^r = \begin{cases} \frac{J_r(k)}{n^r}, & \text{if } k \mid n \\ 0, & \text{if } k \nmid n \end{cases}$$

for each $r \in \mathbb{N}$ or more precisely

$$\Upsilon^r = \begin{bmatrix} \frac{J_r(1)}{1} & 0 & 0 & 0 & 0 & 0 & \cdots \\ \frac{J_r(1)}{2^r} & \frac{J_r(2)}{2^r} & 0 & 0 & 0 & 0 & \cdots \\ \frac{J_r(1)}{3^r} & 0 & \frac{J_r(3)}{3^r} & 0 & 0 & 0 & \cdots \\ \frac{J_r(1)}{4^r} & \frac{J_r(2)}{4^r} & 0 & \frac{J_r(4)}{4^r} & 0 & 0 & \cdots \\ \frac{J_r(1)}{5^r} & 0 & 0 & 0 & \frac{J_r(5)}{5^r} & 0 & \cdots \\ \frac{J_r(1)}{6^r} & \frac{J_r(2)}{6^r} & \frac{J_r(3)}{6^r} & 0 & 0 & \frac{J_r(6)}{6^r} & \cdots \\ \cdots & \cdots & \cdots & \cdots & \cdots & \cdots & \cdots \end{bmatrix}.$$

If $r = 1$, the Euler totient matrix is obtained. It is observed that this special transformation is regular which means it maps c into c and preserves the limit. In fact, the matrix Υ^r satisfies the following three conditions:

1. $\sup_{n\in\mathbb{N}}(\sum_{k=1}^{\infty} |\upsilon_{nk}^r|) < \infty$,
2. $\lim_{n\to\infty} \upsilon_{nk}^r = 0$ for each fixed $k, r \in \mathbb{N}$,
3. $\lim_{n\to\infty} \sum_{k=1}^{\infty} \upsilon_{nk}^r = 1$.

The inverse $(\Upsilon^r)^{-1} = ((\upsilon_{nk}^r)^{-1})$ of the matrix Υ^r is computed as

$$(\upsilon_{nk}^r)^{-1} = \begin{cases} \frac{\mu(\frac{n}{k})}{J_r(n)} k^r, & \text{if } k \mid n \\ 0, & \text{if } k \nmid n \end{cases}$$

or more precisely

$$(\Upsilon^r)^{-1} = \begin{bmatrix} \frac{\mu(1)}{J_r(1)} & 0 & 0 & 0 & 0 & 0 & \cdots \\ \frac{\mu(2)}{J_r(2)} & 2^r\frac{\mu(1)}{J_r(2)} & 0 & 0 & 0 & 0 & \cdots \\ \frac{\mu(3)}{J_r(3)} & 0 & 3^r\frac{\mu(1)}{J_r(3)} & 0 & 0 & 0 & \cdots \\ \frac{\mu(4)}{J_r(4)} & 2^r\frac{\mu(2)}{J_r(4)} & 0 & 4^r\frac{\mu(1)}{J_r(4)} & 0 & 0 & \cdots \\ \frac{\mu(5)}{J_r(5)} & 0 & 0 & 0 & 5^r\frac{\mu(1)}{J_r(5)} & 0 & \cdots \\ \frac{\mu(6)}{J_r(6)} & 2^r\frac{\mu(3)}{J_r(6)} & 3^r\frac{\mu(2)}{J_r(6)} & 0 & 0 & 6^r\frac{\mu(1)}{J_r(6)} & \cdots \\ \cdots & \cdots & \cdots & \cdots & \cdots & \cdots & \cdots \end{bmatrix}.$$

The main purpose of this study is to define matrix domains of the Jordan totient matrix in the spaces c_0, c, ℓ_∞ and give some inclusion theorems related to the resulting spaces. Finally, the dual spaces of the resulting spaces are determined and certain matrix mappings are characterized on these spaces.

5.2 The Domains of the Jordan Totient Matrix in the Spaces c_0, c, ℓ_∞

In this section, we introduce the spaces consisting of sequences whose Υ^r-transforms are in the spaces c_0, c, ℓ_∞. By using usual matrix product, the Υ^r-transform of a sequence $x = (x_n) \in \omega$ is obtained as the sequence

$$y = \Upsilon^r x = ((\Upsilon^r x)_n) = \left(\frac{1}{n^r} \sum_{k|n} J_r(k) x_k \right).$$

Also,

$$x_n = \sum_{k|n} \frac{\mu(\frac{n}{k})}{J_r(n)} k^r y_k \tag{5.2}$$

holds.

The resulting spaces are

$$c_0(\Upsilon^r) = \left\{ x = (x_n) \in \omega : \lim_n \frac{1}{n^r} \sum_{k|n} J_r(k) x_k = 0 \right\}$$

$$c(\Upsilon^r) = \left\{ x = (x_n) \in \omega : \lim_n \frac{1}{n^r} \sum_{k|n} J_r(k) x_k \text{ exists} \right\}$$

$$\ell_\infty(\Upsilon^r) = \left\{ x = (x_n) \in \omega : \sup_n \left| \frac{1}{n^r} \sum_{k|n} J_r(k) x_k \right| < \infty \right\}.$$

Theorem 5.2.1. *The spaces $c_0(\Upsilon^r)$, $c(\Upsilon^r)$, $\ell_\infty(\Upsilon^r)$ are Banach spaces according to the norm $\|x\|_{\Upsilon^r} = \sup_{n \in \mathbb{N}} \{ | \frac{1}{n^r} \sum_{k|n} J_r(k) x_k | \}$.*

Proof. It is a routine verification to prove this result. □

Theorem 5.2.2. *The spaces $c_0(\Upsilon^r)$, $c(\Upsilon^r)$, $\ell_\infty(\Upsilon^r)$ are linearly isomorphic to c_0, c, ℓ_∞, respectively.*

Proof. Let X be c_0, c or ℓ_∞. Then the mapping $L : X(\Upsilon^r) \to X$ defined as $L(x) = \Upsilon^r x$ is linear, injective, surjective and L preserves norms since the equality $\|x\|_{\Upsilon^r} = \|Lx\|_X$ holds. Hence we conclude that $c_0(\Upsilon^r)$, $c(\Upsilon^r)$, $\ell_\infty(\Upsilon^r)$ are linearly isomorphic to c_0, c, ℓ_∞, respectively. □

Theorem 5.2.3. *The following inclusions strictly hold.*

1. $c_0(\Upsilon^r) \subset c(\Upsilon^r) \subset \ell_\infty(\Upsilon^r)$.
2. $c_0 \subset c_0(\Upsilon^r)$.
3. $c \subset c(\Upsilon^r)$.

Proof. One can easily observe that inclusions in (1) strictly holds since the inclusions $c_0 \subset c \subset \ell_\infty$ are strict.

The inclusion in (2) follows from the fact that $\Upsilon^r \in (c_0, c_0)$ since $\sup_{n\in\mathbb{N}}(\sum_{k=1}^{\infty}|\upsilon_{nk}^r|) < \infty$ and $\lim_{n\to\infty}\upsilon_{nk}^r = 0$ for each fixed $k, r \in \mathbb{N}$.

The inclusion in (3) follows from the fact that $\Upsilon^r \in (c, c)$ since $\sup_{n\in\mathbb{N}}(\sum_{k=1}^{\infty}|\upsilon_{nk}^r|) < \infty$, $\lim_{n\to\infty}\upsilon_{nk}^r = 0$ for each fixed $k, r \in \mathbb{N}$ and $\lim_{n\to\infty}\sum_{k=1}^{\infty}\upsilon_{nk}^r = 1$.

Also, the following example shows that the last two inclusions are strict. □

Example 5.2.4. *Consider the sequence (x_n) given by*

$$x_n = \begin{cases} 1, & \text{if } n = 2 \cdot 3 \cdot \ldots \cdot p_l \\ 0, & \text{otherwise}, \end{cases}$$

where p_l is the successive prime number. Clearly, this sequence is not convergent in the usual sense. On the other hand, we have

$$\lim_{n\to\infty} y_n = \lim_{n\to\infty}\frac{1}{n^r}\sum_{k|n} J_r(k)x_k = 0.$$

Let the largest divisor of n, which is product of consecutive primes starting with 2, be $2 \cdot 3 \cdot \ldots \cdot p_i$, and set $n = (2 \cdot 3 \cdot \ldots \cdot p_i)k_n$ and $(p_{i+1} \nmid k_n)$. Then, we write

$$y_n = \frac{1}{k_n^r}\cdot\frac{1}{2^r \cdot 3^r \cdot \ldots \cdot p_i^r}\sum_{k|2\cdot 3\cdot\ldots\cdot p_i} J_r(k)x_k.$$

As $n \to \infty$, $k_n^r + p_i^r \to \infty$. It is therefore sufficient to show that

$$\frac{1}{2^r \cdot 3^r \cdot \ldots \cdot p_i^r}\sum_{k|2\cdot 3\cdot\ldots\cdot p_i} J_r(k)x_k \to 0$$

as $i \to \infty$. This will follow from

$$\sum_{k|2\cdot 3\cdot\ldots\cdot p_i} J_r(k)x_k = J_r(2) + J_r(2\cdot 3) + \ldots + J_r(2\cdot 3\cdot\ldots\cdot p_{i-1}) + J_r(2\cdot 3\cdot\ldots\cdot p_i)$$

$$
\begin{aligned}
&\leq \sum_{k|2\cdot 3\cdot \ldots \cdot p_{i-1}} J_r(k) + J_r(2\cdot 3\cdot \ldots \cdot p_i) \\
&= 2^r \cdot 3^r \cdot \ldots \cdot p_{i-1}^r + J_r(2\cdot 3\cdot \ldots \cdot p_i) \\
&= 2^r \cdot 3^r \cdot \ldots \cdot p_{i-1}^r + 2^r \cdot 3^r \cdot \ldots \cdot p_i^r \prod_{v=1}^{i}\left(1-\frac{1}{p_v^r}\right) \\
&= 2^r \cdot 3^r \cdot \ldots \cdot p_i^r \left(\frac{1}{p_i^r} + \prod_{v=1}^{i}\left(1-\frac{1}{p_v^r}\right)\right) \\
&= o(2^r \cdot 3^r \cdot \ldots \cdot p_i^r).
\end{aligned}
$$

Theorem 5.2.5. *Neither of the spaces ℓ_∞ and $c(\Upsilon^r)$ includes the other one.*

Proof. We have $\Upsilon^r \notin (\ell_\infty, c)$ since $(c,c)\cap(\ell_\infty, c)=\emptyset$. Then, we can find an $x \in \ell_\infty$ such that $x \notin c(\Upsilon^r)$ which means $\ell_\infty \nsubseteq c(\Upsilon^r)$.

On the other hand, it is known that if a regular matrix sums a bounded divergent sequence, then it also sums some unbounded sequence. This means $c(\Upsilon^r) \nsubseteq \ell_\infty$. □

5.3 The α-, β- and γ-Duals

The following conditions will be used to characterize certain classes of matrices. Here and in what follows, $\mathcal{N}$ denotes the family of all finite subsets of $\mathbb{N}$.

$$\sup_{N,M\in\mathcal{N}} \left|\sum_{n\in N}\sum_{k\in M} a_{nk}\right| < \infty. \tag{5.3}$$

$$\sup_{n\in\mathbb{N}} \sum_k |a_{nk}| < \infty. \tag{5.4}$$

$$\lim_{n\to\infty} a_{nk} \text{ exists for each } k\in\mathbb{N}. \tag{5.5}$$

$$\lim_{n\to\infty}\sum_k a_{nk} \text{ exists.} \tag{5.6}$$

$$\sup_{N\in\mathcal{N}}\sum_k\left|\sum_{n\in N} a_{nk}\right| < \infty. \tag{5.7}$$

$$\lim_{n\to\infty}\sum_k |a_{nk}| = \sum_k \left|\lim_{n\to\infty} a_{nk}\right|. \tag{5.8}$$

Lemma 5.3.1. *[29] The necessary and sufficient conditions for $A = (a_{nk}) \in (X, Y)$ with $X \in \{c_0, c, \ell_\infty\}$ and $Y \in \{\ell_1, c, \ell_\infty\}$ can be read from Table 5.1.*

From \ To	ℓ_1	c	ℓ_∞
c_0	**1.**	**2.**	**4.**
c	**1.**	**3.**	**4.**
ℓ_∞	**5.**	**6.**	**4.**

TABLE 5.1: The characterization of the class (X, Y), where $X \in \{c_0, c, \ell_\infty\}$ and $Y \in \{\ell_1, c, \ell_\infty\}$.

1. (5.3) *holds.*
2. (5.4) *and* (5.5) *hold.*
3. (5.4), (5.5) *and* (5.6) *hold.*
4. (5.4) *holds.*
5. (5.7) *holds.*
6. (5.5) *and* (5.8) *hold.*

The multiplier space of the spaces X and Y consists of sequences $t \in \omega$ such that $tx \in Y$ for any $x \in X$ and it is denoted by $M(X, Y)$. If $Y = \ell_1$, $Y = cs$ or $Y = bs$, the multiplier spaces $M(X, \ell_1)$, $M(X, cs)$ and $M(X, bs)$ are called as α-, β- and γ-duals of the space X and denoted by X^α, X^β and X^γ, respectively.

Theorem 5.3.2. *The α-dual of the spaces $c_0(\Upsilon^r)$, $c(\Upsilon^r)$, $\ell_\infty(\Upsilon^r)$ are as follows:*

$$(c_0(\Upsilon^r))^\alpha = (c(\Upsilon^r))^\alpha = \{t = (t_n) \in \omega : \sup_{N,M \in \mathcal{N}} \left| \sum_{n \in N} \sum_{k \in M, k|n} t_n \frac{\mu(\frac{n}{k})}{J_r(n)} k^r \right| < \infty\},$$

$$(\ell_\infty(\Upsilon^r))^\alpha = \{t = (t_n) \in \omega : \sup_{N \in \mathcal{N}} \sum_k \left| \sum_{n \in N, k|n} t_n \frac{\mu(\frac{n}{k})}{J_r(n)} k^r \right| < \infty\}.$$

Proof. Let $t = (t_n) \in \omega$ and $D = (d_{nk})$ be an infinite matrix such that

$$d_{nk} = \begin{cases} \frac{\mu(\frac{n}{k})}{J_r(n)} k^r t_n, & \text{if } k \mid n \\ 0, & \text{if } k \nmid n. \end{cases}$$

From (5.2), we obtain that

$$t_n x_n = \sum_{k|n} \frac{\mu(\frac{n}{k})}{J_r(n)} k^r t_n y_k = (Dy)_n$$

for any $x = (x_n) \in c_0(\Upsilon^r)$, $c(\Upsilon^r)$ or $\ell_\infty(\Upsilon^r)$. This yields $tx \in \ell_1$ for $x = (x_n) \in c_0(\Upsilon^r)$, $c(\Upsilon^r)$ or $\ell_\infty(\Upsilon^r)$ if and only if $Dy \in \ell_1$ for $y \in c_0$, c or ℓ_∞. That is, $t \in (c_0(\Upsilon^r))^\alpha$, $(c(\Upsilon^r))^\alpha$ or $(\ell_\infty(\Upsilon^r))^\alpha$ if and only if $D \in (c_0, \ell_1) = (c, \ell_1)$ or $D \in (\ell_\infty, \ell_1)$. Hence, by Lemma 5.3.1, we deduce that

$$(c_0(\Upsilon^r))^\alpha = (c(\Upsilon^r))^\alpha = \{t = (t_n) \in \omega : \sup_{N,M \in \mathcal{N}} \left| \sum_{n \in N} \sum_{k \in M, k|n} t_n \frac{\mu(\frac{n}{k})}{J_r(n)} k^r \right| < \infty\},$$

$$(\ell_\infty(\Upsilon^r))^\alpha = \{t = (t_n) \in \omega : \sup_{N \in \mathcal{N}} \sum_k \left| \sum_{n \in N, k|n} t_n \frac{\mu(\frac{n}{k})}{J_r(n)} k^r \right| < \infty\}.$$

□

Theorem 5.3.3. *[30, Theorem 3.1] Let $B = (b_{nk})$ be defined via a sequence $t = (t_n) \in \omega$ and the inverse $\mathcal{V} = (v_{nk})$ of the triangle matrix $\mathcal{U} = (u_{nk})$ by*

$$b_{nk} = \begin{cases} \sum_{l=k}^{n} t_l v_{lk}, & \text{if } 1 \leq k \leq n \\ 0, & \text{if } k > n. \end{cases}$$

Then,

$$X_U^\beta = \{t = (t_n) \in \omega : B \in (X, c)\},$$

and

$$X_U^\gamma = \{t = (t_n) \in \omega : B \in (X, \ell_\infty)\}.$$

By applying Theorem 5.3.3 with Lemma 5.3.1, the β- and γ-duals of the spaces $c_0(\Upsilon^r)$, $c(\Upsilon^r)$, $\ell_\infty(\Upsilon^r)$ are established in the following theorem.

Theorem 5.3.4. *Define the sets A_1, A_2, A_3, A_4 and A_5 as follows:*

$$A_1 = \left\{ t = (t_k) \in \omega : \sup_n \sum_k \left| \sum_{j=k,k|j}^{n} t_j \frac{\mu(\frac{j}{k})}{J_r(j)} k^r \right| < \infty \right\},$$

$$A_2 = \left\{ t = (t_k) \in \omega : \lim_n \sum_{j=k,k|j}^{n} t_j \frac{\mu(\frac{j}{k})}{J_r(j)} k^r \text{ exists for each } k \in \mathbb{N} \right\},$$

$$A_3 = \left\{ t = (t_k) \in \omega : \lim_n \sum_k \sum_{j=k,k|j}^{n} t_j \frac{\mu(\frac{j}{k})}{J_r(j)} k^r \text{ exists} \right\},$$

$$A_4 = \left\{ t = (t_k) \in \omega : \sup_n \sum_k \left| \sum_{j=k,k|j}^{n} t_j \frac{\mu(\frac{j}{k})}{J_r(j)} k^r \right|^q < \infty \right\},$$

$$A_5 = \left\{ t = (t_k) \in \omega : \lim_n \sum_k \left| \sum_{j=k,k|j}^{n} t_j \frac{\mu(\frac{j}{k})}{J_r(j)} k^r \right| = \sum_k \left| \sum_{j=k,k|j}^{\infty} t_j \frac{\mu(\frac{j}{k})}{J_r(j)} k^r \right| \right\}$$

Then,

1. $(c_0(\Upsilon^r))^\beta = A_1 \cap A_2$, $(c(\Upsilon^r))^\beta = A_1 \cap A_2 \cap A_3$ *and* $(\ell_\infty(\Upsilon^r))^\beta = A_2 \cap A_5$.
2. $(c_0(\Upsilon^r))^\gamma = (c(\Upsilon^r))^\gamma = A_1$ *and* $(\ell_\infty(\Upsilon^r))^\gamma = A_4$.

5.4 Certain Matrix Transformations

In this section, we give the characterization of certain classes of matrices. In addition to Lemma 5.3.1, we need the following result.

$$\lim_{n\to\infty} a_{nk} = 0 \text{ for each } k \in \mathbb{N}. \tag{5.9}$$

$$\lim_{n\to\infty} \sum_k a_{nk} = 0. \tag{5.10}$$

$$\lim_{n\to\infty} \sum_k |a_{nk}| = 0. \tag{5.11}$$

To / From	c_0
c_0	**1*.**
c	**2*.**
ℓ_∞	**3*.**

TABLE 5.2: The characterization of the class (X, c_0), where $X \in \{c_0, c, \ell_\infty\}$.

Lemma 5.4.1. *[29] The necessary and sufficient conditions for $A = (a_{nk}) \in (X, c_0)$ with $X \in \{c_0, c, \ell_\infty\}$ can be read from Table 5.2.*

1.* (5.4) *and* (5.9) *hold.*

2.* (5.9) *and* (5.10) *hold.*

3.* (5.9) *and* (5.11) *hold.*

We obtain the following result from Theorem 4.1 in [31] and we use this result with Lemma 5.3.1 and Lemma 5.4.1 to characterize the classes of matrices from $c_0(\Upsilon^r)$, $c(\Upsilon^r)$, $\ell_\infty(\Upsilon^r)$ into $\ell_1, c_0, c, \ell_\infty$.

Theorem 5.4.2. *Let $X = c_0$, c or ℓ_∞ and Y be any subset of ω. Then, $A = (a_{nk}) \in (X_{\Upsilon^r}, Y)$ if and only if $T^{(n)} = \left(t_{lk}^{(n)}\right) \in (X, c)$ for each fixed $n \in \mathbb{N}$ and $T = (t_{nk}) \in (X, Y)$, where*

$$t_{lk}^{(n)} = \begin{cases} \sum_{j=k,k|j}^{l} a_{nj} \frac{\mu(\frac{j}{k})}{J_r(j)} k^r, & 1 \le k \le l \\ 0, & k > l \end{cases}$$

and

$$t_{nk} = \sum_{j=k,k|j}^{\infty} a_{nj} \frac{\mu(\frac{j}{k})}{J_r(j)} k^r.$$

Proof. Let $A \in (X_{\Upsilon^r}, Y)$ and $x \in X_{\Upsilon^r}$. Then, we have

$$\begin{aligned} \sum_{k=1}^{l} a_{nk} x_k &= \sum_{k=1}^{l} a_{nk} \left(\sum_{j|k} \frac{\mu(\frac{k}{j})}{J_r(k)} j y_j \right) \\ &= \sum_{k=1}^{l} \left(\sum_{j=k,k|j}^{l} a_{nj} \frac{\mu(\frac{j}{k})}{J_r(j)} k \right) y_k = \sum_{k=1}^{l} t_{lk}^{(n)} y_k. \end{aligned} \tag{5.12}$$

Since Ax exists, this yields that $T^{(n)} \in (X, c)$ for each fixed $n \in \mathbb{N}$. We deduce that $Ax = Ty$ as $l \to \infty$ in (5.12). Hence, $Ax \in Y$ implies that $Ty \in Y$; that is $T \in (X, Y)$.

For the converse, suppose that $T^{(n)} = \left(t_{lk}^{(n)}\right) \in (X, c)$ for each fixed $n \in \mathbb{N}$ and $T = (t_{nk}) \in (X, Y)$. Let $x \in X_{\Upsilon^r}$. Then, $(t_{nk}) \in X^\beta$ for each fixed $n \in \mathbb{N}$ implies that $(a_{nk}) \in X_{\Upsilon^r}^\beta$ for each fixed $n \in \mathbb{N}$. Hence, Ax exists. From equality (5.12), it follows that $Ax = Ty$ as $l \to \infty$. This proves that $A \in (X_{\Upsilon^r}, Y)$. □

Theorem 5.4.3. *Given any infinite matrix $A = (a_{nk})$, we have*

1. *$A \in (c_0(\Upsilon^r), \ell_1)$ if and only if*

$$\sup_{l\in\mathbb{N}} \sum_{k=1}^{l} \left| \sum_{j=k,k|j}^{l} a_{nj} \frac{\mu(\frac{j}{k})}{J_r(j)} k^r \right| < \infty \text{ for each fixed } n \in \mathbb{N}, \tag{5.13}$$

$$\lim_{l\to\infty} \sum_{j=k,k|j}^{l} a_{nj} \frac{\mu(\frac{j}{k})}{J_r(j)} k^r \text{ exists for each fixed } n, k \in \mathbb{N} \tag{5.14}$$

and

$$\sup_{N,M\in\mathcal{N}} \left| \sum_{n\in N} \sum_{k\in M} \sum_{j=k,k|j}^{\infty} a_{nj} \frac{\mu(\frac{j}{k})}{J_r(j)} k^r \right| < \infty. \tag{5.15}$$

2. *$A \in (c_0(\Upsilon^r), c_0)$ if and only if (5.13) and (5.14) hold, and*

$$\sup_{n\in\mathbb{N}} \sum_{k} \left| \sum_{j=k,k|j}^{\infty} a_{nj} \frac{\mu(\frac{j}{k})}{J_r(j)} k^r \right| < \infty \tag{5.16}$$

and

$$\lim_{n\to\infty} \sum_{j=k,k|j}^{\infty} a_{nj} \frac{\mu(\frac{j}{k})}{J_r(j)} k^r = 0 \text{ for each } k \in \mathbb{N} \tag{5.17}$$

also hold.

3. *$A \in (c_0(\Upsilon^r), c)$ if and only if (5.13), (5.14) and (5.16) hold, and*

$$\lim_{n\to\infty} \sum_{j=k,k|j}^{\infty} a_{nj} \frac{\mu(\frac{j}{k})}{J_r(j)} k^r \text{ exists for each } k \in \mathbb{N}. \tag{5.18}$$

4. *$A \in (c_0(\Upsilon^r), \ell_\infty)$ if and only if (5.13), (5.14) and (5.16) hold.*

Proof. The proof follows from Lemma 5.3.1, Lemma 5.4.1 and Theorem 5.4.2. □

Theorem 5.4.4. *Given any infinite matrix* $A = (a_{nk})$, *we have*

1. $A \in (c(\Upsilon^r), \ell_1)$ *if and only if* (5.13), (5.14) *and* (5.15) *hold, and*

$$\lim_{l\to\infty} \sum_{k=1}^{l} \sum_{j=k,k|j}^{l} a_{nj} \frac{\mu(\frac{j}{k})}{J_r(j)} k^r \text{ exists for each } n \in \mathbb{N}. \tag{5.19}$$

2. $A \in (c(\Upsilon^r), c_0)$ *if and only if* (5.13), (5.14), (5.19) *and* (5.17) *hold, and*

$$\lim_{n\to\infty} \sum_{k} \sum_{j=k,k|j}^{\infty} a_{nj} \frac{\mu(\frac{j}{k})}{J_r(j)} k^r = 0. \tag{5.20}$$

3. $A \in (c(\Upsilon^r), c)$ *if and only if* (5.13), (5.14), (5.19), (5.16) *and* (5.18) *hold, and*

$$\lim_{n\to\infty} \sum_{k} \sum_{j=k,k|j}^{\infty} a_{nj} \frac{\mu(\frac{j}{k})}{J_r(j)} k^r \text{ exists.} \tag{5.21}$$

4. $A \in (c(\Upsilon^r), \ell_\infty)$ *if and only if* (5.13), (5.14), (5.19) *and* (5.16) *hold.*

Proof. The proof follows from Lemma 5.3.1, Lemma 5.4.1 and Theorem 5.4.2. □

Theorem 5.4.5. *Given any infinite matrix* $A = (a_{nk})$, *we have*

1. $A \in (\ell_\infty(\Upsilon^r), \ell_1)$ *if and only if* (5.14) *holds, and*

$$\lim_{l\to\infty} \sum_{k} \left| \sum_{j=k,k|j}^{l} a_{nj} \frac{\mu(\frac{j}{k})}{J_r(j)} k^r \right| = \sum_{k} \left| \lim_{l\to\infty} \sum_{j=k,k|j}^{l} a_{nj} \frac{\mu(\frac{j}{k})}{J_r(j)} k^r \right| \tag{5.22}$$

and

$$\sup_{N\in\mathcal{N}} \sum_{k} \left| \sum_{n\in N} \sum_{j=k,k|j}^{\infty} a_{nj} \frac{\mu(\frac{j}{k})}{J_r(j)} k^r \right| < \infty.$$

2. $A \in (\ell_\infty(\Upsilon^r), c_0)$ *if and only if* (5.14), (5.22) *and* (5.17) *hold, and*

$$\lim_{n} \sum_{k} \left| \sum_{j=k,k|j}^{\infty} a_{nj} \frac{\mu(\frac{j}{k})}{J_r(j)} k^r \right| = 0.$$

3. $A \in (\ell_\infty(\Upsilon^r), c)$ *if and only if* (5.14), (5.22) *and* (5.18) *hold, and*

$$\lim_{n} \sum_{k} \left| \sum_{j=k,k|j}^{\infty} a_{nj} \frac{\mu(\frac{j}{k})}{J_r(j)} k^r \right| = \sum_{k} \left| \lim_{n} \sum_{j=k,k|j}^{\infty} a_{nj} \frac{\mu(\frac{j}{k})}{J_r(j)} k^r \right|.$$

4. $A \in (\ell_\infty(\Upsilon^r), \ell_\infty)$ *if and only if* (5.14), (5.22) *and* (5.16) *hold.*

Proof. The proof follows from Lemma 5.3.1, Lemma 5.4.1 and Theorem 5.4.2. □

Corollary 5.4.6. *Given any infinite matrix* $A = (a_{nk})$, *we have*

1. $A \in (c_0(\Upsilon^r), cs_0)$ *if and only if* (5.13) *and* (5.14) *hold, and*

$$\sup_{n\in\mathbb{N}} \sum_k \left| \sum_{l=1}^{n} \sum_{j=k,k|j}^{\infty} a_{lj} \frac{\mu(\frac{j}{k})}{J_r(j)} k^r \right| < \infty \tag{5.23}$$

and

$$\lim_{n\to\infty} \sum_{l=1}^{n} \sum_{j=k,k|j}^{\infty} a_{lj} \frac{\mu(\frac{j}{k})}{J_r(j)} k^r = 0 \text{ for each } k \in \mathbb{N}. \tag{5.24}$$

2. $A \in (c_0(\Upsilon^r), cs)$ *if and only if* (5.13), (5.14) *and* (5.23) *hold, and*

$$\lim_{n\to\infty} \sum_{l=1}^{n} \sum_{j=k,k|j}^{\infty} a_{lj} \frac{\mu(\frac{j}{k})}{J_r(j)} k^r \text{ exists for each } k \in \mathbb{N}. \tag{5.25}$$

3. $A \in (c_0(\Upsilon^r), bs)$ *if and only if* (5.13), (5.14) *and* (5.23) *hold.*

Corollary 5.4.7. *Given any infinite matrix* $A = (a_{nk})$, *we have*

1. $A \in (c(\Upsilon^r), cs_0)$ *if and only if* (5.13), (5.14), (5.19), (5.23) *and* (5.24) *hold, and*

$$\lim_{n\to\infty} \sum_k \sum_{l=1}^{n} \sum_{j=k,k|j}^{\infty} a_{lj} \frac{\mu(\frac{j}{k})}{J_r(j)} k^r = 0. \tag{5.26}$$

2. $A \in (c(\Upsilon^r), cs)$ *if and only if* (5.13), (5.14), (5.19), (5.23) *and* (5.25) *hold, and*

$$\lim_{n\to\infty} \sum_k \sum_{l=1}^{n} \sum_{j=k,k|j}^{\infty} a_{lj} \frac{\mu(\frac{j}{k})}{J_r(k)} k^r \text{ exists.} \tag{5.27}$$

3. $A \in (c(\Upsilon^r), bs)$ *if and only if* (5.13), (5.14), (5.19) *and* (5.23) *hold.*

Corollary 5.4.8. *Given any infinite matrix* $A = (a_{nk})$, *we have*

1. $A \in (\ell_\infty(\Upsilon^r), cs_0)$ *if and only if* (5.14), (5.22) *and* (5.24) *hold, and*

$$\lim_{n} \sum_k \left| \sum_{l=1}^{n} \sum_{j=k,k|j}^{\infty} a_{lj} \frac{\mu(\frac{j}{k})}{J_r(j)} k^r \right| = 0.$$

2. $A \in (\ell_\infty(\Upsilon^r), cs)$ *if and only if* (5.14), (5.22) *and* (5.25) *hold, and*

$$\lim_n \sum_k \left| \sum_{l=1}^{n} \sum_{j=k,k|j}^{\infty} a_{lj} \frac{\mu(\frac{j}{k})}{J_r(j)} k^r \right| = \sum_k \left| \lim_n \sum_{l=1}^{n} \sum_{j=k,k|j}^{\infty} a_{lj} \frac{\mu(\frac{j}{k})}{J_r(j)} k^r \right|.$$

3. $A \in (\ell_\infty(\Upsilon^r), bs)$ *if and only if* (5.14), (5.22) *and* (5.23) *hold.*

Finally, we need the following condition:

$$\sup_{n,k \in \mathbb{N}} |a_{nk}| < \infty. \tag{5.28}$$

Lemma 5.4.9. *[29] The necessary and sufficient conditions for* $A = (a_{nk}) \in (\ell_1, Y)$ *with* $Y \in \{c_0, c, \ell_\infty\}$ *can be read from Table 5.3.*

To / From	c_0	c	ℓ_∞
ℓ_1	**1**.**	**2**.**	**3**.**

TABLE 5.3: The characterization of the class (X, Y), where $X \in \{\ell_1, \ell_\infty\}$ and $Y \in \{c_0, c\}$.

*1**.* (5.9) *and* (5.28) *hold.*

*2**.* (5.5) *and* (5.28) *hold.*

*3**.* (5.28) *holds.*

The following result with Lemma 5.3.1, Lemma 5.4.1 and Lemma 5.4.9 is required to characterize the classes of matrices from $\ell_1, c_0, c, \ell_\infty$ into $c_0(\Upsilon^r)$, $c(\Upsilon^r)$, $\ell_\infty(\Upsilon^r)$.

Theorem 5.4.10. *Let* $X = c_0$, c *or* ℓ_∞ *and* Y *be an arbitrary subset of* ω. *Given any infinite matrix* $A = (a_{nk})$, *define the infinite matrix* $\tilde{A} = (\tilde{a}_{nk})$ *as follows:*

$$\tilde{a}_{nk} = \frac{1}{n^r} \sum_{j|n} J_r(j) a_{jk}.$$

Then, $A = (a_{nk}) \in (Y, X_{\Upsilon^r})$ *if and only if* $\tilde{A} = (\tilde{a}_{nk}) \in (Y, X)$.

Proof. If $x \in Y$, the equality

$$\sum_{k=1}^{\infty} \tilde{a}_{nk} x_k = \sum_{k=1}^{\infty} \left(\frac{1}{n^r} \sum_{j|n} J_r(j) a_{jk} \right) x_k = \frac{1}{n^r} \sum_{j|n} J_r(j) \left(\sum_{k=1}^{\infty} a_{jk} x_k \right)$$

holds. This yields $(\tilde{A}x)_n = (\Upsilon^r(Ax))_n$ for all $n \in \mathbb{N}$. That is, $\tilde{A} = \Upsilon^r \circ A$ and so $Ax \in X_{\Upsilon^r}$ for any $x \in Y$ if and only if $\tilde{A}x \in X$ for any $x \in Y$. □

Theorem 5.4.11. *Given any infinite matrix $A = (a_{nk})$, we have*

1. $A \in (\ell_1, c_0(\Upsilon^r))$ *if and only if*

$$\lim_{n\to\infty} \frac{1}{n^r} \sum_{j|n} J_r(j) a_{jk} = 0 \text{ for each } k \in \mathbb{N} \tag{5.29}$$

and

$$\sup_{n,k\in\mathbb{N}} \left| \frac{1}{n^r} \sum_{j|n} J_r(j) a_{jk} \right| < \infty. \tag{5.30}$$

2. $A \in (c_0, c_0(\Upsilon^r))$ *if and only if* (5.29) *holds and*

$$\sup_{n\in\mathbb{N}} \sum_k \left| \frac{1}{n^r} \sum_{j|n} J_r(j) a_{jk} \right| < \infty. \tag{5.31}$$

3. $A \in (c, c_0(\Upsilon^r))$ *if and only if* (5.29) *holds and*

$$\lim_{n\to\infty} \sum_k \left(\frac{1}{n^r} \sum_{j|n} J_r(j) a_{jk} \right) = 0. \tag{5.32}$$

4. $A \in (\ell_\infty, c_0(\Upsilon^r))$ *if and only if* (5.29) *holds and*

$$\lim_{n\to\infty} \sum_k \left| \frac{1}{n^r} \sum_{j|n} J_r(j) a_{jk} \right| = 0. \tag{5.33}$$

Proof. The proof follows from Lemma 5.3.1, Lemma 5.4.1, Lemma 5.4.9 and Theorem 5.4.10. □

Theorem 5.4.12. *Given any infinite matrix $A = (a_{nk})$, we have*

1. $A \in (\ell_1, c(\Upsilon^r))$ *if and only if* (5.30) *holds and*

$$\lim_{n\to\infty} \frac{1}{n^r} \sum_{j|n} J_r(j) a_{jk} \text{ exists for each } k \in \mathbb{N}. \tag{5.34}$$

2. $A \in (c_0, c(\Upsilon^r))$ *if and only if* (5.31) *and* (5.34) *hold.*
3. $A \in (c, c(\Upsilon^r))$ *if and only if* (5.31) *and* (5.34) *hold, and*

$$\lim_{n\to\infty} \sum_k \left(\frac{1}{n^r} \sum_{j|n} J_r(j) a_{jk} \right) \text{ exists.} \tag{5.35}$$

4. $A \in (\ell_\infty, c(\Upsilon^r))$ *if and only if* (5.34) *holds and*

$$\lim_{n\to\infty} \sum_k \left| \frac{1}{n^r} \sum_{j|n} J_r(j) a_{jk} \right| = \sum_k \left| \lim_{n\to\infty} \frac{1}{n^r} \sum_{j|n} J_r(j) a_{jk} \right|. \quad (5.36)$$

Proof. The proof follows from Lemma 5.3.1, Lemma 5.4.1, Lemma 5.4.9 and Theorem 5.4.10. □

Theorem 5.4.13. *Given any infinite matrix* $A = (a_{nk})$, *we have*

1. $A \in (\ell_1, \ell_\infty(\Upsilon^r))$ *if and only if* (5.30) *holds.*

2. $A \in (c_0, \ell_\infty(\Upsilon^r))$ *if and only if* (5.31) *holds.*

3. $A \in (c, \ell_\infty(\Upsilon^r))$ *if and only if* (5.31) *holds.*

4. $A \in (\ell_\infty, \ell_\infty(\Upsilon^r))$ *if and only if* (5.31) *holds.*

Proof. The proof follows from Lemma 5.3.1, Lemma 5.4.1, Lemma 5.4.9 and Theorem 5.4.10. □

Bibliography

[1] M. Mursaleen, Generalized spaces of difference sequences, J. Math. Anal. Appl. 203(3) (1996) 738–745.

[2] M. Mursaleen, A. K. Noman, On the spaces of λ-convergent and bounded sequences, Thai J. Math. 8(2) (2012) 311–329.

[3] F. Başar, M. Kirişçi, Almost convergence and generalized difference matrix, Comput. Math. Appl. 61 (2011) 602–611.

[4] M. Candan, Almost convergence and double sequential band matrix, ActaMath. Sci. Ser. B Engl. Ed. 34B(2) (2014) 354–366.

[5] M. Candan, Domain of the double sequential band matrix in the spaces of convergent and null sequences, Adv. Difference Equ. 2014 (2014), Article 163.

[6] E. E. Kara, M. İlkhan, Some properties of generalized Fibonacci sequence spaces, Linear Multilinear Algebra 64(11) (2016) 2208–2223.

[7] S. Ercan, Ç. A. Bektaş, Some geometrical and topological properties of a new BK-space derived by using regular matrix of Fibonacci numbers, Linear Multilinear Algebra 65(5) (2017) 909–921.

[8] A. Das, B. Hazarika, Some new Fibonacci difference spaces of non-absolute type and compact operators, Linear Multilinear Algebra 65(12) (2017) 2551–2573.

[9] S. Demiriz, O. Duyar, Domain of the generalized double Cesaro matrix in some paranormed spaces of double sequences, Tbilisi Math. J. 10(2) (2017) 43–56.

[10] O. Duyar, S. Demiriz, On vector-valued operator Riesz sequence spaces, Commun. Fac. Sci. Univ. Ank. Ser. A1 Math. Stat. 68(1) (2019) 236–247.

[11] H. Roopaei, T. Yaying, Quasi-Cesaro matrix and associated sequence spaces, Turkish J. Math. 45(1) (2021) 153–166.

[12] T. Yaying, B. Hazarika, On sequence spaces defined by the domain of a regular tribonacci matrix, Math. Slovaca 70(3) (2020) 697–706.

[13] M. İlkhan, S. Demiriz, E. E. Kara, A new paranormed sequence space defined by Euler totient matrix, Karaelmas Sci. Eng. J. 9(2) (2019) 277–282.

[14] S. Demiriz, S. Erdem, Domain of Euler totient matrix operator in the space $\mathcal{L}_p$, Korean J. Math. 28(2) (2020) 361–378.

[15] M. A. Sarıgöl, Spaces of series summable by absolute Cesàro and matrix operators, Comm. Math. Appl. 7(1) (2016) 11–22.

[16] G. C. Hazar, M. A. Sarıgöl, On absolute Nörlund spaces and matrix operators, Acta Math. Sin. (Engl. Ser.) 34(5) (2018) 812–826.

[17] F. Gökçe, M. A. Sarıgöl, Generalization of the space $\ell(p)$ derived by absolute Euler summability and matrix operators, J. Inequal. Appl. 2018 (2018), Article 133.

[18] J. Meng, L. Mei, The matrix domain and the spectra of a generalized difference operator, J. Math. Anal. Appl. 470(2) (2019) 1095–1107.

[19] G. C. Hazar, M. A. Sarıgöl, Compact and matrix operators on the space $|C,-1|_k$, J. Comput. Anal. Appl. 25(6) (2018) 1014–1024.

[20] T. Yaying, M. İlkhan Kara, On sequence spaces defined by the domain of tribonacci matrix in c_0 and c, Korean J. Math. 29(1) (2021) 25–40.

[21] T. Yaying, B. Hazarika, S.A. Mohiuddine, M. Mursaleen, K.J. Ansari, Sequence spaces derived by the triple band generalized Fibonacci difference operator, Adv. Difference Equ. 2020 (2020) 639.

[22] T. Yaying, On the paranormed Nörlund difference sequence space of fractional order and geometric properties, Math. Slovaca 71(1) (2021) 155–170.

[23] S. Erdem, S. Demiriz, 4-dimensional Euler totient matrix operator and some double sequence spaces, Math. Sci. Appl. E-Notes 8(2) (2020) 110–122.

[24] S. Demiriz, M. İlkhan, E. E. Kara, Almost convergence and Euler totient matrix, Ann. Funct. Anal. 11(3) (2020) 604–616.

[25] F. Başar, Summability Theory and Its Applications, Bentham Science Publishers, İstanbul (2012).

[26] M. İlkhan, E. E. Kara, A new Banach space defined by Euler totient matrix operator, Oper. Matrices 13(2) (2019) 527–544.

[27] M. İlkhan, Matrix domain of a regular matrix derived by Euler totient function in the spaces c_0 and c, Mediterr. J. Math. 17(1) (2020) 1–21.

[28] M. İlkhan, N. Şimşek, E. E. Kara, A new regular infinite matrix defined by Jordan totient function and its matrix domain in ℓ_p, Math. Meth. Appl. Sci. 44(9) (2021) 7622–7633.

[29] M. Stieglitz, H. Tietz, Matrix transformationen von folgenraumen eine ergebnisbersicht, Math. Z. 154 (1977) 1–16.

[30] B. Altay, F. Başar, Certain topological properties and duals of the matrix domain of a triangle matrix in a sequence space, J. Math. Anal. Appl. 336 (2007) 632–645.

[31] M. Kirişçi, F. Başar, Some new sequence spaces derived by the domain of generalized difference matrix, Comput. Math. Appl. 60 (2010) 1299–1309.

Chapter 6

A Study of Fibonacci Difference I–Convergent Sequence Spaces

Vakeel A. Khan

Aligarh Muslim University, vakhanmaths@gmail.com

Kamal M. A. S. Alshlool

Aligarh Muslim University, k_moto@yahoo.com

Sameera A. A. Abdullah

Sultan Qaboos University, sameera173a@gmail.com

6.1 Introduction and Preliminaries

FIGURE 6.1: Statue of Fibonacci in Pisa

DOI: 10.1201/9781003178200-6

Leonardo has left an irrevocable impact on the history of mathematics, therefore, he is known to the latter Italian mathematicians as Fibonacci. Leonardo was born in Pisa, Italy (1175–1250). He was assisted by his father who stayed by his side preparing him to become the way he is. In 1192, Leonardo went to North Africa as his father arranged some lessons in calculation techniques for him to become a merchant. Through traveling many different places around the world, Leonardo learned the Hindu-Arabic numerals from one of his Arab instructors and extended the boundaries of his mathematical knowledge. In 1200, Fibonacci returned to Pisa. In 1202, his book entitled Liber Abaci, which contains a compilation of mathematics known since the Greeks, was published. This book provided him to be consultant on accounting and educational matters for citizens in the city. In 1863, in recognition of his mathematical achievements and to honor him, a statue of Fibonacci was set in Pisa, that is located in the western gallery of Camposanto, the historical cemetery on the Piazza dei Miracolithe, Republic of Pisa. There are many mathematical concepts named after Fibonacci referring to the connection they have with the Fibonacci numbers. For example, the Brahmagupta–Fibonacci identity and the Pisano period. Beyond Mathematics, namesakes of Fibonacci include the asteroid 6765 Fibonacci. He wrote some books such as, Liber Abaci (The Book of Calculating) in 1202, Practica Geometriae (Practical Geometry) in 1220, Flos in 1225, and Liber Quadratorum (The Book of Squares) in 1225.

6.1.1 Fibonacci sequence

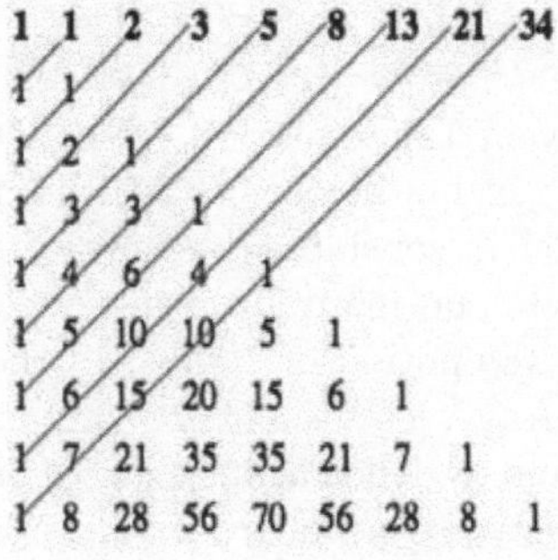

FIGURE 6.2: Fibonacci sequence

Although Fibonacci effectively contributed to the field of mathematics, Fibonacci is not remembered for many reasons, but rather for a single sequence of numbers that provided the solution to an equation included in his book (Liber Abaci). If we follow the origin of the sequence, we will find the question is, how many pairs of rabbits come out from one pair in a year. Then, Fibonacci started to use the mathematical operations. He learned from all the Mediterranean countries to produce his famous sequence, which was built on the basis of the third number in the sequence, which is equal to the sum of the two numbers before it, as represented by linear recurrence relation

$$f_n = f_{n-1} + f_{n-2}, n \geq 2.$$

Putting the values $f_1 = 0$, $f_2 = 1$ or the sequence starts with $(n = 3)$ because the sequence as mentioned will start with the third number equal to the sum of the numbers before it. For clarification, see the following table:

f_1	f_2	f_3	f_4	f_5	f_6	f_7	f_8	f_9	f_{10}	f_{11}	f_{12}	f_{13}	f_{14}	f_{15}
0	1	1	2	3	5	8	13	21	34	55	89	144	233	377

The sequence $\{f_n\}_{n=0}^{\infty}$ of Fibonacci numbers has some fundamental properties (see, Koshy [25]):

$$\lim_{n\to\infty} \frac{f_{n+1}}{f_n} = \frac{1+\sqrt{5}}{2} = \alpha, \quad \text{(Golden Ratio)}$$

$$\sum_{k=0}^{n} f_k = f_{n+2} - 1 \quad (n \in \mathbb{N}),$$

$$\sum_k \frac{1}{f_k} \quad \text{converges,}$$

$$f_{n-1}f_{n+1} - f_n^2 = (-1)^{n+1} \quad n \geq 1 \text{ (Cassini formula).}$$

It can be obtained $f_{n-1}^2 + f_n f_{n-1} - f_n^2 = (-1)^{n+1}$, if we can substitute for f_{n+1} in Cassini's formula.

Fibonacci did not invent this equation by himself, but his solution has forever immortalized him in mathematics. So, the arithmetic sequence that takes the name Fibonacci is considered as one of the most famous in the financial fields and markets and many branches of science.

It is well known that the notion of ω denotes the vector space of all real (or complex) sequences. Any vector subspace of ω is called a sequence space such as the spaces ℓ_∞, c and c_0 denoting the classes of all bounded, convergent and null sequences, respectively, normed by $\|x\|_\infty = \sup_k |x_k|$, where $k \in \mathbb{N}$. Let λ and μ be two sequence spaces and $A = (a_{nk})$ be an infinite matrix of real numbers a_{nk}, where $n, k \in \mathbb{N}$. Then, we say that A defines a *matrix transformation* from λ into μ and we denote it by writing $A : \lambda \longrightarrow \mu$ if for every sequence $x = (x_k) \in \lambda$ the sequence $Ax = \{A_n(x)\}$, the A–transform of x, is in μ where

$$A_n(x) = \sum_{k=0}^{\infty} a_{nk}x_k, \quad \text{for each } n \in \mathbb{N}. \tag{6.1}$$

By (λ, μ), we denote the class of all matrices A. Thus, $A \in (\lambda, \mu)$ if and only if the series on the right side of (6.1) converges for each $n \in \mathbb{N}$ and every $x \in \lambda$ and we have $Ax \in \mu$ for all $x \in \lambda$, where $A_n = (a_{nk})_{k \in \mathbb{N}}$ denotes the sequence in the n–th row of A. The matrix domain of an infinite matrix A in a sequence space λ is defined by

$$\lambda_A := \{x = (x_k) \in \omega : Ax \in \lambda\} \tag{6.2}$$

which is a sequence space. If we take $\lambda = c$, then c_A is called convergence domain of A, and we write the limit of Ax as $\lim_A x = \lim_{n \to \infty} \sum_k a_{nk} x_k$.

If $A = \triangle$, where $\triangle$ is the backward difference matrix defined by

$$\triangle = \triangle_{nk} = \begin{cases} (-1)^{n-k}, & n-1 \leq k \leq n \\ 0, & 0 \leq k < n-1 \text{ or } k > n \end{cases}$$

for all $n, k \in \mathbb{N}$, then $\lambda_\triangle$ is called the *difference sequence space* defined by the domain of a triangle matrix A whenever λ is a normed linear space or paranormed sequence space. The notion of difference sequence spaces was introduced by Kizmaz [24] as follows:

$$\lambda(\triangle) := \{x = (x_n) \in \omega : (x_n - x_{n+1}) \in \lambda\}$$

for $\lambda \in \{\ell_\infty, c, c_0\}$. The theory of matrix transformations which plays the major role in generalizing the notions of limit and convergence, has always been of great interest in the theory of sequence spaces. This theory was the focus of attention of many researchers, see for instance [2, 7, 9, 11, 19, 35, 53]. Recently, by using the sequence of Fibonacci numbers, Kara [10] defined the Fibonacci difference matrix $\hat{F} = (\hat{f}_{nk})$ as follows:

$$\hat{f}_{nk} = \begin{cases} -\frac{f_{n+1}}{f_n}, & k = n-1, \\ \frac{f_n}{f_{n+1}}, & k = n, \\ 0, & 0 \leq k < n-1 \text{ or } k > n, \end{cases} \tag{6.3}$$

for all $n, k \in \mathbb{N}$. One can derive by a straightforward calculation that the inverse $\hat{F}^{-1} = (\hat{g}_{nk})$ of the Fibonacci difference matrix is given by

$$\hat{g}_{nk} = \begin{cases} \frac{f_{n+1}^2}{f_k f_{k+1}}, & 0 \leq k \leq n, \\ 0, & k > n, \end{cases} \tag{6.4}$$

for all $n, k \in \mathbb{N}$. And then, Kara [10] introduced the difference sequence spaces

$$\ell_\infty(\hat{F}) = \left\{ x = (x_n) \in \omega : \sup_{n \in \mathbb{N}} \left| \frac{f_n}{f_{n+1}} x_n - \frac{f_{n+1}}{f_n} x_{n-1} \right| < \infty \right\},$$

and

$$\ell_p(\hat{F}) = \left\{ x = (x_n) \in \omega : \sum_n \left| \frac{f_n}{f_{n+1}} x_n - \frac{f_{n+1}}{f_n} x_{n-1} \right|^p < \infty, 1 \leq p < \infty \right\},$$

Afterwards, by using the same infinite Fibonacci matrix $\hat{F}$ and the same technique, Başarir et al. [1] have introduced the Fibonacci difference sequence spaces $c_0(\hat{F})$ and $c(\hat{F})$ as the sets of all sequences whose $\hat{F}$–transforms are in the spaces c_0 and c, respectively, that is,

$$c_0(\hat{F}) := \left\{x = (x_n) \in \omega : \lim_{n\to\infty} \hat{F}_n(x) = 0\right\},$$

and

$$c(\hat{F}) := \left\{x = (x_n) \in \omega : \exists \ell \in \mathbb{R} \ni \lim_{n\to\infty} \hat{F}_n(x) = \ell\right\},$$

where the sequence $\hat{F}_n(x)$ is the $\hat{F}$–transform of a sequence $x = (x_n) \in \omega$ defined as follows:

$$\hat{F}_n(x) = \begin{cases} \frac{f_0}{f_1}x_0 = x_0, & n = 0, \\ \frac{f_n}{f_{n+1}}x_n - \frac{f_{n+1}}{f_n}x_{n-1}, & n \geq 1 \end{cases} \tag{6.5}$$

for all $n \in \mathbb{N}$. For a more detailed information about Fibonacci sequence spaces, we refer to [3, 4, 12, 23, 34].

Definition 6.1.1. *[26] Let X be a non–empty set. Then a family I of subsets of X is said to be an ideal in X if and only if*

(i) $\emptyset \in I$,

(ii) for each $A, B \in I$, *we have* $A \cup B \in I$,

(iii) for each $A \in I$, $B \subset A$, *we have* $B \in I$.

- *An ideal I is called non–trivial, if $I \neq X$.*
- *A non–trivial ideal I is said to be an admissible in X if and only if it contains all singletons. i.e., $I \supset \{\{x\} : x \in X\}$.*
- *A non–trivial ideal I is maximal, if there cannot exist any non–trivial ideal $J \neq I$ containing I as a subset.*

Definition 6.1.2. *[26] Let X be a non–empty set. Then a non–empty family $\mathcal{F}$ of subsets of X is said to be a filter on X if and only if*

(i) $\emptyset \notin \mathcal{F}$,

(ii) for each $A, B \in \mathcal{F}$, *we have* $A \cap B \in \mathcal{F}$,

(iii) for each $A \in \mathcal{F}$ *with* $A \subset B$, *we have* $B \in \mathcal{F}$.

Remark 6.1.3. *[26] For each ideal I the class*

$$\mathcal{F}(I) = \{K \subseteq X : K^c \in I\} \text{ where } K^c = \mathbb{N} \setminus K$$

is a filter on X, called the filter associated with ideal I.

Depending on the structure of admissible ideals of subsets of the natural numbers $\mathbb{N}$, Kostyrko et al. [26] defined the notion of ideal convergence (or I–convergence) for real sequences as a generalization of both usual and statistical convergence which was introduced by Fast [6] and Steinhaus [45] in 1951. Later on, the notion of I–convergence was further investigated from the sequence space point of view and linked with the summability theory by Salát et al. [41], Tripathy and Hazarika [47–49], Kumar and Kumar [28, 29], Hazarika and Savas [8, 15], Khan and Ebdullah [14], Khan et al. [16], Khan and Tabassum [21], Das et al. [5], and many other authors. Salát et al. [42] came up with a new notion that is the notion of I–summable by using the infinite matrix of operators A and the notion of I–convergence and introduced some new sequence spaces c_A^I and m_A^I, the I–convergence field and bounded I–convergence field of an infinite matrix A, respectively. For further details on ideal convergence, we refer to [18, 51].

Now, we recall some definitions, lemmas and remarks that will be used throughout the chapter.

Definition 6.1.4. *[38] Let A be a subset of $\mathbb{N}$. The natural density $d(A)$ is defined as*

$$d(A) = \lim_{n\to\infty} \frac{1}{n}|\{k \in A : k \le n\}|$$

provided the limit exists. The vertical bars denote the order of the enclosed set.

Definition 6.1.5. *[6, 45] A sequence $x = (x_n) \in \omega$ is said to be statistically convergent to a number $\ell \in \mathbb{R}$ if, for every $\epsilon > 0$, the natural density of the set $\{n \in \mathbb{N} : |x_n - \ell| \ge \epsilon\}$ equal zero. We write $st\text{-}\lim x_n = \ell$. In case, $\ell = 0$ then (x_n) is said to be st–null.*

Definition 6.1.6. *[41] A sequence $x = (x_n) \in \omega$ is said to be I–Cauchy if, for every $\epsilon > 0$, there exists a number $N = N(\epsilon) \in \mathbb{N}$ such that the set $\{n \in \mathbb{N} : |x_n - x_N| \ge \epsilon\} \in I$.*

Definition 6.1.7. *[26] A sequence $x = (x_n) \in \omega$ is said to be I–convergent to a number $\ell \in \mathbb{R}$ if, for every $\epsilon > 0$, the set $\{n \in \mathbb{N} : |x_n - \ell| \ge \epsilon\} \in I$. We write $I\text{-}\lim x_n = \ell$. In case, $\ell = 0$ then (x_n) is said to be I–null.*

$$c^I := \text{The space of all } I\text{–convergent sequences,}$$
$$c_0^I := \text{The space of all } I\text{–null sequences.}$$

Definition 6.1.8. *[21] A sequence $x = (x_n) \in \omega$ is said to be I–bounded if there exists $K > 0$, such that, the set $\{n \in \mathbb{N} : |x_n| > K\} \in I$.*

$$\ell_\infty^I := \text{The space of all } I\text{–bounded sequences.}$$

Definition 6.1.9. *[21] Let $x = (x_n)$ and $z = (z_n)$ be two sequences. We say that $x_n = z_n$ for almost all n relative to I (in short a.a.n.r.I) if the set $\{n \in \mathbb{N} : x_n \ne z_n\} \in I$.*

Definition 6.1.10. *[41] A sequence space E is said to be convergence free, if $(y_n) \in E$ whenever $(x_n) \in E$ and $x_n = 0$ implies $y_n = 0$.*

Definition 6.1.11. *[41] A sequence space E is said to be solid or normal, if $(\alpha_n x_n) \in E$ whenever $(x_n) \in E$ and for any sequence of scalars $(\alpha_n) \in \omega$ with $|\alpha_n| < 1$, for every $n \in \mathbb{N}$.*

Definition 6.1.12. *[21] A sequence space E is said to be sequence algebra, if $(x_n) * (z_n) = (x_n \cdot z_n) \in E$ whenever $(x_n), (z_n) \in E$.*

Definition 6.1.13. *[41] Let $K = \{n_i \in \mathbb{N} : n_1 < n_2 < \dots\} \subseteq \mathbb{N}$ and E be a sequence space. A K–step space of E is a sequence space*

$$\lambda_K^E = \{(x_{n_i}) \in \omega : (x_n) \in E\}.$$

A canonical pre–image of a sequence $(x_{n_i}) \in \lambda_K^E$ is a sequence $(y_n) \in \omega$ defined as follows:

$$y_n = \begin{cases} x_n, & \text{if } n \in K \\ 0, & \text{otherwise.} \end{cases}$$

A canonical pre–image of a step space λ_K^E is a set of canonical pre–images of all elements in λ_K^E. i.e., y is in canonical pre–image of λ_K^E iff y is canonical pre–image of some element $x \in \lambda_K^E$.

Definition 6.1.14. *[41] A sequence space E is said to be monotone, if it is contains the canonical pre–images of it is step space. (i.e., if for all infinite $K \subseteq \mathbb{N}$ and $(x_n) \in E$ the sequence $(\alpha_n x_n)$, where $\alpha_n = 1$ for $n \in K$ and $\alpha_n = 0$ otherwise, belongs to E).*

Definition 6.1.15. *A map h defined on a domain $D \subset X$, i.e., $h : D \subset X \longrightarrow \mathbb{R}$ is said to satisfy Lipschitz condition if $|h(x) - h(y)| \leq K|x - y|$ where K is known as the Lipschitz constant.*

Remark 6.1.16. *[41] A convergence field of I–convergence is a set*

$$\mathcal{F}(I) = \{x = (x_k) \in \ell_\infty : \text{ there exists } I\text{–}\lim x \in \mathbb{R}\}.$$

Definition 6.1.17. *[26] The convergence field $\mathcal{F}(I)$ is a closed linear subspace of ℓ_∞ with respect to the supremum norm, $\mathcal{F}(I) = \ell_\infty \cap c^I$.*

Definition 6.1.18. *[41] The function $h : D \subset X \longrightarrow \mathbb{R}$ defined by $h(x) = I\text{–}\lim x$, for all $x \in \mathcal{F}(I)$ is a Lipschitz function.*

Lemma 6.1.19. *[41] Every solid space is monotone.*

Lemma 6.1.20. *[42] Let $K \in \mathcal{F}(I)$ and $M \subseteq \mathbb{N}$. If $M \notin I$, then $M \cap K \notin I$.*

6.2 Fibonacci Difference Sequence Spaces

Combining the two definitions, those of Fibonacci difference matrix and ideal convergence, Khan et al. [20] introduced some new sequence spaces as a domain of Fibonacci difference matrix $\hat{F}$ and studying some inclusion theorems and some topological and algebraic properties on these resulting. Throughout this chapter, the sequences $x = (x_n) \in \omega$ and $\hat{F}_n(x)$ are connected with the relation (6.5), I being an admissible ideal of subset of $\mathbb{N}$. We define

$$c_0^I(\hat{F}) := \Big\{x = (x_n) \in \omega : \Big\{n \in \mathbb{N} : \Big|\hat{F}_n(x)\Big| \geq \epsilon\Big\} \in I\Big\}, \tag{6.6}$$

$$c^I(\hat{F}) := \Big\{x = (x_n) \in \omega : \Big\{n \in \mathbb{N} : \Big|\hat{F}_n(x) - L\Big| \geq \epsilon, \text{ for some } L \in \mathbb{R}\Big\} \in I\Big\}, \tag{6.7}$$

$$\ell_\infty^I(\hat{F}) := \Big\{x = (x_n) \in \omega : \exists K > 0 \text{ s.t } \Big\{n \in \mathbb{N} : \Big|\hat{F}_n(x)\Big| > K\Big\} \in I\Big\}, \tag{6.8}$$

and write

$$m_0^I(\hat{F}) := c_0^I(\hat{F}) \cap \ell_\infty(\hat{F}), \tag{6.9}$$

and

$$m^I(\hat{F}) := c^I(\hat{F}) \cap \ell_\infty(\hat{F}). \tag{6.10}$$

With the notation of (6.2), the spaces $c_0^I(\hat{F})$, $c^I(\hat{F})$, $\ell_\infty^I(\hat{F})$, $m^I(\hat{F})$ and $m_0^I(\hat{F})$ can be redefined as follows:

$$c_0^I(\hat{F}) = (c_0^I)_{\hat{F}}, \quad c^I(\hat{F}) = (c^I)_{\hat{F}}, \quad \ell_\infty^I(\hat{F}) = (\ell_\infty^I)_{\hat{F}},$$

$$m^I(\hat{F}) = (m^I)_{\hat{F}} \text{ and } m_0^I(\hat{F}) = (m_0^I)_{\hat{F}}.$$

Also, Khan et al. [20] defined the notion of Fibonacci I–Cauchy as follows:

Definition 6.2.1. *Let I be an admissible ideal of subset of $\mathbb{N}$. If for each $\epsilon > 0$ there exists a number $N = N(\epsilon) \in \mathbb{N}$ such that $\Big\{n \in \mathbb{N} : |\hat{F}_n(x) - \hat{F}_N(x)| \geq \epsilon\Big\} \in I$ then a sequence $x = (x_n) \in \omega$ is called Fibonacci I–Cauchy.*

Example 6.2.2. *Define $I_f = \{A \subseteq \mathbb{N} : A \text{ is finite}\}$. I_f is an admissible ideal in $\mathbb{N}$ and $c^{I_f}(\hat{F}) = c(\hat{F})$.*

Example 6.2.3. *Define the non–trivial ideal as $I_d = \{A \subseteq \mathbb{N} : d(A) = 0\}$, where $d(A)$ is the natural density of the set A. In this case, $c^{I_d}(\hat{F}) = S(\hat{F})$. Where $S(\hat{F})$ is the space of Fibonacci difference statistically convergent sequence we defined as follows:*

$$S(\hat{F}) := \Big\{x = (x_n) \in \omega : d\big(\{n \in \mathbb{N} : |\hat{F}_n(x) - L| \geq \epsilon\}\big) = 0, \text{ for some } L \in \mathbb{R}\Big\}. \tag{6.11}$$

Theorem 6.2.4. *The spaces $X(\hat{F})$ are normed linear spaces with the norm*

$$\|x\|_{X(\hat{F})} = \sup_n |\hat{F}_n(x)| \text{ where } X \in \{m^I, m_0^I\}. \tag{6.12}$$

Proof. The proof of the result is easy in view of existing techniques and hence omitted. □

Theorem 6.2.5. *Let $I \subseteq 2^{\mathbb{N}}$ be a non–trivial ideal. Then the inclusion $c(\hat{F}) \subset c^I(\hat{F})$ strictly holds.*

Proof. We know that $c \subseteq c^I$ and for any X and Y spaces, if $X \subseteq Y$ then $X(\hat{F}) \subseteq Y(\hat{F})$ (see [23] Lemma 2.1). Hence, it is easy to see that $c(\hat{F}) \subset c^I(\hat{F})$. Now, we constructing the following example to show the strictness of the inclusion.

Example 6.2.6. *Define the sequence $x = (x_n) \in \omega$ such that*

$$\hat{F}_n(x) = \begin{cases} \sqrt{n}, & \text{if } n \text{ is a square} \\ 0, & \text{otherwise.} \end{cases}$$

Then $x \in c^{I_d}(\hat{F})$ but $x \notin c(\hat{F})$.

Example 6.2.7. *Define the ideal I such that*

$$A \in I \Longleftrightarrow A \text{ eventually only contains even natural numbers}$$

then I is a non–trivial ideal in $\mathbb{N}$. When

$$\hat{F}_n(x) = (1, 1, 1, 0, 1, 1, 1, 0, 0, 1, 0, 1, 0, \dots)$$

we have

$$A_\epsilon = \{n \in \mathbb{N} : \hat{F}_n(x) \neq 0\} = \{1, 2, 3, 5, 6, 7, 10, 12, 14, 16, 18, \dots\}$$

and $(x_n) \in c_0^I(\hat{F})$. Hence, $A_\epsilon \in I$ and $\hat{F}_n(x) \in c^I$. Now lets look at the statistical convergence of the sequence.

$$\lim_{n\to\infty} \frac{1}{n}|A_\epsilon| = \lim_{n\to\infty} \frac{1}{n}\left|B + \frac{n}{2}\right| = \frac{1}{2}$$

where B is a finite number and $|A_\epsilon|$ is the cardinality of A_ϵ. Hence, $\hat{F}_n(x) \notin S$. □

Theorem 6.2.8. *A sequence $x = (x_n) \in \omega$ is Fibonacci I–convergent if and only if for every $\epsilon > 0$, there exists $N = N(\epsilon) \in \mathbb{N}$, such that*

$$\left\{n \in \mathbb{N} : \left|\hat{F}_n(x) - \hat{F}_N(x)\right| < \epsilon\right\} \in \mathcal{F}(I). \tag{6.13}$$

Proof. Suppose that the sequence $x = (x_n) \in \omega$ is Fibonacci I–convergent to some number $L \in \mathbb{R}$, then for a given $\epsilon > 0$, the set

$$B_\epsilon = \left\{n \in \mathbb{N} : \left|\hat{F}_n(x) - L\right| < \frac{\epsilon}{2}\right\} \in \mathcal{F}(I).$$

Fix an integer $N = N(\epsilon) \in B_\epsilon$. Then we have

$$\left|\hat{F}_n(x) - \hat{F}_N(x)\right| \leq \left|\hat{F}_n(x) - L\right| + \left|L - \hat{F}_N(x)\right| < \frac{\epsilon}{2} + \frac{\epsilon}{2} = \epsilon,$$

which holds for all $n \in B_\epsilon$. Hence the set (6.13) holds.

Conversely, suppose that (6.13) holds for all $\epsilon > 0$. Then the set

$$C_\epsilon = \left\{n \in \mathbb{N} : \hat{F}_n(x) \in \left[\hat{F}_n(x) - \epsilon, \hat{F}_n(x) + \epsilon\right]\right\} \in \mathcal{F}(I), \text{ for all } \epsilon > 0.$$

Let $J_\epsilon = \left[\hat{F}_n(x) - \epsilon, \hat{F}_n(x) + \epsilon\right]$. Fixing $\epsilon > 0$, then we have $C_\epsilon \in \mathcal{F}(I)$ and $C_{\frac{\epsilon}{2}} \in \mathcal{F}(I)$. Hence $C_\epsilon \cap C_{\frac{\epsilon}{2}} \in \mathcal{F}(I)$. This implies that

$$J = J_\epsilon \cap J_{\frac{\epsilon}{2}} \neq \emptyset,$$

that is the set

$$\left\{n \in \mathbb{N} : \hat{F}_n(x) \in J\right\} \in \mathcal{F}(I),$$

and thus

$$\text{diam } (J) \leq \frac{1}{2} \text{ diam } (J_\epsilon),$$

where the diam (J) denote the length of interval J. Proceeding in this way, by induction, we get a sequence of closed intervals

$$J_\epsilon = I_0 \supseteq I_1 \supseteq \cdots \supseteq I_n \supseteq \cdots$$

such that

$$\text{diam } (I_n) \leq \frac{1}{2} \text{ diam } (I_{n-1}), \text{ for } n = 2, 3, \ldots$$

and

$$\left\{n \in \mathbb{N} : \hat{F}_n(x) \in I_n\right\} \in \mathcal{F}(I).$$

Then there exists a number $L \in \bigcap_{n \in \mathbb{N}} I_n$, and it is a routine work to verify that $L = I\text{–}\lim \hat{F}_n(x)$ showing that $x = (x_n) \in \omega$ is Fibonacci I–convergent. Hence the result. □

Theorem 6.2.9. *Let I be an admissible ideal. Then the following are equivalent:*

(i) $(x_n) \in c^I(\hat{F})$,

(ii) there exists $(y_n) \in c(\hat{F})$ *such that* $x_n = y_n$, *for a.a.n.r.I,*

(iii) *there exists* $(y_n) \in c(\hat{F})$ *and* $(z_n) \in c_0^I(\hat{F})$ *such that* $x_n = y_n + z_n$ *for all* $n \in \mathbb{N}$
and $\left\{n \in \mathbb{N} : \left|\hat{F}_n(x) - L\right| \geq \epsilon\right\} \in I$,

(iv) *there exists a subset* $K = \{n_i : i \in \mathbb{N}, n_1 < n_2 < n_3 < \dots\}$ *of* $\mathbb{N}$, *such that* $K \in \mathcal{F}(I)$
and $\lim_{i\to\infty} \left|\hat{F}_{n_i}(x) - L\right| = 0$.

Proof. (i) implies (ii). Let $x = (x_n) \in c^I(\hat{F})$, then for any $\epsilon > 0$, there exists a number $L \in \mathbb{R}$ such that

$$\left\{n \in \mathbb{N} : \left|\hat{F}_n(x) - L\right| \geq \epsilon\right\} \in I.$$

Let (m_t) be an increasing sequence with $m_t \in \mathbb{N}$ such that

$$\left\{n \leq m_t : \left|\hat{F}_n(x) - L\right| \geq t^{-1}\right\} \in I.$$

Define a sequence $y = (y_n)$ as $y_n = z_n$ for all $n \leq m_1$. For $m_t < n < m_{t+1}$, for $t \in \mathbb{N}$, as

$$y_n = \begin{cases} x_n, & \text{if } \left|\hat{F}_n(x) - L\right| < t^{-1} \\ L, & \text{otherwise.} \end{cases}$$

Then $y_n \in c(\hat{F})$, and from the following inclusion

$$\{n \leq m_t : x_n \neq y_n\} \subseteq \left\{n \in \mathbb{N} : \left|\hat{F}_n(x) - L\right| \geq \epsilon\right\} \in I$$

we get $x_n = y_n$ for *a.a.n.r.I.*
(ii) implies (iii). For $x = (x_n) \in c^I(\hat{F})$ there exists $y = (y_n) \in c(\hat{F})$ such that $x_n = y_n$ for *a.a.n.r.I.* Let $K = \{n \in \mathbb{N} : x_n \neq y_n\}$. Then $K \in I$. Define a sequence $z = (z_n)$ as

$$z_n = \begin{cases} x_n - y_n, & \text{if } n \in K \\ 0, & \text{otherwise.} \end{cases}$$

Then $(z_n) \in c_0^I(\hat{F})$ and $(y_n) \in c(\hat{F})$.
(iii) implies (iv). Let $P = \left\{n \in \mathbb{N} : |\hat{F}_n(x)| \geq \epsilon\right\} \in I$, and

$$K = P^c = \{(n_i \in \mathbb{N} : i \in \mathbb{N}, n_1 < n_2 < n_3 < \dots\} \in \mathcal{F}(I).$$

Then we have

$$\lim_{i\to\infty} \left|\hat{F}_{n_i}(x) - L\right| = 0.$$

(iv) implies (i). Let $\epsilon > 0$ be given and suppose that (iii) holds. Then for any $\epsilon > 0$, and by Lemma 6.1.20 we have

$$\left\{n \in \mathbb{N} : \left|\hat{F}_n(x) - L\right| \geq \epsilon\right\} \subseteq K^c \cup \left\{n \in K : \left|\hat{F}_n(x) - L\right| \geq \epsilon\right\}.$$

Thus $(x_n) \in c^I(\hat{F})$. □

Theorem 6.2.10. *The inclusions $c_0^I(\hat{F}) \subset c^I(\hat{F}) \subset \ell_\infty^I(\hat{F})$ are strict.*

Proof. The inclusion $c_0^I(\hat{F}) \subset c^I(\hat{F})$ is obvious. Now, to show the strictness, consider the sequence $x = (x_n) \in \omega$ such that $\hat{F}_n(x) = 1$. It easy to see that the sequence $\hat{F}_n(x) \in c^I$ but $\hat{F}_n(x) \notin c_0^I$, that is to say that $x \in c^I(\hat{F}) \backslash c_0^I(\hat{F})$. Next, let $x = (x_n) \in c^I(\hat{F})$. Then there exists $L \in \mathbb{R}$ such that $I\text{–}\lim |\hat{F}_n(x) - L| = 0$, that is,

$$\left\{n \in \mathbb{N} : \left|\hat{F}_n(x) - L\right| \geq \epsilon\right\} \in I.$$

We have

$$\left|\hat{F}_n(x)\right| = \left|\hat{F}_n(x) - L + L\right| \leq \left|\hat{F}_n(x) - L\right| + |L|.$$

From this, it follows that the sequence (x_n) must belongs to $\ell_\infty^I(\hat{F})$. Further, we show the strictness of the inclusion $c^I(\hat{F}) \subset \ell_\infty^I(\hat{F})$ by constructing the following example.

Example 6.2.11. *Consider the sequence $x = (x_n) \in \omega$ such that*

$$\hat{F}_n(x) = \begin{cases} \sqrt{n}, & \text{if } n \text{ is square} \\ 1, & \text{if } n \text{ is odd non–square} \\ 0, & \text{if } n \text{ is even non–square.} \end{cases}$$

Then the sequence $\hat{F}_n(x) \in \ell_\infty^I$ but $\hat{F}_n(x) \notin c^I$, which implies that the sequence $x \in \ell_\infty^I(\hat{F}) \backslash c^I(\hat{F})$.

Thus, the inclusion $c_0^I(\hat{F}) \subset c^I(\hat{F}) \subset \ell_\infty^I(\hat{F})$ strictly holds. □

Remark 6.2.12. *A Fibonacci bounded sequence is obviously Fibonacci I–bounded as the empty set belongs to the ideal I. However, the converse is not true. For example, we consider the sequence*

$$\hat{F}_n(x) = \begin{cases} n, & \text{if } n \text{ is sequere} \\ 0, & \text{otherwise.} \end{cases}$$

Clearly, $\hat{F}_n(x)$ is not a bounded sequence. However, $\{n \in \mathbb{N} : \left|\hat{F}_n(x)\right| > \frac{1}{2}\} \in I$. Hence (x_n) is Fibonacci I–bounded.

Theorem 6.2.13. *The spaces $m^I(\hat{F})$ and $m_0^I(\hat{F})$ are Banach spaces normed by* (6.12).

Proof. Let $(x_n^{(i)})$ be a Cauchy sequence in $m^I(\hat{F}) \subset \ell_\infty(\hat{F})$. Then $(x_n^{(i)})$ converges in $\ell_\infty(\hat{F})$ and $\lim_{i\to\infty} \hat{F}_n^{(i)}(x) = \hat{F}_n(x)$. Let $I\text{–}\lim \hat{F}_n^{(i)}(x) = L_i$ for each $i \in \mathbb{N}$. Then, we have to show that

(i) (L_i) is convergent say to L,

(ii) $I\text{–}\lim \hat{F}_n(x) = L$.

Proof of (i). Since $(x_n^{(i)})$ is a Cauchy, so for each $\epsilon > 0$ there exists $n_0 \in \mathbb{N}$ such that

$$\left|\hat{F}_n^{(i)}(x) - \hat{F}_n^{(j)}(x)\right| < \frac{\epsilon}{3}, \text{ for all } i, j \geq n_0. \tag{6.14}$$

Now, let E_i and E_j be the following sets in I

$$E_i = \left\{n \in \mathbb{N} : \left|\hat{F}_n^{(i)}(x) - L_i\right| \geq \frac{\epsilon}{3}\right\} \tag{6.15}$$

and

$$E_j = \left\{n \in \mathbb{N} : \left|\hat{F}_n^{(j)}(x) - L_j\right| \geq \frac{\epsilon}{3}\right\}. \tag{6.16}$$

Consider $i, j \geq n_0$ and $n \notin E_i \cap E_j$. Then we have

$$|L_i - L_j| \leq \left|\hat{F}_n^{(i)}(x) - L_i\right| + \left|\hat{F}_n^{(j)}(x) - L_j\right| + \left|\hat{F}_n^{(i)}(x) - \hat{F}_n^{(j)}(x)\right| < \epsilon$$

by (6.14), (6.15) and (6.16). Thus (L_i) is a Cauchy sequence of $\mathbb{R}$, thus converges say to L, i.e., $\lim_{i\to\infty} L_i = L$.

Proof of (ii). Let $\delta > 0$ be given. Then we can fined m_0 such that

$$|L_i - L| < \frac{\delta}{3}, \text{ for each } i > m_0. \tag{6.17}$$

We have $(x_n^{(i)}) \to x_n$ as $i \to \infty$. Thus

$$\left|\hat{F}_n^{(i)}(x) - \hat{F}_n(x)\right| < \frac{\delta}{3}, \text{ for each } i > m_0. \tag{6.18}$$

Since $(\hat{F}_n^{(j)}(x))$ is I–convergent to L_j, there exists $D \in I$ such that, for each $n \notin D$, we have

$$\left|\hat{F}_n^{(j)}(x) - L_j\right| < \frac{\delta}{3}. \tag{6.19}$$

Without loss of generality, let $j > m_0$ then for all $n \notin D$, we have by (6.17), (6.18) and (6.19) that

$$\left|\hat{F}_n(x) - L\right| \leq \left|\hat{F}_n(x) - \hat{F}_n^{(j)}(x)\right| + \left|\hat{F}_n^{(j)}(x) - L_j\right| + |L_j - L| < \delta.$$

Hence, (x_n) is Fibonacci I–convergent to L. Thus $m^I(\hat{F})$ is a Banach space. The other cases can be similarly established. □

The following results is a consequence of the above Theorem 6.2.13.

Theorem 6.2.14. *The spaces $m^I(\hat{F})$ and $m_0^I(\hat{F})$ are K–spaces.*

Theorem 6.2.15. *The set $m^I(\hat{F})$ is closed subspace of $\ell_\infty(\hat{F})$.*

Since the inclusions $m^I(\hat{F}) \subset \ell_\infty(\hat{F})$ and $m_0^I(\hat{F}) \subset \ell_\infty(\hat{F})$ are strict, in view of Theorem 6.2.15, we have the following result.

Theorem 6.2.16. *The spaces $m^I(\hat{F})$ and $m_0^I(\hat{F})$ are nowhere dense subset of $\ell_\infty(\hat{F})$.*

Theorem 6.2.17. *The spaces $c_0^I(\hat{F})$ and $m_0^I(\hat{F})$ are solid and monotone.*

Proof. We will prove the result for $c_0^I(\hat{F})$ and for $m_0^I(\hat{F})$ the result can be established similarly. Let $x = (x_n) \in c_0^I(\hat{F})$ for $\epsilon > 0$, the set

$$\left\{n \in \mathbb{N} : \left|\hat{F}_n(x)\right| \geq \epsilon\right\} \in I. \tag{6.20}$$

Let $\alpha = (\alpha_n)$ be a sequence of scalars with $|\alpha| \leq 1$ for all $n \in \mathbb{N}$. Then,

$$\left|\hat{F}_n(\alpha x)\right| = \left|\alpha \hat{F}_n(x)\right| \leq |\alpha| \left|\hat{F}_n(x)\right| \leq \left|\hat{F}_n(x)\right|, \text{ for all } n \in \mathbb{N}.$$

From this inequality and (6.20), we have that

$$\left\{n \in \mathbb{N} : \left|\hat{F}_n(\alpha x)\right| \geq \epsilon\right\} \subseteq \left\{n \in \mathbb{N} : \left|\hat{F}_n(x)\right| \geq \epsilon\right\} \in I$$

implies

$$\left\{n \in \mathbb{N} : \left|\hat{F}_n(\alpha x)\right| \geq \epsilon\right\} \in I$$

Therefore, $(\alpha x_n) \in c_0^I(\hat{F})$. Hence the space $c_0^I(\hat{F})$ is solid and hence by Lemma 1, the space $c_0^I(\hat{F})$ is monotone. □

Theorem 6.2.18. *The spaces $c_0^I(\hat{F})$ and $c^I(\hat{F})$ are sequence algebra.*

Proof. Let $x = (x_n), y = (y_n) \in c_0^I(\hat{F})$. Then

$$I\text{-}\lim|\hat{F}_n(x)| = 0 \text{ and } I\text{-}\lim|\hat{F}_n(y)| = 0. \tag{6.21}$$

Therefore, from (6.21), we have $I\text{-}\lim|\hat{F}_n(x \cdot y)| = 0$. This implies that $\left\{n \in \mathbb{N} : |\hat{F}_n(x \cdot y)| \geq \epsilon\right\} \in I$. Thus, $(x \cdot y) \in c_0^I(\hat{F})$. Hence, $c_0^I(\hat{F})$ is sequence algebra. Similarly, we can prove that $c^I(\hat{F})$ is sequence algebra. □

Theorem 6.2.19. *The function $h : m^I(\hat{F}) \to \mathbb{R}$ defined by $h(x) = \left|I\text{-}\lim \hat{F}_n(x)\right|$, where $m^I(\hat{F}) = \ell_\infty(\hat{F}) \cap c^I(\hat{F})$, is a Lipschitz function and hence uniformly continuous.*

Proof. First of all, we show that the function is well defined. Let $x, y \in m^I(\hat{F})$, such that

$$\begin{aligned} x = y &\Rightarrow I\text{-}\lim \hat{F}_n(x) = I\text{-}\lim \hat{F}_n(y) \\ &\Rightarrow |I\text{-}\lim \hat{F}_n(x)| = |I\text{-}\lim \hat{F}_n(y)| \Rightarrow h(x) = h(y). \end{aligned}$$

Thus, h is well defined. Next, let $x = (x_n), y = (y_n) \in m^I(\hat{F})$, $x \neq y$. Then

$$A_x = \left\{n \in \mathbb{N} : \left|\hat{F}_n(x) - h(x)\right| \geq |x - y|_*\right\} \in I$$

and

$$A_y = \left\{n \in \mathbb{N} : \left|\hat{F}_n(y) - h(y)\right| \geq |x-y|_*\right\} \in I,$$

where $|x-y|_* = \sup_n \left|\hat{F}_n(x) - \hat{F}_n(y)\right|$. Thus

$$B_x = \left\{n \in \mathbb{N} : \left|\hat{F}_n(x) - h(x)\right| < |x-y|_*\right\} \in \mathcal{F}(I)$$

and

$$B_y = \left\{n \in \mathbb{N} : \left|\hat{F}_n(y) - h(y)\right| < |x-y|_*\right\} \in \mathcal{F}(I).$$

Hence $B = B_x \cap B_y \in \mathcal{F}(I)$, so that B is non–empty set. Therefore, choosing $n \in B$, therefore, we have

$$\begin{aligned} |h(x) - h(y)| &\leq \left|h(x) - \hat{F}_n(x)\right| + \left|\hat{F}_n(x) - \hat{F}_n(y)\right| + \left|\hat{F}_n(y) - h(y)\right| \\ &\leq 3|x-y|_*. \end{aligned}$$

Thus, h is Lipschitz function and hence uniformly continuous. □

Theorem 6.2.20. *If $x = (x_n), y = (y_n) \in m^I(\hat{F})$ with $\hat{F}_n(x \cdot y) = \hat{F}_n(x) \cdot \hat{F}_n(y)$, then $(x \cdot y) \in m^I(\hat{F})$ and $h(x \cdot y) = h(x) \cdot h(y)$, where $h : m^I(\hat{F}) \longrightarrow \mathbb{R}$ is defined by $h(x) = \left|I\text{–}\lim \hat{F}_n(x)\right|$.*

Proof. For $\epsilon > 0$,

$$B_x = \left\{n \in \mathbb{N} : \left|\hat{F}_n(x) - h(x)\right| < \epsilon\right\} \in \mathcal{F}(I) \tag{6.22}$$

and

$$B_y = \left\{n \in \mathbb{N} : \left|\hat{F}_n(y) - h(y)\right| < \epsilon\right\} \in \mathcal{F}(I), \tag{6.23}$$

where $\epsilon = |x-y|_* = \sup_n \left|\hat{F}_n(x) - \hat{F}_n(y)\right|$. Now, we have

$$\begin{aligned} \left|\hat{F}_n(x \cdot y) - h(x)h(y)\right| &= \left|\hat{F}_n(x)\hat{F}_n(y) - \hat{F}_n(x)h(y) + \hat{F}_n(x)h(y) - h(x)h(y)\right| \\ &\leq \left|\hat{F}_n(x)\right| \left|\hat{F}_n(y) - h(y)\right| + |h(y)| \left|\hat{F}_n(x) - h(x)\right|. \end{aligned} \tag{6.24}$$

As $m^I(\hat{F}) \subseteq \ell_\infty(\hat{F})$, there exists an $M \in \mathbb{R}$ such that $|\hat{F}_n(x)| < M$. Therefore, from the equations (6.22), (6.23) and (6.24) we have

$$\begin{aligned} \left|\hat{F}_n(xy) - h(x)h(y)\right| &= \left|\hat{F}_n(x) \cdot \hat{F}_n(y) - h(x)h(y)\right| \\ &\leq M\epsilon + |h(y)|\epsilon = \epsilon_1, \text{ (say)} \end{aligned}$$

for all $n \in B_x \cap B_y \in \mathcal{F}(I)$. Hence $(x \cdot y) \in m^I(\hat{F})$ and $h(x \cdot y) = h(x) \cdot h(y)$. □

6.3 Orlicz Fibonacci Difference Sequence Spaces

Recall in [27] that an Orlicz function is a function $M : [0, \infty) \longrightarrow [0, \infty)$, which is continuous, convex and non–decreasing with $M(0) = 0$, $M(x) > 0$ and $M(x) \to \infty$ as $x \to \infty$. If the convexity of an Orlicz function is replaced by $M(x + y) \leq M(x) + M(y)$ then this function is called Modulus function, introduced by Nakano [37] and it was further investigated with applications to sequences by Maddox [46]. It is well known if M is a convex function and $M(0) = 0$, then $M(\lambda x) \leq \lambda M(x)$ for all λ with $0 < \lambda < 1$.

An Orlicz function M is said to satisfy Δ_2–condition for all values of u if there exists a constant $K > 0$ such that $M(Lu) \leq KLM(u)$ for all values $L > 1$ (see [27]). Lindenstrauss and Tzafriri [30] used the idea of Orlicz function to define the sequence space

$$l_M = \left\{ x = (x_n) \in \omega : \sum_{n=1}^{\infty} M\left(\frac{|x_n|}{\rho}\right) < \infty \text{ for some } \rho > 0 \right\}$$

which is called an Orlicz sequence space. This space is a Banach space with the norm

$$\|x\| = \inf\left\{ \rho > 0, \sum_{n=1}^{\infty} M\left(\frac{|x_n|}{\rho}\right) \leq 1 \right\}$$

and the space l_M is closely related to the space l_p which is an Orlicz sequence space with $M(t) = t^p$ for $1 \leq p < \infty$. Orlicz function was used to define sequence spaces by Parashar and Choudhary [39], Khan et al. [13]. Later on, with the help of the notion of I–convergence, Tripathy and Hazarika [48] introduced some new sequence spaces defined by Orlicz function and further studied by Khan et al. [17, 21] and many others. Recently, by using the idea of Orlicz function and the Fibonacci matrix $\hat{F}$, Khan et al. [50] introduced some new Fibonacci difference sequence spaces and presented some inclusion relations concerning these spaces.

$$c^I(M, \hat{F}) := \left\{ x = (x_n) \in \omega : \left\{ n \in \mathbb{N} : \sum_{n=1}^{\infty} M\left(\frac{|\hat{F}_n(x) - \ell|}{\rho}\right) \geq \epsilon \right\} \in I \right\}, \tag{6.25}$$

$$c_0^I(M, \hat{F}) := \left\{ x = (x_n) \in \omega : \left\{ n \in \mathbb{N} : \sum_{n=1}^{\infty} M\left(\frac{|\hat{F}_n(x)|}{\rho}\right) \geq \epsilon, \text{ for some } \ell \in \mathbb{R} \right\} \in I \right\}, \tag{6.26}$$

$$\ell^I_\infty(M,\hat{F}) := \left\{x=(x_n)\in\omega : \exists \quad K>0 \text{ s.t} \left\{n\in\mathbb{N} \right.\right.$$
$$\left.\left. : \sum_{n=1}^{\infty} M\left(\frac{|\hat{F}_n(x)|}{\rho}\right) > K\right\}\in I\right\}, \tag{6.27}$$

$$\ell_\infty(M,\hat{F}) := \left\{x=(x_n)\in\omega : \sup_n \sum_{n=1}^{\infty} M\left(\frac{|\hat{F}_n(x)|}{\rho}\right) < \infty\right\}. \tag{6.28}$$

We write

$$m^I_0(M,\hat{F}) := c^I_0(M,\hat{F}) \cap \ell_\infty(M,\hat{F}), \tag{6.29}$$

and

$$m^I(M,\hat{F}) := c^I(M,\hat{F}) \cap \ell_\infty(M,\hat{F}). \tag{6.30}$$

Theorem 6.3.1. *The spaces $m^I_0(M,\hat{F})$ and $m^I(M,\hat{F})$ are Banach spaces normed by*

$$\|\hat{F}_n(x)\| = \inf\left\{\rho>0 : \sup_n \sum_{n=1}^{\infty} M\left(\frac{|\hat{F}_n(x)|}{\rho}\right) \le 1\right\} \tag{6.31}$$

Proof. The proof of this result is easy, so omitted. □

Theorem 6.3.2. *Let M_1, M_2 be Orlicz functions that satisfy the Δ_2-condition. Then*

(i) $W(M_2,\hat{F}) \subseteq W(M_1\cdot M_2,\hat{F})$,

(ii) $W(M_1,\hat{F}) \cap W(M_2,\hat{F}) = W(M_1+M_2,\hat{F})$ for $W = c^I, c^I_0, m^I_0, m^I$.

Proof. (i) Let $(x_n)\in c^I_0(M_2,\hat{F})$. Then there exists $\rho>0$ such that

$$\left\{n\in\mathbb{N} : \sum_{n=1}^{\infty} M_2\left(\frac{|\hat{F}_n(x)|}{\rho}\right) \ge \epsilon\right\}\in I. \tag{6.32}$$

Let $\epsilon>0$ and choose δ with $0<\delta<1$ such that $M_1(t)<\epsilon$ for $0\le t\le\delta$. Write

$$\hat{F}_n(y) = \sum_{n=1}^{\infty} M_2\left(\frac{|\hat{F}_n(x)|}{\rho}\right)$$

and consider

$$\lim_{\substack{n\in\mathbb{N}\\ 0\le\hat{F}_n(y)\le\delta}} M_1(|\hat{F}_n(y)|) = \lim_{\substack{n\in\mathbb{N}\\ \hat{F}_n(y)\le\delta,}} M_1(|\hat{F}_n(y)|) + \lim_{\substack{n\in\mathbb{N}\\ \hat{F}_n(y)>\delta}} M_1(|\hat{F}_n(y)|).$$

By the convexity of Orlicz function, we have

$$\lim_{\substack{n\in\mathbb{N}\\ \hat{F}_n(y)\le\delta}} M_1(|\hat{F}_n(y)|) \le M_1(2) \lim_{\substack{n\in\mathbb{N}\\ \hat{F}_n(y)\le\delta}} M_1(|\hat{F}_n(y)|). \tag{6.33}$$

For $\hat{F}_n(y) > \delta$, we have

$$\hat{F}_n(y) < \frac{\hat{F}_n(y)}{\delta} < 1 + \frac{\hat{F}_n(y)}{\delta}.$$

Since M_1 is non–decreasing and convex, it follows that

$$M_1(\hat{F}_n(y)) < M_1\left(1 + \frac{\hat{F}_n(y)}{\delta}\right) < \frac{1}{2}M_1(2) + \frac{1}{2}M_1\left(\frac{2\hat{F}_n(y)}{\delta}\right).$$

Since M_1 satisfies Δ_2–condition, we have

$$M_1(\hat{F}_n(y)) < \frac{1}{2}K\frac{\hat{F}_n(y)}{\delta}M_1(2) + \frac{1}{2}K\frac{\hat{F}_n(y)}{\delta}M_1(2) = K\frac{\hat{F}_n(y)}{\delta}M_1(2).$$

Hence,

$$\lim_{\substack{n\in\mathbb{N}\\ \hat{F}_n(y)>\delta}} M_1(\hat{F}_n(y)) \le \max(1, K\delta^{-1}M_1(2)) \lim_{\substack{n\in\mathbb{N}\\ \hat{F}_n(y)>\delta}} \hat{F}_n(y). \tag{6.34}$$

From above equations (8.12), (6.33) and (6.34), we have $(x_n) \in c_0^I(M_1 \cdot M_2, \hat{F})$. Thus $c_0^I(M_2, \hat{F}) \subseteq c_0^I(M_1 \cdot M_2, \hat{F})$. The other cases can be proved similarly.
(ii) Let $(x_n) \in c_0^I(M_1, \hat{F}) \cap c_0^I(M_2, \hat{F})$. Then there exists $\rho > 0$ such that

$$\left\{n \in \mathbb{N} : \sum_{n=1}^{\infty} M_1\left(\frac{|\hat{F}_n(x)|}{\rho}\right) \ge \epsilon\right\} \in I$$

and

$$\left\{n \in \mathbb{N} : \sum_{n=1}^{\infty} M_2\left(\frac{|\hat{F}_n(x)|}{\rho}\right) \ge \epsilon\right\} \in I.$$

The rest of the proof follows from the following equality

$$\lim_{n\in\mathbb{N}}(M_1 + M_2)\left(\frac{|\hat{F}_n(x)|}{\rho}\right) = \lim_{n\in\mathbb{N}} M_1\left(\frac{|\hat{F}_n(x)|}{\rho}\right) + \lim_{n\in\mathbb{N}} M_2\left(\frac{|\hat{F}_n(x)|}{\rho}\right).$$

□

Let us take $M_2(x) = x$ and $M_1(x) = x$ for all $x \in [0, \infty)$ in the Theorem 6.3.2(i). Then, we have the following result:

Corollary 6.3.3. *$W(\hat{F}) \subseteq W(M, \hat{F})$ for $W = c_0^I$, c^I, m_0^I, m^I.*

Proof. The proof of the result is easy, so omitted. □

Proposition 6.3.4. *The space $c^I(\hat{F}, M)$ and $c_0^I(\hat{F}, M)$ are not convergent free in general.*

Proof. The proof of this result follows from the following example.

Example 6.3.5. *Let $I = I_f$ and $M(x) = x^3$ for all $x \in [0, \infty)$. Consider the sequences (x_n) and (y_n) defined by $x_n = \frac{1}{n}$ and $y_n = n$ for all $n \in \mathbb{N}$. Then (x_n) belongs to $c^I(M, \hat{F})$ and $c_0^I(M, \hat{F})$, but (y_n) does not belong to $c^I(M, \hat{F})$ and $c_0^I(M, \hat{F})$. Hence the spaces are not convergence free.*

□

6.4 Paranormed Fibonacci Difference Sequence Spaces

Let X be a linear space. A function $g : X \to \mathbb{R}$ is called paranorm if for all $x, y \in X$, (1) $g(x) \geq 0$, (2) $g(-x) = g(x)$, (3) $g(x + y) \leq g(x) + g(y)$, (4) (c_n) is a sequence of scalars with $c_n \to c$ as $(n \to \infty)$ and (x_n) is a sequence of vectors with $g(x_n - x) \to 0$ as $(n \to \infty)$, $g(c_n x_n - cx) \to 0$ as $(n \to \infty)$. A paranorm g which $g(x) = 0$ implies that $x = 0$ is called a total paranorm and the pair (X, g) is called a totally paranormed space. The concept of paranorm is related to the linear metric spaces given by some total paranorm [33,52]. The notion of paranormed sequence was studied at the initial stage by Nakano [36] and Simons [44]. Later on it was investigated by Maddox [31, 32] and many others. Tripathy and Hazarika [49] generalized the sequence spaces of Maddox to introduce the new idea of paranorm I–convergent sequence spaces $c_0^I(p)$, $c^I(p)$, $\ell_\infty^I(p)$ and $\ell_\infty(p)$ where $p = (p_n)$ is the sequence of strictly positive real numbers. In 2013, Kara and Demiriz [10] used the concept of paranorm and Fibonacci matrix domain to introduce the paranormed sequence spaces $c_0(\hat{F}, p)$, $c(\hat{F}, p)$ and $\ell_\infty(\hat{F}, p)$, i.e.,

$$c_0(\hat{F}, p) = \left\{x = (x_n) \in \omega : \lim_{n \to \infty} \left|\frac{f_n}{f_{n+1}} x_n - \frac{f_{n+1}}{f_n} x_{n-1}\right|^{p_n} = 0\right\},$$

$$c(\hat{F}, p) = \left\{x = (x_n) \in \omega : \exists L \in \mathbb{C} : \lim_{n \to \infty} \left|\frac{f_n}{f_{n+1}} x_n - \frac{f_{n+1}}{f_n} x_{n-1}\right|^{p_n} = L\right\},$$

$$\ell_\infty(\hat{F}, p) = \left\{x = (x_n) \in \omega : \sup_{n \in \mathbb{N}} \left|\frac{f_n}{f_{n+1}} x_n - \frac{f_{n+1}}{f_n} x_{n-1}\right|^{p_n} < \infty\right\}.$$

Afterwards, by using the notion of ideal convergence and same technique, Khan et al. [22] introduced the paranorm Fibonacci difference ideal convergent sequence spaces $c_0^I(\hat{F}, p)$, $c^I(\hat{F}, p)$, $\ell_\infty^I(\hat{F}, p)$ and $\ell_\infty(\hat{F}, p)$ related to the matrix domain of $\hat{F}$ in the sequence spaces $c_0^I(p)$, $c^I(p)$, $\ell_\infty^I(p)$ and $\ell^I(p)$ introduced in [49]. Further, Khan et al. [22] studied some topological and algebraic properties and examined some inclusion relations concerning these spaces.

$$c_0^I(\hat{F}, p) := \left\{x = (x_n) \in \omega : \left\{n \in \mathbb{N} : \left|\hat{F}_n(x)\right|^{p_n} \geq \epsilon\right\} \in I\right\}, \tag{6.35}$$

$$c^I(\hat{F},p) := \Big\{x=(x_n)\in\omega : \Big\{n\in\mathbb{N} : \Big|\hat{F}_n(x)-L\Big|^{p_n} \geq \epsilon,$$
$$\text{for some } L\in\mathbb{R}\Big\}\in I\Big\}, \quad (6.36)$$

$$\ell^I_\infty(\hat{F},p) := \Big\{x=(x_n)\in\omega : \exists K>0 \text{ s.t } \Big\{n\in\mathbb{N} : \Big|\hat{F}_n(x)\Big|^{p_n} \geq K\Big\}\in I\Big\}, \quad (6.37)$$

and write

$$m^I_0(\hat{F},p) := c^I_0(\hat{F},p)\cap\ell_\infty(\hat{F},p), \quad (6.38)$$

and

$$m^I(\hat{F},p) := c^I(\hat{F},p)\cap\ell_\infty(\hat{F},p). \quad (6.39)$$

Theorem 6.4.1. *The classes of sequences $m^I(\hat{F},p)$ and $m^I_0(\hat{F},p)$ are paranormed spaces, paranormed by $g(x_n) = \sup_n(\|x_n\|)^{\frac{p_n}{M}}$, where $M = \max\{1,\sup_n p_n\}$.*

Proof. The proof of the result is easy, so we leave for reader. □

Theorem 6.4.2. *The space $c^I(\hat{F},p)$ and $m^I(\hat{F},p)$ are neither solid nor monotone in general.*

Proof. Here we give a counter example for establishment of this result.

Example 6.4.3. *Let $I = I = \{A\subseteq\mathbb{N} : A$ is finite$\}$. Let $p_n = 1$ if n is even and $p_n = 2$ if n is odd. Consider the K–step spaces $E_K(\hat{F},p)$ of $E(\hat{F},p)$ defined as follows: Let $x=(x_n)\in E(\hat{F},p)$ and $y=(y_n)\in E_K(\hat{F},p)$ be such that*

$$\hat{F}_n(y) = \begin{cases} \hat{F}_n(x), & \text{if } n \text{ is even} \\ 0, & \text{otherwise.} \end{cases} \quad (6.40)$$

Consider the sequence $x=(x_n)\in\omega$ such that $\hat{F}_n(x) = \frac{1}{2}$, for all $n\in\mathbb{N}$. Then $(x_n)\in E(\hat{F},p)$, but its Kth–step space pre-image does not belong to $E(\hat{F},p)$, where $E = c^I$ and m^I. Thus $c^I(\hat{F},p)$ and $m^I(\hat{F},p)$ are not monotone and hence by Lemma 1, the spaces $c^I(\hat{F},p)$ and $m^I(\hat{F},p)$ are not solid.

□

Theorem 6.4.4. *Let (p_n) and (q_n) be two sequences of positive real numbers. Then $m^I_0(\hat{F},p)\supseteq m^I_0(\hat{F},q)$, if and only if $\lim_{n\in N}\inf\frac{p_n}{q_n}>0$, where $N\subseteq\mathbb{N}$ such that $N\in\mathcal{F}(I)$.*

Proof. Let $\lim_{n\in N} inf\frac{p_n}{q_n}>0$ and $(x_n)\in m^I_0(\hat{F},q)$. Then there exists $\beta>0$ such that $p_n>\beta q_n$, for all sufficiently large $n\in N$. Since $(x_n)\in m^I_0(\hat{F},q)$, for a given $\epsilon>0$, we have

$$B = \{n\in\mathbb{N} : |x_n|^{q_n}\geq\epsilon\}\in I. \quad (6.41)$$

Let $G = N \cup B$. Then $G \in I$. Then for all sufficiently large $n \in G$,

$$\{n \in \mathbb{N} : |x_n|^{p_n} \geq \epsilon\} \subseteq \{n \in \mathbb{N} : |x_n|^{\beta q_n} \geq \epsilon\} \in I.$$

Therefore $(x_n) \in m_0^I(\hat{F}, p)$. The converse part of the result follows obviously. □

Corollary 6.4.5. *Let (p_n) and (q_n) be two sequences of positive real numbers. Then $m_0^I(\hat{F}, p) = m_0^I(\hat{F}, q)$ if and only if $\lim_{n \in N} \inf p_n q_n > 0$, where $N \subseteq \mathbb{N}$ such that $N \in \mathcal{F}(I)$.*

Theorem 6.4.6. *If I is neither maximal nor $I = I_f$, then the spaces $H(\hat{F}, p)$ are not symmetric, where $H = c_0^I, c^I, m_0^I,$ and m^I.*

Proof. We prove the result with the help of the following example. □

Example 6.4.7. *Let $I = I_c = \{A \subseteq \mathbb{N} : \sum_{n \in A} n^{-1} < \infty\}$, (see [40]). Let*

$$\begin{aligned} A &= \{n : n = s^2 \text{ or } t^3, \text{ for } s, t \in \mathbb{N}\} \\ &= \{n \in \mathbb{N} : n = s^2 \text{ for } n \in \mathbb{N}\} \cup \{n \in \mathbb{N} : n = t^3, t \in \mathbb{N}\}, \end{aligned}$$

then $\sum_{n \in A} n^{-1} < \infty$. Let $p_n = 1$ if n is even and $p_n = 2$, if n is odd. Consider the sequence $x = (x_n)$ such that

$$\hat{F}_n(x) = \begin{cases} n^{-1}, & \text{if } n = t^3, t \in \mathbb{N}, \\ 0, & \text{otherwise.} \end{cases}$$

Define the rearrangement $\hat{F}_n(y)$ of $\hat{F}_n(x)$ by

$$\hat{F}_n(y) = (\hat{F}_1(x), \hat{F}_3(x), \hat{F}_3(x), \hat{F}_8(x), \hat{F}_4(x), \hat{F}_5(x), \hat{F}_{27}(x), \hat{F}_6(x), \hat{F}_7(x), \hat{F}_{64}(x), \hat{F}_8(x), \hat{F}_9(x), \ldots).$$

Then $(y_n) \notin H(\hat{F}, p)$ but $(x_n) \in H(\hat{F}, p)$, where $H = c_0^I$, c^I, m_0^I and m^I.

Theorem 6.4.8. *Let (p_n) and (q_n) be two sequences of positive real numbers. Then $m_0^I(\hat{F}, q) \supseteq m_0^I(\hat{F}, p)$, if and only if $\lim_{n \in N} \inf \frac{p_n}{q_n} > 0$, where $N^c \subseteq \mathbb{N}$ such that $N \in I$.*

Proof. The proof follows similarly as that of the proof of theorem (6.4.4). □

Theorem 6.4.9. *Let (p_n) and (q_n) be two sequences of positive real numbers. Then $m_0^I(\hat{F}, p) = m_0^I(\hat{F}, q)$ and only if $\lim_{n \in N} \inf \frac{p_n}{q_n} > 0$, and $\lim_{n \in N} \inf \frac{q_n}{p_n} > 0$ where $N \subseteq \mathbb{N}$ such that $N^c \in I$.*

Proof. On combining Theorems (6.4.4) and (6.4.8), we get the desired result. □

Theorem 6.4.10. *The spaces $m_0^I(\hat{F},p)$ and $m^I(\hat{F},p)$ are not separable.*

Proof. Let $M = \{m_1 < m_{2<\ldots}\}$ be an infinite subset of $\mathbb{N}$ such that $M \in I$. Let

$$\hat{F}_n(x) = \begin{cases} 1, & \text{if } n \in M; \\ 2, & \text{otherwise.} \end{cases}$$

Since M is infinite, so

$$\hat{F}_n(0) = \begin{cases} x_n, & \text{if } x_n = 0; \\ 1, & \text{if } n = m_j, j \in \mathbb{N}, \\ 0, & \text{otherwise.} \end{cases}$$

Since M is infinite, so $\hat{F}_n(0)$ is uncountable. Consider the class of open balls $B_1 = \{B(z, \frac{1}{2}) : z \in m_0^I(\hat{F},p)\}$. Let C_1 be an open cover of $m_0^I(\hat{F},p)$ or $m^I(\hat{F},p)$ containing B_1. Since B_1 is uncountable, so C_1 cannot be reduced to a countable subcover for $m_0^I(\hat{F},p)$ as well as $m^I(\hat{F},p)$. Thus $m_0^I(\hat{F},p)$ and $m^I(\hat{F},p)$ are not separable. □

Bibliography

[1] M. Başarir, F. Başar, E. E. Kara, On the spaces of fibonacci difference absolutely p-summable, null and convergent sequences, Sarajevo J. Math. 12(25) (2016) 167–182

[2] M. Candan, K. Kayaduman, Almost convergent sequence space derived by generalized Fibonacci matrix and Fibonacci core, Brithish J. Math. Comput. Sci. 7(2) (2015) 150–167.

[3] A. Das, B. Hazarika, Some new Fibonacci difference spaces of non-absolute type and compact operators, Linear Multilinear Algebra 65(12) (2017) 2551–2573.

[4] A. Das, B. Hazarika, Matrix transformation of Fibonacci band matrix on generalized bv-space and its dual spaces, Bol. Soc. Parana. Mat. 36(3) (2018) 41–52.

[5] P. Das, P. Kostyrko, W. Wilczyński, P. Malik, I and $I^\star$-convergence of double sequences, Math. Slovaca 58(5) (2008) 605–620.

[6] H. Fast, Sur la convergence statistique, Colloq. Math. 2 (1951) 241–244.

[7] B. Hazarika, A. Das, Some properties of generalized Fibonacci difference bounded and p-absolutely convergent sequences, Bol. Soc. Parana. Mat. 36(1) (2018) 37–50.

[8] B. Hazarika, E. Savaş, Some I-convergent lambda-summable difference sequence spaces of fuzzy real numbers defined by a sequence of Orlicz functions, Math. Comp. Model. 54(11) (2011) 2986–2998.

[9] B. Hazarika, K. Tamanag, On Zweier generalized difference ideal convergent sequences in a locally convex space defined by Musielak-Orlicz function, Bol. Soc. Parana. Mat. 35(2) (2016) 19–37.

[10] E. E. Kara, S. Demiriz, Some new paranormed difference sequence spaces derived by Fibonacci numbers, arXiv preprint arXiv:1309.0154 (2013).

[11] E. E. Kara, M. İlkhan, On some Banach sequence spaces derived by a new band matrix, British J. Math. Comput. Sci. 9(2) (2015) 141–159.

[12] E. E. Kara, M. İlkhan, Some properties of generalized Fibonacci sequence spaces, Linear Multilinear Algebra 64(11) (2016) 2208–2223.

[13] V. A. Khan, On a new sequence space defined by Orlicz functions, Commun. Fac. Sci. Univ. Ank. Ser. A1 Math. Stat. 57(2) (2008) 25–33.

[14] V. A. Khan, K. Ebadullah, On some I-convergent sequence spaces defined by a modulus function, Theory Appl. Math. Comput. Sci. 1(2) (2011) 22–30.

[15] V. A. Khan, K. Ebadullah, I-Convergent difference sequence spaces defined by a sequence of moduli, J. Math. Comput. Sci. 2(2) (2012) 265–273.

[16] V. A. Khan, K. Ebadullah, A. Ahmad, I-Pre-Cauchy sequences and Orlicz function, J. Math. Anal. 3(1) (2012) 21–26.

[17] V. A. Khan, H. Fatima, S. A. A. Abdullah, K. M. A. S. Alshlool, On paranorm BV_σ I-convergent double sequence spaces defined by an Orlicz function, Analysis 37(3) (2017) 157–167.

[18] V. A. Khan, H. Fatima, A. Esi, S. A. A. Abdullah, K. M. A. S. Alshlool. On some new I-convergent double sequence spaces defined by a compact operator, Int. J. Adv. Appl. Sci. 4(4) (2017) 43–48.

[19] V. A. Khan, N. Khan, Y. Khan, On Zweier paranorm I-convergent double sequence spaces, Cogent Math. 3(1) (2016) 1122257.

[20] V. A. Khan, R. K. A. Rababah, K. M. A. S. Alshlool, S. A. A. Abdullah, A. Ahmad, On ideal convergence Fibonacci difference sequence spaces, Adv. Difference Equ. 2018 (20181), Article 199.

[21] V. A. Khan, S. Tabassum, On ideal convergent difference double sequence spaces in n-normed spaces defined by Orlicz function, Theory Appl. Math. Comput. Sci. 3(1) (2013) 90–98.

[22] V. A. Khan, S. A. A. Abdullah, K. M. A. S. Alshlool, Paranorm ideal convergent Fibonacci difference sequence spaces, Commun. Adv. Math. Sci. 2(4) (2019) 293–302.

[23] M. Kirişci, A. Karaisa, Fibonacci statistical convergence and Korovkin type approximation theorems, J. Inequal. Appl. 2017 (2017), Article 229.

[24] H. Kizmaz, Certain sequence spaces, Can. Math. Bull. 24(2) (1981) 169–176.

[25] T. Koshy, Fibonacci and Lucas numbers with applications, John Wiley & Sons (2019).

[26] P. Kostyrko, T. Šalát, W. Wilczyński, I-Convergence, Real Anal. Exchange 26(2) (2000/2001) 669–686.

[27] M. A. Krasnoselskii, Y. B. Rutitsky. Convex functions and Orlicz functions, Groningen, Netherlands (1961).

[28] V. Kumar, On I and $I^{\star}$-Convergence of double sequences, Math. Commun. 12(2) (2007) 171–181.

[29] V. Kumar, K. Kumar, On the ideal convergence of sequences of fuzzy numbers, Inform. Sci. 178 (2008) 4670–4678.

[30] J. Lindenstrauss, L. Tzafriri, On Orlicz sequence spaces, Israel J. Math. 10(3) (1971) 379–390.

[31] I. J. Maddox, Spaces of strongly summable sequences, Quart. J. Math. 18(1) (1967) 345–355.

[32] I. J. Maddox, Paranormed sequence spaces generated by infinite matrices, Math. Proc. Cambridge Philos. Soc. 64 (1968) 335–340.

[33] J. E. Maddux, Self-efficacy, adaptation, and adjustment, Springer (1995).

[34] S. J. Miller, Y. Wang, From Fibonacci numbers to central limit type theorems, J. Comb. Theory Ser. A 119(7) (2012) 1398–1413.

[35] M. Mursaleen, A. K. Noman, On some new sequence spaces of non-absolute type related to the spaces ℓ_p and ℓ_∞, Filomat 25(2) (2011) 33–51.

[36] H. Nakano, Modulared sequence spaces, Proc. Japan Acad. 27(9) (1951) 508–512.

[37] H. Nakano, Concave modulars, J. Math. Soc. Japan 5(1) (1953) 29–49.

[38] I. Niven, H. S. Zuckerman, H. L. Montgomery, An introduction to the theory of numbers, John Wiley & Sons (1991).

[39] S. D. Parashar, B. Choudhary, Sequence spaces defined by Orlicz functions, Indian J. Pure Appl. Math. 25 (1994) 419–428.

[40] G. M. Petersen, Regular matrix transformations, McGraw-Hill (1966).

[41] T. Šalát, B. C. Tripathy, M. Ziman, On some properties of I-convergence, Tatra Mt. Math. Publ. 28(2) (2004) 274–286.

[42] T. Šalát, B. C. Tripathy, M. Ziman, On I-convergence field, Ital. J. Pure Appl. Math. 17(5) (2005) 1–8.

[43] B. Schweizer, A. Sklar, Statistical metric spaces, Pacific J. Math. 10 (1960) 313–334.

[44] S. Simons, The sequence spaces $\ell(pv)$ and $m(pv)$, Proc. London Math. Soc. 15(3) (1965) 422–436.

[45] H. Steinhaus, Sur la convergence ordinaire et la convergence asymptotique, Colloq. Math. 2 (1951) 73–74.

[46] I. J. Maddox, Elements of functional analysis, Cambridge University Press (1970).

[47] B. C. Tripathy, B. Hazarika, I-Convergent sequence spaces associated with multiplier sequences, Math. Inequal. Appl. 11(3) (2008) 543–548.

[48] B. C. Tripathy, B. Hazarika, Some I-convergent sequence spaces defined by Orlicz functions, Acta Math. Appl. Sin. Engl. Ser. 27(1) (2011) 149–154.

[49] B. C. Tripathy, B. Hazarika, Paranorm I-convergent sequence spaces, Math. slovaca 59 (4) (2009) 485–494.

[50] V. A. Khan, K. M. A. S. Alshlool, A. A. H. Makharesh, S. A. A. Abdullah, On spaces of ideal convergent Fibonacci difference sequence defined by Orlicz function, Sigma J. Eng. Nat. Sci. 37(1) (2019) 143–154.

[51] A. Esi, S. A. A. Abdullah, V. A. Khan, R. K. A. Rababah, K. M. A. S. Alshlool, Some new spaces of ideal convergent double sequences by using compact operator, J. Appl. Sci. 17(9) (2017) 467–474.

[52] A. Wilansky, Summability through functional analysis, North-Holland Mathematics Studies 85, Amsterdam-New York-Oxford (1984).

[53] P. Z. Alp, M. İlkhan. On the difference sequence space $\ell_p(\hat{T}^q)$, Math. Sci. Appl. E-Notes 7(2) (2019) 161–173.

Chapter 7

Theory of Approximation for Operators in Intuitionistic Fuzzy Normed Linear Spaces

Nabanita Konwar

Birjhora Mahavidyalaya, nabnitakonwar@gmail.com

Pradip Debnath

Assam University, debnath.pradip@yahoo.com

7.1 Introduction

In nonlinear analysis, a significant number of problems are associated with a class of objects, many of which turn out to be linear spaces or vector spaces. If the limit process is imposed in such classes of objects then these areas are turned as the branch of metric or topology. When the induced metric fulfill the translation invariance property the "property of the norm" can be defined on it. The new structure is then converted to normed linear space. In Mathematics, there are some situations where the crisp norm is unable to measure the length of a vector accurately then for such type of situations the concept of fuzzy norm seems to be helpful. There has been a systematic improvement of fuzzy normed linear spaces (FNLSs) and one of the such important improvement of FNLS is the idea of intuitionistic fuzzy normed

DOI: 10.1201/9781003178200-7

linear space (IFNLS). The investigation and evaluation of analytic properties, topological structures and systematic generalizations of IFNLSs remain well motivated areas of research.

7.1.1 Background

The theory of approximation property for operators takes an important role in the structure theory of complete normed linear spaces (i.e. Banach spaces). Initially, the systematic study of this property in Banach space was investigated by Grothendick [18] in 1935. We say that a Banach space Y satisfy the approximation property (AP) if, for each compact set C and $\epsilon > 0$, there exist a bounded finite rank operator $F : Y \longrightarrow Y$ such that $||F(\alpha) - \alpha|| < \epsilon$, $\forall\ \alpha \in C$, i.e. the identity operator $I(\alpha)$ on Y on a compact set can be approximated uniformly by the finite rank operators. And Y satisfy the bounded approximation property (BAP) if for each compact set C and $\epsilon > 0$, there exist a bounded finite rank operator $F : \alpha \longrightarrow \alpha$ with $||F|| \leq \lambda$ for some $\lambda > 0$ such that $||F(\alpha) - \alpha|| < \epsilon\ \forall\ \alpha \in C$.

The APs take paramount importance in the development of infinite dimensional Banach space theory and also in the study of Schauder basis. A Banach space satisfy the BAPs if and only if it is a complemented subspace of a Banach space with a basis. In this branch of mathematics, the concept of basis implies the Schauder basis. The idea of Schauder basis was introduced by Schauder in 1927. Initially, when Grothendick started to study the approximation property he used the concept of the topology of uniform convergence on compact sets. In 1972, Enflo [15] provide the first example of a complete normed linear space which fails to satisfy the AP. Also there exist some Banach spaces with the BAP which fail to have bases. However, a complemented subspace of L_p for $1 < p < \infty$ has a basis. In 1992, Lusky [30] has established that the Disk Algebra A has a basis without using the results from the complex function theory, but using the results from the AP. Some of the important references which are related to the present work is found in [8,14,24,34,43].

The concept of a fuzzy norm on a linear space was established by Katsaras [23]. Felbin [17] put forward the concept of a fuzzy norm whose associated metric is of Kaleva and Seikkala [22] type. Another notion fuzzy norm was introduced by Cheng and Mordeson [7] on a linear space whose associated metric of this new space is of Kramosil and Michalek [25] type. Following the concept of Cheng and Mordeson, one more systematic development of fuzzy normed linear space was provided by Bag and Samanta [3]. The idea of intuitionistic fuzzy set (IFS) was introduced by Atanassov [2]. After this development Atanassov has triggered for some debate (for details, see [6,13,19]) regarding the use of the terminology "intuitionistic". In response to this debate, Atanassov justified the terminology in [1]. This new field of mathematics provides a well motivated application base area for the branch of science and engineering.

With the help of the concept of arbitrary continuous t-norm and continuous t-conorm, Saadati and Park [41] put forward the notion of IFNLS. Further there has been more systematic improvement over the field IFNLS. One of such improvement is the study of topological structure of an intuitionistic fuzzy 2-normed space which was provided and established by Mursaleen and Lohani in [35]. Simultaneously, a large number of interesting and analytic properties of IFNLS have been established by Mursaleen and Mohiuddine [37–40]. Furthermore, Vijayabalaji et al. [45] generalized the concept of Saadati and Park, and established the notion of an intuitionistic fuzzy n-normed linear space (IFnNLS). Various analytical properties of IFnNLS have been introduced by N. Thillaigovindan, S. Anita Shanti and Y. B. Jun in [44]. Several recent work in similar context can be found in [10–12, 16, 20, 21, 31–33, 36, 42].

In fuzzy normed spaces the concept of AP was introduced by Yilmaz [46] and established some more interesting and analytical results on it. Recently, regarding the continuity of fuzzy operators, Keun Young Lee [28] established some limitations in Yilmaz's definitions. He modified Yilmaz's definitions and set up a researchable environment to study the approximation property (AP) and bounded approximation property (BAP) on fuzzy normed spaces.

7.1.2 Main goal

The main goal of this chapter is to investigate and discuss some important concepts on AP and BAP in the settings of complete intuitionistic fuzzy normed linear spaces (shortly, CIFNLS). In order to study such properties, we introduce some important operators called strong intuitionstic fuzzy continuous operators and strong intuitionistic fuzzy bounded operators in CIFNLSs. With the help of these operator, we develop the existence of approximation properties in CIFNLS.

7.2 Basic Definitions

This section contains some basic and useful definitions which are useful for the whole chapter.

Definition 7.2.1. *[27] Consider 5-tuple $(X, \varpi, \upsilon, \otimes, \odot)$ where X is a linear space, $\otimes$ is a continuous t-norm, $\odot$ is a continuous t-conorm, and ϖ, υ fuzzy sets on $X \times (0, \infty)$ then $(X, \varpi, \upsilon, \otimes, \odot)$ is said to be an IFNLS if it satisfy the following conditions for every $a, y \in X$ and $s, t > 0$:*

- $\varpi(a, t) > 0$,
- $\varpi(a, t) = 1$ *if and only if* $a = 0$,
- $\varpi(\alpha a, t) = \varpi(a, \frac{t}{|\alpha|})$ *for each* $\alpha \neq 0$,
- $\varpi(a + y, t + s) \geq \min \varpi(a, t), \varpi(y, s)$,

- $\lim_{t\to\infty} \varpi(a,t) = 1$ *and* $\lim_{t\to 0} \varpi(a,t) = 0$,
- $\upsilon(a,t) < 1$,
- $\upsilon(a,t) = 0$ *if and only if* $a = 0$,
- $\upsilon(\alpha a,t) = \upsilon(a, \frac{t}{|\alpha|})$ *for each* $\alpha \neq 0$,
- $\upsilon(a+y, t+s) \leq \max \upsilon(a,t), \upsilon(y,s)$,
- $\lim_{t\to\infty} \upsilon(a,t) = 0$ *and* $\lim_{t\to 0} \upsilon(a,t) = 1$.

The pair (ϖ, υ) *is called an intuitionistic fuzzy norm (shortly IFN).*

Definition 7.2.2. *[9] Consider an IFNLS* $(X, \varpi, \upsilon, \otimes, \odot)$ *and a sequence* $x = \{x_k\}$ *in* X. *Then* $x = \{x_k\}$ *is called convergent to* $\xi \in X$ *with respect to the intuitionistic fuzzy norm* (ϖ, υ) *if, for every* $\varepsilon \in (0,1)$ *and* $t > 0$, $\exists\ k_0 \in \mathbb{N}$ *such that* $\varpi(x_k - \xi, t) > 1 - \varepsilon$ *and* $\upsilon(x_k - \xi, t) < \varepsilon\ \forall\ k \geq k_0$. *It is denoted by* $(\varpi, \upsilon) - \lim x_k = \xi$.

Definition 7.2.3. *[41] Consider an IFNLS* $(X, \varpi, \upsilon, \otimes, \odot)$ *and a sequence* $x = \{x_k\}$ *in* X. *Then* $x = \{x_k\}$ *is called a Cauchy sequence with respect to the intuitionistic fuzzy norm* (ϖ, υ) *if, for every* $\alpha \in (0,1)$ *and* $t > 0$, $\exists\ k_0 \in \mathbb{N}$ *such that* $\varpi(x_k - x_m, t) > 1 - \alpha$ *and* $\upsilon(x_k - x_m, t) < \alpha$ *for all* $k, m \geq k_0$.

Definition 7.2.4. *[11] An IFNLS* $(X, \varpi, \upsilon, \otimes, \odot)$ *is complete iff every Cauchy sequence of* $(X, \varpi, \upsilon, \otimes, \odot)$ *is convergent.*

Definition 7.2.5. *[26] Suppose* $(X, \varpi, \upsilon, \otimes, \odot)$ *is an IFNLS and* S *is a subset* S *in* $(X, \varpi, \upsilon, \otimes, \odot)$. *Then* S *is called compact if every sequence of elements of* S *has a convergent subsequence.*

Definition 7.2.6. *[9] Suppose* $(X, \varpi, \upsilon, \otimes, \odot)$ *is an IFNLS. Then for* $t > 0$ $B(x, r, t)$ *with center at* $x \in X$ *and radius* $0 < r < 1$ *is called an open ball if*

$$B(x,r,t) = \{y \in X : \varpi(x-y,t) > 1 - r, \upsilon(x-y,t) < r\}.$$

Theorem 7.2.7. *Consider an IFNLS* $(X, \varpi, \upsilon, \otimes, \odot)$ *with the condition*

$$\varpi(a,t) > 0 \text{ and } \upsilon(a,t) < 1 \text{ implies } x = 0, \text{ for all } t \in \mathbb{R}^+. \tag{7.1}$$

Suppose $\|x\|_\alpha = inf\{t \in \mathbb{R}^+ : \varpi(a,t) > \alpha$ *and* $\upsilon(a,t) < 1 - \alpha\}$ *for each* $\alpha \in (0,1)$. *Then* $\{\|\cdot\|_\alpha : \alpha \in (0,1)\}$ *is an ascending class of norms on* X *and these norms are called* α*-norms on the IFN* (ϖ, υ).

Proof of this theorem is similar to [4].

Definition 7.2.8. *[36] Suppose* (x_n) *is a sequence in an IFNLS* $(X, \varpi, \upsilon, \otimes, \odot)$. *Then* (x_n) *is a basis of* X *if for every* $x \in X$ $\exists$ *a unique sequence* (a_n) *of scalars such that*

$$(\varpi, \upsilon) - \lim \textstyle\sum_{k=1}^{n} a_k x_k = x.$$

that is, for each $\alpha \in (0,1)$ *and* $\epsilon > 0$, $\exists\ n_0 = n_0(\alpha, \epsilon) \in N$ *such that* $n \geq n_0$ *implies,*

$$\varpi(x - \sum_{k=1}^{n} a_k x_k, \epsilon) > 1 - \alpha \text{ and } \upsilon(x - \sum_{k=1}^{n} a_k x_k, \epsilon) < \alpha, \text{ where } x = \sum_{k=1}^{\infty} a_k x_k.$$

7.3 Definitions and Main Results

This section includes the discussion of the main results for the chapter. Initially, we establish some essential definitions in connection with AP in the area of IFNLS.

7.3.1 Essential definitions

Definition 7.3.1. *Consider a complete IFNLS* $(X, \varpi, \upsilon, \otimes, \odot)$. *Then* $(X, \varpi, \upsilon, \otimes, \odot)$ *is known to satisfy the approximation property (shortly AP), if for each compact set* K *in* X, $\exists$ *an operator* T *of finite rank, for each* $\alpha > (0,1)$ *and* $\epsilon > 0$ *such that* $\varpi(T_\alpha(x) - x, \epsilon) > 1 - \alpha$ *and* $\upsilon(T_\alpha(x) - x, \epsilon) < \alpha$ *for every* $x \in K$.

Definition 7.3.2. *Suppose* λ *is a real number. A complete IFNLS* $(X, \varpi, \upsilon, \otimes, \odot)$ *is known to satisfy the* λ*-bounded approximation property (shortly,* λ*-BAP), if for each compact set* K *in* X, $\exists$ *an operator* $T \in F(X, X, \lambda)$ *for each* $\alpha \in (0,1)$ *and* $\epsilon > 0$ *such that* $\varpi(T(x) - x, \epsilon) > 1 - \alpha$ *and* $\upsilon(T(x) - x, \epsilon) < \alpha$ *for every* $x \in K$.

Definition 7.3.3. *Let* (x_n) *be a basis of a complete IFNLS* $(X, \varpi, \upsilon, \otimes, \odot)$. *Then* m^{th} *natural projection* P_m *for* x_m, *for every positive integer* m, *is the function*

$$\sum_{n=1}^{\infty} a_n x_n \longrightarrow \sum_{n=1}^{m} a_n x_n \text{ from } (X, \varpi, \upsilon, \otimes, \odot) \text{ to } (X, \varpi, \upsilon, \otimes, \odot).$$

Definition 7.3.4. *Consider two complete IFNLS* $(X, \varpi, \upsilon, \otimes, \odot)$ *and* $(Y, \varpi', \upsilon', \otimes, \odot)$ *and* $T : X \longrightarrow Y$ *is a linear operator where* (ϖ, υ) *and* (ϖ', υ') *are intuitionistic fuzzy normed. Then*

1. *Operator* T *is known as strongly intuitionistic fuzzy (shortly sif) continuous at* $a \in X$ *if, for any* $\epsilon > 0$, $\exists\ \delta > 0$ *such that* $\varpi'(T(x) - T(a), \epsilon) \geq \varpi(x - a, \delta)$ *and* $\upsilon'(T(x) - T(a), \epsilon) \leq \upsilon(x - a, \delta)$ *for all* $x \in X$. *If at each point of* X, T *is sif-continuous, then* T *is known as sif-continuous on* X.

2. *Operator* T *is known as strongly intuitionistic fuzzy (shortly sif) bounded on* X *if* $\exists$ *a positive real number* m *such that* $\varpi'(T(x), t) \geq \varpi(x, \frac{t}{m})$

and $\upsilon'(T(x),t) \leq \upsilon(x,\frac{t}{m})$ $\forall$ $x \in X$ *and* $t \in \mathbb{R}$. *The set of all sif-bounded operators form* X *to* Y *are denoted by* $F(X,Y)$. *Furthermore,* $F(X,Y)$ *is a vector space. Next* $F(X,Y,M)$, $\forall$ $M > 0$ *is denoted by* $\{T \in F(X,Y) : \varpi'(T(x),t) \geq \varpi(x,\frac{t}{m}), \upsilon'(T(x),t) \leq \upsilon(x,\frac{t}{m}), \forall x \in X, \forall t \in \mathbb{R}\}$, *where* m *is a positive real number. If* $\mathbb{S} = F(X,Y,M)$, *for some* $m > 0$ *then* $\mathbb{S}$ *is known as a bounded subset of* $F(X,Y)$. *And* $\bar{F}(X,Y)$ *implies the set of all finite rank sif-bounded operators from* X *to* Y. *Then* $\bar{F}(X,Y)$ *is subspace of* $F(X,Y)$. *Additionally, for some* $m > 0$, $\bar{F}(X,Y,m)$ *is also a subspace of* $F(X,Y,m)$.

If $(X,\varpi,\upsilon,\otimes,\odot)$ and $(Y,\varpi',\upsilon',\otimes,\odot)$ be two IFNLSs satisfying the conditions: $\varpi(a,t) > 0$ and $\upsilon(a,t) < 1$ implies $x = 0$ $\forall$ $t \in \mathbb{R}^+$ and for $x \neq 0$, $\varpi(a,t)$ is continuous and strictly increasing on $\{t : 0 < \varpi(a,t) < 1\}$, while $\upsilon(a,t)$ is continuous and strictly decreasing on $\{t : 0 < \varpi(a,t) < 1\}$ and $M > 0$. Then we obtain

$$F(X,Y,M) = \{T \in F(X,Y) : ||T(x)||_\alpha \leq M||x||_\alpha, \forall x \in X, \forall \alpha \in (0,1)\}.$$

Therefore, $F(X,Y,M)$ and $\bar{F}(X,Y,M)$ are bounded convex subsets of $F(X,Y)$.

7.3.2 Main results

Theorem 7.3.5. *Let* $(X,\varpi,\upsilon,\otimes,\odot)$ *and* $(Y,\varpi',\upsilon',\otimes,\odot)$ *be two IFNLSs satisfying condition 12.1 and* $T : X \longrightarrow Y$ *be a linear operator. Then* T *is sif-bounded if and only if it is uniformly bounded with respect to* α*-norms of* (ϖ,υ) *and* (ϖ',υ'). *That is,* $\exists$ *some* $M > 0$, *independent of* α, *such that* $||T(x)||_\alpha \leq M||x||_\alpha$, $\forall$ $\alpha \in (0,1)$.

Proof. Proof of this theorem is similar to [5]. □

Theorem 7.3.6. *Consider a complete IFNLS* $(X,\varpi,\upsilon,\otimes,\odot)$ *and a Schauder basis* (x_n) *in* X. *Then* (x_n) *is a basis where*

$$\varpi(a,t) = \begin{cases} \frac{t-||x||}{t+||x||} & if\ t > ||x|| \\ 0, & if\ t \leq ||x|| \end{cases}$$

and

$$\upsilon(a,t) = \begin{cases} 1 - \frac{t-||x||}{t+||x||}, & if\ t > ||x|| \\ 1, & if\ t \leq ||x||. \end{cases}$$

Also, each natural projection is sif-continuous.

Proof. Suppose that (x_n) is a basis for an IFNLS $(X,\varpi,\upsilon,\otimes,\odot)$. Then, we have to establish that natural projection $P_n : (X,\varpi,\upsilon,\otimes,\odot) \rightarrow$

$(X, \varpi, \upsilon, \otimes, \odot)$ is sif-bounded for every $x \in N$. Let $n \in N$, $t \in R$, $x \in X$. Consider $M = ||P_n||$. When $t \leq 0$, the result is trivial. Consider that $t > 0$. Then, we have to establish that

$$\varpi\left(P_n\left(x\right), t\right) \geq \varpi\left(x, \frac{t}{M}\right) \text{ and } \upsilon\left(P_n\left(x\right), t\right) \leq \upsilon\left(x, \frac{t}{M}\right).$$

By using the similar process of the Proposition 3.4 of [28], we can obtain the proof of $\varpi\left(P_n\left(x\right), t\right) \geq \varpi\left(x, \frac{t}{M}\right)$. Next, for υ, we have $t > M||x||$. Then

$$\upsilon\left(x, \frac{t}{M}\right) = 1 - \frac{\frac{t}{M} - ||x||}{\frac{t}{M} + ||x||}.$$

By the assumption, $t > M||x|| = ||P_n||||x|| \geq ||P_n\left(x\right)||$ and

$$\frac{t - ||P_n\left(x\right)||}{t + ||P_n\left(x\right)||} \geq \frac{\frac{t}{M} - ||x||}{\frac{t}{M} + ||x||}.$$

Therefore, we have

$$\upsilon\left(P_n\left(x\right), t\right) = 1 - \frac{t - ||P_n\left(x\right)||}{t + ||P_n\left(x\right)||} \leq 1 - \frac{\frac{t}{M} - ||x||}{\frac{t}{M} + ||x||} = \upsilon\left(x, \frac{t}{M}\right).$$

Hence $\upsilon\left(P_n\left(x\right), t\right) \leq \upsilon\left(x, \frac{t}{M}\right)$. Secondly, take $t \leq ||Mx||$. Then $\upsilon\left(Mx, t\right) = 1$. Thus, $\upsilon\left(P_n\left(x\right), t\right) \leq \upsilon\left(x, \frac{t}{M}\right)$. Hence, we have obtained the existence of a complete IFNLS having a basis such that each natural projection is sif-continuous. □

7.3.3 Modified version of definitions of AP and BAP

Next, we establish a modified version of definitions of APs and BAPs in a complete IFNLSs by introducing the concept of continuity of approximating operators.

Definition 7.3.7. *Consider a complete IFNLS $(X, \varpi, \upsilon, \otimes, \odot)$. Then $(X, \varpi, \upsilon, \otimes, \odot)$ satisfy the approximation property, if for each compact set K in $(X, \varpi, \upsilon, \otimes, \odot)$ and for every $\alpha \in (0, 1)$ and $\epsilon > 0$, $\exists$ an operator $T \in \bar{F}(X, X)$ such that $\varpi\left(T\left(x\right) - x, \epsilon\right) > 1 - \alpha$ and $\upsilon(T(x) - x, \epsilon) < \alpha$ for every $x \in K$.*

Definition 7.3.8. *Consider a complete IFNLS $(X, \varpi, \upsilon, \otimes, \odot)$ and a positive real number λ. Then $(X, \varpi, \upsilon, \otimes, \odot)$ satisfy the λ-BAP, if for each compact set K in $(X, \varpi, \upsilon, \otimes, \odot)$ and for every $\alpha \in (0, 1)$ and $\epsilon > 0$, $\exists$ an operator $T \in \bar{F}(X, X, \lambda)$ such that $\varpi\left(T\left(x\right) - x, \epsilon\right) > 1 - \alpha$ and $\upsilon(T\left(x\right) - x, \epsilon) < \alpha$, $\forall\ x \in K$. One can also conclude that $(X, \varpi, \upsilon, \otimes, \odot)$ satisfy the BAP if for some $\lambda > 0$, $(X, \varpi, \upsilon, \otimes, \odot)$ satisfy the λ-BAP.*

7.3.4 Certain related results and examples

Theorem 7.3.9. *Consider a complete IFNLS* $(X, \varpi, \upsilon, \otimes, \odot)$*. Then the below statements are equivalent.*

(i) $(X, \varpi, \upsilon, \otimes, \odot)$ *satisfy the AP.*

(ii) If $(Y, \varpi', \upsilon', \otimes, \odot)$ *is an IFNLS, then for each* $T \in F(X, Y)$*, every compact set* K *in* $(X, \varpi, \upsilon, \otimes, \odot)$ *and for every* $\alpha \in (0,1)$ *and* $t > 0$, $\exists$ *an operator* $S \in \bar{F}(X, Y)$ *such that* $\varpi'(S(x) - T(x), t) > 1 - \alpha$ *and* $\upsilon'(S(x) - T(x), t) < \alpha$ *for each* $x \in K$.

(iii) If $(Y, \varpi', \upsilon', \otimes, \odot)$ *is an IFNLS, then for each* $T \in F(Y, X)$*, every compact set* K *in* $(Y, \varpi', \upsilon', \otimes, \odot)$ *and for every* $\alpha \in (0,1)$ *and* $t > 0$, $\exists$ *an operator* $S \in \bar{F}(Y, X)$ *such that* $\varpi(S(y) - T(y), t) > 1 - \alpha$ *and* $\upsilon(S(y) - T(y), t) < \alpha$ *for each* $y \in K$.

Proof. (*i*) $\Rightarrow$ (*ii*) Let us consider a compact set K in $(X, \varpi, \upsilon, \otimes, \odot)$, $T \in F(X, Y)$, $\alpha \in (0,1)$ and $t > 0$ for $t \in \mathbb{R}$. Then $\exists$ a positive real number M such that

$$\varpi'(T(x), t) \geq \varpi\left(x, \frac{t}{M}\right) \text{ and } \upsilon'(T(x), t) \leq \upsilon\left(x, \frac{t}{M}\right), \ \forall x \in X.$$

As $(X, \varpi, \upsilon, \otimes, \odot)$ satisfy the AP, therefore $\exists$ an operator $R \in F(X, X)$ such that $\varpi\left(R(x) - x, \frac{t}{M}\right) > 1 - \alpha$ and $\upsilon\left(R(x) - x, \frac{t}{M}\right) < \alpha$ for each $x \in K$. Next we consider $S = TR$. As T and R both are sif-bounded operators, therefore S is also a sif-bounded operator. Thus

$$\varpi'(S(x) - T(x), t) = \varpi'(TR(x) - T(x), t) \geq \varpi\left(R(x) - x, \frac{t}{M}\right) > 1 - \alpha$$

and

$$\upsilon'(S(x) - T(x), t) = \upsilon'(TR(x) - T(x), t) \leq \upsilon\left(R(x) - x, \frac{t}{M}\right) < \alpha.$$

for every $x \in K$.

(*i*) $\Rightarrow$ (*iii*) Suppose K is a compact set in $\left(Y, \varpi', \upsilon', \otimes, \odot\right)$, $\alpha \in (0,1)$, $T \in F(Y, X)$ and $t > 0$ for some $t \in R$. As $(X, \varpi, \upsilon, \otimes, \odot)$ satisfy the AP and $T(K)$ is compact set in $(X, \varpi, \upsilon, \otimes, \odot)$, $\exists$ an operator $R \in \bar{F}(X, X)$ such that $\varpi(R(x) - x, t) > 1 - \alpha$ and $\upsilon(R(x) - x, t) < \alpha$ for each $x \in T(K)$. Next, we consider $S = RT \in \bar{F}(Y, X)$. Then

$$\varpi(S(y) - T(y), t) = \varpi(RT(y) - T(y), t) > 1 - \alpha$$

and

$$\upsilon(S(y) - T(y), t) = \upsilon(RT(y) - T(y), t) < \alpha$$

for each $y \in K$. Hence, (*i*) implies both (*ii*) and (*iii*). We conclude that (*i*), (*ii*) and (*iii*) are equivalent. □

Theorem 7.3.10. *Consider a subset K in a complete IFNLS $(X, \varpi, \upsilon, \otimes, \odot)$. If K is a compact set in $(X, \varpi, \upsilon, \otimes, \odot)$, then for each $\alpha \in (0,1)$ and $t > 0$, $\exists$ a finite set $\{x_1, x_2, \ldots, x_n\}$ in K such that for every $x \in K$, we have for some x_i, $x \in B(x_i, \alpha, t)$.*

Proof. Proof of this theorem is similar to Theorem 4.2 of [28]. □

Theorem 7.3.11. *Consider a complete IFNLS $(X, \varpi, \upsilon, \otimes, \odot)$ with intuitionistic fuzzy norm (ϖ, υ) and $M > 0$. Assume that for every $x \in X$, $\exists$ a sequence $(T_n) \in \bar{F}(X, X, M)$ such that $T_n(x) \longrightarrow x$, then the IFNLS $(X, \varpi, \upsilon, \otimes, \odot)$ satisfy AP.*

Proof. Assume that (T_n) is a sequence in $\bar{F}(X, X, M)$ such that $T_n(x) \longrightarrow x$ for all $x \in X$. Let $\alpha \in (0,1)$ and $t > 0$, and K be a compact set in $(X, \varpi, \upsilon, \otimes, \odot)$. Then, from the Theorem 7.3.10, $\exists$ a finite set $\{x_1, x_2, ..., x_n\} \subset K$ such that for $x \in K$, we have $x \in B(x_i, \alpha, t)$ for some x_i. Therefore $\exists$ $N_1, N_2 \in \mathbb{N}$ such that for every i, if $n \geq N_1, N_2$ we have, $\varpi(T_n(x_i) - x_i, t) > 1 - \alpha$ and $\upsilon(T_n(x_i) - x_i, t) < \alpha$. Consider $x \in K$ and choose i such that $x \in B(x_i, \alpha, t)$, that is, $\varpi(x_i - x, t) > 1 - \alpha$ and $\upsilon(x_i - x, t) < \alpha$. Then, for $n \geq N_1, N_2$, we obtain

$$\begin{aligned}
&\varpi(T_n(x) - x, t) \\
&= \varpi(T_n(x) + (-T_n(x_i)) + (T_n(x_i)) + (-x_i) + x_i + (-x), t) \\
&\geq \min\left\{\varpi\left(T_n(x - x_i), \frac{t}{3}\right), \varpi\left(T_n(x_i) - x_i, \frac{t}{3}\right), \varpi\left(x_i - x, \frac{t}{3}\right)\right\} \\
&\geq \min\left\{\varpi\left(x - x_i, \frac{t}{3M}\right), \varpi\left(T_n(x_i) - x_i, \frac{t}{3}\right), \varpi\left(x_i - x, \frac{t}{3}\right)\right\} \\
&> 1 - \alpha
\end{aligned}$$

and

$$\begin{aligned}
&\upsilon(T_n(x) - x, t) \\
&= \upsilon(T_n(x) + (-T_n(x_i)) + (T_n(x_i)) + (-x_i) + x_i + (-x), t) \\
&\leq \max\left\{\varpi\left(T_n(x - x_i), \frac{t}{3}\right), \varpi\left(T_n(x_i) - x_i, \frac{t}{3}\right), \varpi\left(x_i - x, \frac{t}{3}\right)\right\} \\
&\leq \max\left\{\varpi\left(x - x_i, \frac{t}{3M}\right), \varpi\left(T_n(x_i) - x_i, \frac{t}{3}\right), \varpi\left(x_i - x, \frac{t}{3}\right)\right\} < \alpha.
\end{aligned}$$

Hence, $(X, \varpi, \upsilon, \otimes, \odot)$ satisfy AP. □

With the help of above theorem, we establish the following result:

Theorem 7.3.12. *Consider a basis $\{x_n\}$ of $(X, \varpi, \upsilon, \otimes, \odot)$ and every natural projection $P_n : (X, (\varpi, \upsilon)) \longrightarrow (X, (\varpi, \upsilon))$ is sif-continuous. Then $(X, \varpi, \upsilon, \otimes, \odot)$ satisfy the AP but the converse is not necessarily true.*

Theorem 7.3.13. *A complete IFNLS $(X, \varpi, \upsilon, \otimes, \odot)$ satisfying condition 12.1 has satisfy the AP iff for every $\alpha \in (0,1)$, $\epsilon > 0$ and for each compact set K in $(X, \varpi, \upsilon, \otimes, \odot)$, $\exists$ an operator $T \in \bar{F}(X, X)$ for every $x \in K$ such that $||T(x) - x||_\alpha < \epsilon$.*

Theorem 7.3.14. *Consider a complete IFNLS $(X, \varpi, \upsilon, \otimes, \odot)$ satisfying the condition 12.1 and $\lambda > 0$. Then $(X, \varpi, \upsilon, \otimes, \odot)$ satisfy λ-BAP iff for every $\alpha \in (0,1)$, $\epsilon > 0$ and for every compact set K in $(X, \varpi, \upsilon, \otimes, \odot)$, there exists an operator $T \in \bar{F}(X, X, \lambda)$ for each $x \in K$ such that $||T(x) - x||_\alpha < \epsilon$.*

The proof of above results can be easily obtained from [47].

Now we are going to establish some examples of IFNLS which satisfy or fails to satisfy the AP and BAP. Next we are going to present an example which fails to have the AP.

Example 7.3.15. *In this example, we consider the Banach space $(X, ||\cdot||)$ which fails to have the approximation property and we are going to show that the complete IFNLS $(X, \varpi, \upsilon, \otimes, \odot)$ also fails to satisfy the AP.*

Consider a function $\varpi, \upsilon : X \times \mathbb{R} \longrightarrow [0,1]$ defined by

$$\varpi(a,t) = \begin{cases} 1 & if \;\; t > ||x|| \\ 0 & if \;\; t \leq ||x|| \end{cases} \quad and \quad \upsilon(a,t) = \begin{cases} 0 & if \;\; t > ||x|| \\ 1 & if \;\; t \leq ||x|| \end{cases},$$

where (ϖ, υ) is the intuitionistic fuzzy norm and $||x||_\alpha = ||x||$, for each $\alpha \in (0,1)$. Next, consider that $(X, \varpi, \upsilon, \otimes, \odot)$ satisfy the AP. Suppose $\alpha \in (0,1)$ and $\epsilon > 0$ and K is a compact set in X. Since $||x||_\alpha = ||x||$ for every $\alpha \in (0,1)$, K is compact in $(X, \varpi, \upsilon, \otimes, \odot)$. Then from the Theorem 7.3.13, $\exists$ an operator $T_\alpha \in \bar{F}(X, X)$ for each $x \in K$ such that $||T(x) - x||_\alpha < \epsilon$. Therefore, for each $x \in K$, we get $||T(x) - x|| = ||T(x) - x||_\alpha < \epsilon$ which is a contradiction as $(X, ||\cdot||)$ fails to satisfy the approximation property. Here $(X, \varpi, \upsilon, \otimes, \odot)$ fails to satisfy the AP.

We mow provide an example of the existence of a complete IFNLS which satisfy the AP but fails to satisfy the BAP (as shown in Example 4.9 of [28]).

Example 7.3.16. *In classical mathematics there exist a a Banach Space X_0 which satisfy the metric approximation property but its dual space X_0^* fails to satisfy the approximation property. This example was established by Enflo and Lidenstrauss [14, 29].*

For this example, we consider a sequence $(||\cdot||_n)$ of equivalent norms on X_0 so that $(X_0, ||\cdot||_n)$ fails to satisfy the n-BAP. Now we consider $X_n = (X_0, ||\cdot||_n)$. Then we have $(\sum \oplus X_n)_{l_2}$ which fails to satisfy the BAP where $(\sum \oplus X_n)_{l_2}$ is a Banach space and all its elements are the sequence of the form $(x_1, x_2, ...)$, where $\sum_{n=1}^{\infty} ||x_n||_n^2 < \infty$ and $x_n \in X_n$. Suppose that $X = (\sum \oplus X_n)_{l_2}$, $||x|| = (\sum_{n=1}^{\infty} ||x_n||_n^2)^{\frac{1}{2}}$ and $||x||_1 = sup_n ||x_n||$ for all $x = (x_1, x_2, ...) \in X$.

Next, considering the intuitionistic fuzzy norm (ϖ, υ) *we define a function* $\varpi, \upsilon : X \times \mathbb{R} \longrightarrow [0,1]$ *by*

$$\varpi(a,t) = \begin{cases} 1 & if \ \ t > ||x|| \\ \frac{1}{2} & if \ \ ||x||_1 < t \leq ||x|| \\ 0 & if \ \ t \leq ||x||_1 \end{cases} \ and \ \upsilon(a,t) = \begin{cases} 0 & if \ \ t > ||x|| \\ \frac{1}{2} & if \ \ ||x||_1 < t \leq ||x|| \\ 1 & if \ \ t \leq ||x||_1. \end{cases}$$

Assume the α*-norms as*

$$||x||_\alpha = \begin{cases} ||x|| & if \ \ 1 > \alpha > \frac{1}{2} \\ ||x||_1 & if \ \ 0 < \alpha \leq \frac{1}{2}. \end{cases}$$

Consider that $(X, \varpi, \upsilon, \otimes, \odot)$ *satisfy the BAP and* K *is a compact set in* $(X, || \cdot ||)$. *Then we have to show that* K *is a compact set in* $(X, \varpi, \upsilon, \otimes, \odot)$.

Suppose $\epsilon > 0$ *and* (x_n) *is a sequence in* K. *Since* K *is compact subset in* $(X, || \cdot ||)$, $\exists$ *a subsequence* (x_{n_k}) *in* $(X, || \cdot ||)$. *Hence for some integers* $\varpi, \upsilon > 0$, $\exists$ *an* $x \in X$ *such that, for* $k \geq \varpi, \upsilon$, $||x_{n_k} - x|| < \epsilon$. *Since* $||x||_1 \leq ||x||$, $\forall$ $x \in X$, *then, for all* $\alpha \in (0,1)$ *and for some* $k \geq \varpi, \upsilon$, *we have* $||x_{n_k} - x||_\alpha < \epsilon$. *Therefore,* K *is a compact set in* $(X, \varpi, \upsilon, \otimes, \odot)$.

Again, consider $\alpha \in \left(\frac{1}{2}, 1\right)$ *and* $\epsilon > 0$. *Since* K *is a compact set in* $(X, \varpi, \upsilon, \otimes, \odot)$ *and using* $\alpha \in \left(\frac{1}{2}, 1\right)$ *and* $\epsilon > 0$ *we have* $\lambda > 0$ *and* $T_{\alpha,\epsilon} \in \bar{F}(X, X, \lambda)$ *such that* $||T_{\alpha,\epsilon}(x) - x||_\alpha < \epsilon$ *for every* $x \in K$. *Hence, we have* $||T_{\alpha,\epsilon}(x) - x|| < \epsilon$ *and* $||T_{\alpha,\epsilon}(x)|| \leq \lambda ||x||$, *which is a contradiction. Therefore,* $(X, \varpi, \upsilon, \otimes, \odot)$ *has fails to satisfy the BAP.*

Lastly, we are going to show that $(X, \varpi, \upsilon, \otimes, \odot)$ *satisfy the AP. Consider* $\epsilon > 0$ *and* K *is a compact subset in* $(X, \varpi, \upsilon, \otimes, \odot)$. *Also consider* $P_j : X \longrightarrow \left(\sum_{n=1}^{j} \oplus X_n\right)_{l_2}$ *as the projection, defined by* $P((x)) = (x_1, x_2, ..., x_j)$. *As* K *is a compact set in* X, *therefore from the Theorem 2.4 of [8],* $\exists$ *a natural number* $m \in N$ *and a finite rank operator* $T^{'} : (\sum_{n=1}^{m} \oplus X_n)_{l_2} \longrightarrow (\sum_{n=1}^{m} \oplus X_n)_{l_2}$ *such that, for each* $x \in K$, $||kT^{'} P_m(x) - x|| < \epsilon$, *where* k *is the function defined by* $k : (\sum_{n=1}^{m} \oplus X_n)_{l_2} \longrightarrow X$ *such that* $k(x_1, x_2, ..., x_m) = (x_1, x_2, ..., x_m, 0, ...)$.

We consider $T = kT^{'} P_m$. *Since* T *is a finite rank operator such that* $T : X \longrightarrow X$ *and* $||x||_1 \leq ||x|| \ \forall \ x \in X$, *we have* $||T(x) - x||_1 < \epsilon$, *i.e., for each* $\alpha \in (0,1)$, *we have* $||T(x) - x||_\alpha < \epsilon$.

Now, we are going to show that T *is sif-bounded on* X. *Since* $(\sum_{n=1}^{m} \oplus X_n)_{l_2}$ *and* $(\sum_{n=1}^{m} \oplus X_n)_{l_\infty}$ *are equivalent,* $\exists$ $M^{'} > 1$ *such that*

$$\left(\sum_{n=1}^{m} ||x_n||_n^2\right)^{\frac{1}{2}} \leq M^{'} sup_{1 \leq n \leq m} ||x_n||_n.$$

Then,

$$\begin{aligned} \|T(x)\|_1 \leq ||T(x)|| = \|kT^{'} P_m(x)\| & \\ & \leq ||kT^{'}||\left(\sum_{n=1}^{m} \|x_n\|_n^2\right)^{\frac{1}{2}} \\ & \leq ||kT^{'}||M^{'} sup_{1 \leq n \leq m} ||x_n||_n \\ & \leq ||kT^{'}||M^{'}||x||_1. \end{aligned}$$

Considering $M = max\left\{||T||, ||kT^{'}||, M^{'}\right\}$ *and we have to show that* $\forall$ $x \in X$ *and* $t \in \mathbb{R}$

$$\varpi\left(T\left(x\right), t\right) \geq \varpi\left(x, \frac{t}{M}\right) \text{ and } \upsilon\left(T\left(x\right), t\right) \leq \upsilon\left(x, \frac{t}{M}\right).$$

If $t \leq 0$, *the result is trivial. Consider that* $t > 0$. *Then it is enough to show that* $\varpi\left(T\left(x\right), t\right) \geq \varpi\left(Mx, t\right)$ *and* $\upsilon\left(T\left(x\right), t\right) \leq \upsilon\left(Mx, t\right)$ *for all* $x \in X$ *and* $t \in \mathbb{R}$. *For the first condition, that is,* $t > M||x||$, *we have* $\varpi\left(Mx, t\right) = 1$ *and* $\upsilon\left(Mx, t\right) = 0$. *By the given assumption as* $t > M||x|| \geq ||T||||x|| \geq ||T\left(x\right)||$, *we have* $\varpi\left(T\left(x\right), t\right) = 1$ *and* $\upsilon\left(T\left(x\right), t\right) = 0$. *Hence* $\varpi\left(T\left(x\right), t\right) \geq \varpi\left(Mx, t\right)$ *and* $\upsilon\left(T\left(x\right), t\right) \leq \upsilon\left(Mx, t\right)$.

For the second condition $||Mx||_1 < t \leq ||Mx||$, *we can obtain* $\varpi\left(Mx, t\right) = \frac{1}{2}$ *and* $\upsilon\left(Mx, t\right) = \frac{1}{2}$. *By the assumption* $t > M||x||_1 \geq ||kT^{'}||M^{'}||x||_1 \geq ||T\left(x\right)||_1$, *we have* $\varpi\left(T\left(x\right), t\right) \geq \frac{1}{2}$ *and* $\upsilon\left(T\left(x\right), t\right) \leq \frac{1}{2}$. *Hence* $\varpi\left(T\left(x\right), t\right) \geq \varpi\left(Mx, t\right)$ *and* $\upsilon\left(T\left(x\right), t\right) \leq \upsilon\left(Mx, t\right)$.

For the third condition $t \leq M||x||$, *we have* $\varpi\left(Mx, t\right) = 0$ *and* $\upsilon\left(Mx, t\right) = 1$. *Then, by the given assumption, trivially we obtain* $\varpi\left(T\left(x\right), t\right) \geq \varpi\left(Mx, t\right)$ *and* $\upsilon\left(T\left(x\right), t\right) \leq \upsilon\left(Mx, t\right)$. *Therefore,* $(X, \varpi, \upsilon, \otimes, \odot)$ *satisfy the AP.*

7.4 Conclusion

Throughout the chapter, we have established and discussed the idea and properties of AP and BAP along with some examples in the context of a complete IFNLS. Along with the discussion we also provide some lemmas and theorems in the settings of complete IFNLS. Hence, the results of this chapter provide a better way to understand the analytical structure of operators in an IFNLS. This work will provide an interesting topic for future research which may be applied in the study of dynamical systems and particle physics.

Bibliography

[1] K. Atanassov, Answer to D. Dubois, S. Gottwald, P. Hajek, J. Kacprzyk and H. Prade's paper "Terminological difficulties in fuzzy set theory the case of "Intuitionistic Fuzzy Sets"", Fuzzy Sets Syst. 156 (2005) 496–499.

[2] K. T. Atanassov, Intuitionistic fuzzy sets, Fuzzy Sets Syst. 20 (1986) 87–96.

[3] T. Bag, S. K. Samanta, Finite dimensional fuzzy normed linear spaces, J. Fuzzy Math. 11 (2003) 687–705.

[4] T. Bag, S. K. Samanta, Finite dimensional fuzzy normed linear spaces, J. Fuzzy Math. 11 (2003) 687–705.

[5] T. Bag, S. K. Samanta, Fuzzy bounded linear operators, Fuzzy Sets Syst. 151 (2005) 513–547.

[6] G. Cattaneo, D. Ciucci, Basic intuitionistic principle in fuzzy set theories and its extensions (A terminological debate on Atanassov IFS), Fuzzy Sets Syst. 157 (2006) 3198–3219.

[7] S. C. Cheng, J. N. Mordeson, Fuzzy linear operator and fuzzy normed linear space, Bull. Cal. Math. Soc. 86 (1994) 429–436.

[8] C. Choi, J. M. Kim, K. Y. Lee, Right and left weak approximation properties in Banach spaces, Canad. Math. Bull. 52 (2009) 28–38.

[9] P. Debnath, Lacunary ideal convergence in intuitionistic fuzzy normed linear spaces, Comput. Math. Appl. 63 (2012) 708–715.

[10] P. Debnath, Lacunary difference ideal convergence in intuitionistic fuzzy normed linear spaces, J. Intell. Fuzzy Syst. 28 (2015) 1299–1306.

[11] P. Debnath, M. Sen, Some completeness results in terms of infinite series and quotient spaces in intuitionistic fuzzy n-normed linear spaces, J. Intell. Fuzzy Syst. 26(6) (2014) 975–982.

[12] P. Debnath, M. Sen, Some results of calculus for functions having values in an intuitionistic fuzzy n-normed linear space, J. Intell. Fuzzy Syst. 26(2)(2014) 2983–2991.

[13] D. Dubois, S. Gottwald, P. Hajek, J. Kacprzk, H. Prade, Terminological difficulities in fuzzy set theory-the case of "intuitionistic fuzzy sets", Fuzzy Sets Syst. 156 (2005) 485–491.

[14] P. Enflo, A counter example to the approximation problem, Acta Math. 130 (1973) 309–317.

[15] P. Enflo. A counterexample to the approximation property for banach spaces, Acta Math. 130 (1973) 309–317.

[16] A. Esi, B. Hazarika, λ-ideal convergence in intuitionistic fuzzy 2-normed linear space, J. Intell. Fuzzy Syst. 24 (2013) 725–732.

[17] C. Felbin, Finite dimensional fuzzy normed linear spaces, Fuzzy Sets Syst. 48 (1992) 239–248.

[18] A. Grothendiek, Produits tensoriels topologiques et espaces nucleires, Mem. Amer. Math. Soc. 16 (1955) 1–140.

[19] P. Grzegorzewski and E. Mrowka, Some notes on (atanassov's) intuitionistic fuzzy sets, Fuzzy Sets Syst. 156 (2005) 492–495.

[20] B. Hazarika, A. Alotaibi, S. A. Mohiuddine, Statistical convergence in measure for double sequences of fuzzy-valued functions, Soft Computing 24 (2020) 6613–6622.

[21] U. Kadak, M. Mursaleen, S. A. Mohiuddine, Statistical weighted matrix summability of fuzzy mappings and associated approximation results, J. Intell. Fuzzy Syst. 36 (2019) 3483–3494.

[22] O. Kaleva, S. Seikkala, On fuzzy metric spaces, Fuzzy Sets Syst. 12 (1984) 215–229.

[23] A. K. Katsaras, Fuzzy topological vector spaces, Fuzzy Sets Syst. 12 (1984) 143–154.

[24] J. M. Kim, Characterizations of bounded approximation properties, Taiwanese J. Math. 12 (2008) 179–190.

[25] I. Kormosil, J. Michalek, Fuzzy metric and statistical metric spaces, Kybernetica 11 (1975) 326–334.

[26] F. Lael, K. Nourouzi, Fuzzy compact linear operators, Chaos Solitions Fractals. 34 (2007) 1584–1589.

[27] F. Lael, K. Nourouzi, Some results on the if-normed spaces, Chaos Solitions Fractals 37 (2008) 931–939.

[28] K. Y. Lee, Approximation properties in fuzzy normed spaces, Fuzzy Sets Syst. 283 (2015) 115–130.

[29] J. Lindenstrauss, On Jame's paper "separable conjugate spaces", Israel J. Math. 9 (1971) 279–284.

[30] W. Lusky, On banach spaces with the commuting bounded approximation property, Arch. Math. (Basel) 58(6) (1992) 568–574.

[31] S. A. Mohiuddine, A. Asiri, B. Hazarika, Weighted statistical convergence through difference operator of sequences of fuzzy numbers with application to fuzzy approximation theorems, Int. J. Gen. Syst. 48(5) (2019) 492–506.

[32] S. A. Mohiuddine, B. Hazarika, A. Alotaibi, On statistical convergence of double sequences of fuzzy valued functions, J. Intell. Fuzzy Syst. 32 (2017) 4331–4342.

[33] S. A. Mohiuddine, B. Hazarika, A. Alotaibi, On statistical convergence of double sequences of fuzzy valued functions, J. Intell. Fuzzy Syst. 32 (2017) 4331–4342.

[34] M. Mursaleen, V. Karakaya, S. A. Mohiuddine, Schauder basis, separability and approximation property in intuitionistic fuzzy normed space, Abstr. Appl. Anal. 2010, Article ID 131868, 14 pages (2010).

[35] M. Mursaleen, Q. M. D. Lohani, Intuitionistic fuzzy 2-normed space and some related concepts, Chaos Solitons Fractals 42 (2009) 224–234.

[36] M. Mursaleen, S. A. Mohiuddine, H. H. Edely, On the ideal convergence of double sequences in intuitionistic fuzzy normed spaces, Comput. Math. Appl. 59 (2010) 603–611.

[37] M. Mursaleen, S. A. Mohiuddine, Nonlinear operators between intuitionistic fuzzy normed spaces and fréchet derivative, Chaos Solitions Fractals 42 (2009) 1010–1015

[38] M. Mursaleen, S. A. Mohiuddine, On lacunary statistical convergence with respect to the intuitionistic fuzzy normed space, J. Comput. Appl. Math. 233(2) (2009) 142–149.

[39] M. Mursaleen, S. A. Mohiuddine, On stability of a cubic functional equation in intuitionistic fuzzy normed spaces, Chaos Solitions Fractals 42 (2009) 2997–3005.

[40] M. Mursaleen, S. A. Mohiuddine, Statistical convergent of double sequences in intuitionistic fuzzy normed, Chaos Solitions Fractals 41 (2009) 2414–2421.

[41] R. Saadati, J. H. Park, On the intuitionistic fuzzy topological spaces, Chaos Solitons Fractals 27 (2006) 331–344.

[42] M. Sen, P. Debnath, Lacunary statistical convergence in intuitionistic fuzzy n-normed linear spaces, Math. Comput. Modell. 54 (2011) 2978–2985.

[43] S. J. Szarek, A Banach space without a basis which has the bounded approximation property, Acta Math. 159 (1987) 81–98.

[44] S. Vijayabalaji, N. Thillaigovindam, Y. B. Jun, Intuitionistic fuzzy n-normed linear space, Bull. Korean. Math. Soc. 44 (2007) 291–308.

[45] S. Vijayabalaji, N. Thillaigovindan, Y. B. Jun, Intuitionistic fuzzy n-normed linear space, Bull. Korean. Math. Soc. 44 (2007) 291–308.

[46] Y. Yilmaz, On some basic properties of differentiation in intuitionistic fuzzy normed spaces, Math. Comp. Modell. 52 (2010) 448–458.

[47] Y. Yilmaz, Schauder bases and approximation property in fuzzy normed spaces, Comput. Math. Appl. 59 (2010) 1957–1964.

Chapter 8

Solution of Volterra Integral Equations in Banach Algebras using Measure of Noncompactness

Hemant Kumar Nashine

Vellore Institute of Technology, hemant.nashine@vit.ac.in

Anupam Das

Cotton University, math.anupam@gmail.com

8.1 Introduction and Preliminaries

Denote $\mathbb{R}$:= the set of real numbers, $\mathbb{R}_+ := [0, +\infty)$, $\mathbb{N}$:= the set of natural numbers, and $\mathbb{N}^* := \mathbb{N} \cup \{0\}$.

In 1930, Kuratowski [10] discussed a new direction of research with the notion of measure of noncompactness (MNC, in short) that combines with some algebraic arguments which are useful for studying the mathematical formulations, particularly for solving the existence of solutions of some nonlinear problems under certain conditions. The Kuratowskii and Hausdorff MNC in a metric space are well-known in the literature. Fixed point theory has two main branches: Constructive fixed point theorems in the line of Banach Contraction Principle, and nonconstructive fixed point theorems, where results are obtained by using topological properties in the direction of Brouwer's/ Schauder's/Darbo's fixed point theorem. Schauder discussed the convexity of domains and the compactness of operators whereas Darbo replaced the strong condition of compactness of operators with the MNC and defined appropriate classes of operators [6] which is extended and generalized by many authors [1, 2, 4, 11–21].

Let $(\mathcal{X}, \|\cdot\|)$ be a real Banach space and θ be its zero element. $\mathcal{B}(\vartheta, \zeta)$ will denote the closed ball with center ϑ are radius ζ and $\mathcal{B}_\zeta$ will stand for $\mathcal{B}(\theta, \zeta)$.

DOI: 10.1201/9781003178200-8

Moreover, $\mathfrak{M}_{\mathcal{X}}$ will denote the family of nonempty bounded subsets of $\mathcal{X}$ and $\mathfrak{N}_{\mathcal{X}}$ its subfamily consisting of all relatively compact sets.

Definition 8.1.1. *[5] A mapping $\chi : \mathfrak{M}_{\mathcal{X}} \to \mathbb{R}^+$ is said to be a MNC in $\mathcal{X}$ if it satisfies the following conditions ($\Omega, \Omega_1, \Omega_2 \in \mathfrak{M}_{\mathcal{X}}$):*

(1°) $ker\,\chi := \{\Omega \in \mathfrak{M}_{\mathcal{X}} : \chi(\Omega) = 0\} \neq \emptyset$ *and* $ker\,\chi \subset \mathfrak{N}_{\mathcal{X}}$,

(2°) $\Omega_1 \subseteq \Omega_2 \Rightarrow \chi(\Omega_1) \leq \chi(\Omega_2)$,

(3°) $\chi(\overline{\Omega}) = \chi(\Omega)$,

(4°) $\chi(\overline{conv\,\Omega}) = \chi(\Omega)$,

(5°) $\chi(\lambda\Omega_1 + (1-\lambda)\Omega_2) \leq \lambda\chi(\Omega_1) + (1-\lambda)\chi(\Omega_2)$ *for* $\lambda \in [0,1]$,

(6°) $\chi(\Omega_1 \cup \Omega_2) = \max\{\chi(\Omega_1), \chi(\Omega_2)\}$,

(7°) *If* (Ω_n) *is a decreasing sequence of non-empty closed sets in* $\mathfrak{M}_{\mathcal{X}}$ *and if* $\lim_{n\to\infty} \chi(\Omega_n) = 0$, *then the set* $\Omega_\infty = \bigcap_{n=1}^{\infty} \Omega_n$ *is non-empty and compact.*

The family $ker\chi$ defined in axiom (1^0) is called the kernel of the MNC χ. One of the properties of the MNC is $\mathcal{Y}_\infty \in ker\chi$. Indeed, from the inequality $\chi(\mathcal{Y}_\infty) \leq \chi(\mathcal{Y}_n)$ for $n = 1, 2, 3, \ldots$, we infer that $\chi(\mathcal{Y}_\infty) = 0$.

A map $\alpha : \mathfrak{M}_{\mathcal{X}} \to \mathbb{R}^+$ is said to be a Kuratowski MNC [10] if

$$\alpha(\Omega) = \inf\left\{\epsilon > 0 : \Omega \subset \bigcup_{k=1}^{n} \mathcal{U}_k, \mathcal{U}_k \subset \mathcal{X}, diam(\mathcal{U}_k) < \epsilon\,(k \in \mathbb{N})\right\}. \tag{8.1}$$

For all $\Omega \in \mathfrak{M}_E$, the *Hausdorff measure of noncompactness*

$$\chi(\Omega) = \inf\{\epsilon > 0 : \Omega \text{ can be covered by a finite number of balls of radii } \epsilon\}. \tag{8.2}$$

A continuous mapping $\mathcal{T} : \Omega \to \Omega$ is called a *densifying map* if, for any bounded set Ω with $\mu(\Omega) > 0$, we have $\mu(\mathcal{T}(\Omega)) < \mu(\Omega)$. If $\mu(\mathcal{T}(\Omega) \leq k\mu(\Omega)$, $0 < k < 1$, then $\mathcal{T}$ is a $k-$set contraction.

If $\mu(\mathcal{T}(\Omega) \leq \mu(\Omega)$, then $\mathcal{T}$ is said to be $1-$*set contraction.*

A nonexpansive map is an example of $1-$set contraction. A contraction map is densifying and so is a compact mapping. In the history, there are results in fixed point theory that dealt with combination of two maps for instance, $\mathcal{T}_1 + \mathcal{T}_2$, where $\mathcal{T}_1$ is a contraction map and $\mathcal{T}_2$ is a compact map.

If $\mathcal{T}_1$ and $\mathcal{T}_2$ both are continuous functions, then $\mathcal{T}_1 + \mathcal{T}_2$ is also a continuous map and the fixed point theorem for continuous map is applicable for $\mathcal{T}_1 + \mathcal{T}_2$. However, if $\mathcal{T}_1$ is a contraction map, then Banach fixed point theorem is applied and, if $\mathcal{T}_2$ is a compact map, then Schauder fixed point theorem is applicable. If $\mathcal{T}_1$ is densifying and $\mathcal{T}_2$ is densifying, then $\mathcal{T}_1 + \mathcal{T}_2$ is also densifying.

We denote $\Lambda(\mathcal{X})$ a nonempty, bounded, closed and convex set on Banach space $\mathcal{X}$.

The following extensions of topological Schauder fixed point Theorem 8.1.2 and classical Banach fixed point theorem were proved by Darbo (resp. Sadovskii) in 1955 (resp. 1972).

Theorem 8.1.2. *Let $\Omega \in \Lambda$ be a unbounded subset of a Banach space $\mathcal{X}$, then every compact, continuous map $\mathcal{T} : \Omega \to \Omega$ has at least one fixed point.*

Theorem 8.1.3. *[6] Let $\mathcal{X}$ be a Banach space, $\Omega \in \Lambda(\mathcal{X})$ and $\mathcal{T} : \Omega \to \Omega$ be a continuous operator such that there exists a $\lambda \in [0, 1)$ with*

$$\chi(\mathcal{T}(\mathcal{U})) \leq \lambda \chi(\mathcal{U})$$

for any $\emptyset \neq \mathcal{U} \subset \Omega$, where χ is the Kuratowski MNC on $\mathcal{X}$. Then $\mathcal{T}$ has a fixed point.

Theorem 8.1.4. *[23] Let $\mathcal{X}$ be a Banach space, $\Omega \in \Lambda(\mathcal{X})$ and $\mathcal{T} : \Omega \to \Omega$ be a continuous operator such that*

$$\chi(\mathcal{U}) > 0 \implies \chi(\mathcal{T}(\mathcal{U})) < \chi(\mathcal{U}),$$

for any $\emptyset \neq \mathcal{U} \subset \Omega$, where χ is the Kuratowski MNC on $\mathcal{X}$. Then $\mathcal{T}$ has a fixed point.

8.2 Fixed Point Results

In the papers [18, 21], authors used the notion of F defined by Wardowski [24] and proved the Darbo fixed point results. In this chapter, we extend their work by defining new (F, χ)-set contraction condition and derive some new Darbo type fixed point and n-coupled fixed point results.

Definition 8.2.1. *[21] Let $\mathbb{F} = \{F : \mathbb{R}^+ \to \mathbb{R}\}$ be a family of mappings satisfying:*

(F_1) *F is strictly increasing; that is, for all $\xi, \zeta \in \mathbb{R}^+$ such that $\zeta < \xi$ implies that $F(\zeta) < F(\xi)$;*

(F_2) *For every sequence $\{\zeta_n\}_{n \in \mathbb{N}}$ of positive real numbers $\lim_{n\to\infty} \zeta_n = 0$ if and only if $\lim_{n\to\infty} F(\zeta_n) = -\infty$.*

Definition 8.2.2. *A self-mapping $\mathcal{T}$ on $\Omega \in \Lambda$ is called an (F, χ)-set contraction if there exist $F \in \mathbb{F}$, $\tau \in \mathbb{R}^+$, and a continuous and strictly increasing mapping $\varphi : \mathbb{R}^+ \to \mathbb{R}^+$ such that*

$$\tau(\chi(\mathcal{U}) + \varphi(\chi(\mathcal{U}))) + F(\chi(\mathcal{T}\mathcal{U}) + \varphi(\chi(\mathcal{T}\mathcal{U}))) \leq F(\chi(\mathcal{U}) + \varphi(\chi(\mathcal{U}))), \quad (8.3)$$

for all $\emptyset \neq \mathcal{U} \subseteq \Omega$ with $\chi(\mathcal{T}\mathcal{U}) > 0$, where χ is an arbitrary MNC .

Example 8.2.3. *Let $F : \mathbb{R}^+ \to \mathbb{R}$ given by $F(s) = ln(s)$. Then $F \in \mathbb{F}$. Suppose that $\mathcal{T} : \Omega \to \Omega$ is an (F, χ)-set contraction. From (8.3), we have $\tau(\chi(\mathcal{U}) + \varphi(\chi(\mathcal{U}))) + ln(\chi(\mathcal{T}\mathcal{U}) + \varphi(\chi(\mathcal{T}\mathcal{U}))) \leq ln(\chi(\mathcal{U}) + \varphi(\chi(\mathcal{U})))$ for all $\mathcal{U} \subseteq \Omega$ with $\chi(\mathcal{T}\mathcal{U}) > 0$, which implies that*

$$\chi(\mathcal{T}\mathcal{U}) + \varphi(\chi(\mathcal{T}\mathcal{U})) \leq e^{-\tau(\chi(\mathcal{U})+\varphi(\chi(\mathcal{U})))}[\chi(\mathcal{U}) + \varphi(\chi(\mathcal{U}))].$$

Example 8.2.4. *Let $F : \mathbb{R}^+ \to \mathbb{R}$ given by $F(s) = -\frac{1}{\sqrt{s}}$, $s > 0$. Then $F \in \mathbb{F}$. Here each (F, χ)-set contraction $\mathcal{T} : \Omega \to \Omega$ satisfies*

$$\begin{aligned} &\chi(\mathcal{T}\mathcal{U}) + \varphi(\chi(\mathcal{T}\mathcal{U})) \\ \leq\ &\frac{1}{(1 + \tau(\chi(\mathcal{U}) + \varphi(\chi(\mathcal{U})))\sqrt{\chi(\mathcal{U}) + \varphi(\chi(\mathcal{U}))})^2}[\chi(\mathcal{U}) + \varphi(\chi(\mathcal{U})))] \end{aligned}$$

for all $\emptyset \neq \mathcal{U} \subseteq \Omega$ with $\chi(\mathcal{T}\mathcal{U}) > 0$.

As a result, we state the existence of a fixed point for a continuous (but not necessarily compact) operator satisfying a (F, χ)-contraction condition.

Theorem 8.2.5. *Let $\Omega \in \Lambda$ and $\mathcal{T} : \Omega \longrightarrow \Omega$ be a continuous and (F, χ)-set contraction. Then $\mathcal{T}$ has at least one fixed point in Ω.*

Proof. Define a sequence $\{\Omega_n\}$ with $\Omega_0 = \Omega$ and $\Omega_n = Conv(\mathcal{T}\Omega_{n-1})$, $n \geq 1$. $\mathcal{T}\Omega_0 = \mathcal{T}\Omega \subseteq \Omega = \Omega_0, \Omega_1 = Conv(\mathcal{T}\Omega_0) \subseteq \Omega = \Omega_0$, therefore by continuing this process, we have

$$\Omega_0 \supseteq \Omega_1 \supseteq \Omega_2 \supseteq \cdots .$$

If there exists an integer $N \geq 0$ such that $\chi(\Omega_N) + \varphi(\chi(\Omega_N)) = 0$, that is, $\chi(\Omega_N) = 0$, then Ω_N is relatively compact and since $\mathcal{T}\Omega_N \subseteq Conv(\mathcal{T}\Omega_N) = \Omega_{N+1} \subseteq \Omega_N$. Thus, Theorem 11.2.3 implies that $\mathcal{T}$ has a fixed point. So we assume that $\chi(\Omega_n) + \varphi(\chi(\Omega_n)) \neq 0$ for $n \geq 0$. Also by (8.3), we have

$$\begin{aligned} F(\chi(\Omega_n) + \varphi(\chi(\Omega_n))) &= F(\chi(Conv(\mathcal{T}\Omega_{n-1})) + \varphi(\chi(Conv(\mathcal{T}\Omega_{n-1})))) \\ &= F(\chi(\mathcal{T}\Omega_{n-1}) + \varphi(\chi(\mathcal{T}\Omega_{n-1}))) \\ &\leq F(\chi(\Omega_{n-1}) + \varphi(\chi(\Omega_{n-1}))) \\ &\quad -\tau(\chi(\Omega_{n-1}) + \varphi(\chi(\Omega_{n-1}))) \\ &\leq F(\chi(\Omega_{n-2}) + \varphi(\chi(\Omega_{n-2}))) - \tau(\chi(\Omega_{n-1}) \\ &\quad +\varphi(\chi(\Omega_{n-1}))) - \tau(\chi(\Omega_{n-2}) + \varphi(\chi(\Omega_{n-2}))) \\ &\ \ \vdots \\ &\leq F(\chi(\Omega_0) + \varphi(\chi(\Omega_0))) \\ &\quad -\sum_{i=0}^{n-1} \tau(\chi(\Omega_i) + \varphi(\chi(\Omega_0))). \end{aligned} \tag{8.4}$$

From (8.4), we obtain $\lim_{n\to\infty} F(\chi(\Omega_n) + \varphi(\chi(\Omega_n))) = -\infty$. Now, using (F_2)

we get $\lim_{n\to\infty}\chi(\Omega_n)+\varphi(\chi(\Omega_n))=0$, therefore $\lim_{n\to\infty}\chi(\Omega_n)=0$. Since $\Omega_{n+1}\subseteq\Omega_n$ and $\mathcal{T}\Omega_n\subseteq\Omega_n$ for all $n\geq 1$, then from (7^0) of Definition 8.1.1, $\Omega_\infty=\bigcap_{n=1}^{\infty}\Omega_n$ is a nonempty convex closed set, invariant under $\mathcal{T}$ and belongs to Ω_χ. Therefore Theorem 11.2.3 gives the desired conclusion. □

Proposition 8.2.6. *Let $\Omega\in\Lambda$ and $\mathcal{T}:\Omega\longrightarrow\Omega$ be a continuous and $\mu(\mathcal{T}\mathcal{M})>0$ implies that*

$$\tau(diam(\mathcal{M})+\varphi(diam(\mathcal{M})))+F(diam(\mathcal{T}\mathcal{M})+\varphi(diam(\mathcal{T}\mathcal{M})))$$
$$\leq F(diam(\mathcal{M})+\varphi(diam(\mathcal{M}))), \quad (8.5)$$

for all $\mathcal{M}\subseteq\mathcal{X}$, where $F\in\mathbb{F}$, $\tau\in\mathbb{R}^+$, and a continuous and strictly increasing mapping $\varphi:\mathbb{R}^+\to\mathbb{R}^+$. Then $\mathcal{T}$ has a unique fixed point in $\mathcal{X}$.

Proof. Following Theorem 8.2.5 and argument of Proposition 3.2 [7], $\mathcal{T}$ has a fixed point in $\mathcal{X}$.
To prove uniqueness, we suppose that there exist two distinct fixed points $u,v\in\mathcal{X}$, then we may define the set $M:=\{u,v\}$. In this case $diam(M)=diam(\mathcal{T}(M))=\|u-v\|>0$. Then using (8.5), we get $diam(\mathcal{T}(\Upsilon))>0$ implies that

$$\tau(diam(M)+\varphi(diam(M)))+F(diam(\mathcal{T}(M))+\varphi(diam(\mathcal{T}(M))))$$
$$\leq F(diam(M)+\varphi(diam(M))),$$

a contradiction and hence the result. □

Next we derived classical fixed point result from Proposition 8.2.6.

Corollary 8.2.7. *Let $\Omega\in\Lambda$ and $\mathcal{T}:\Omega\longrightarrow\Omega$ be a continuous and $\|\mathcal{T}u-\mathcal{T}v\|>0$ implies that*

$$\tau(\|u-v\|+\varphi(\|u-v\|))+F(\|\mathcal{T}u-\mathcal{T}v\|+\varphi(\|\mathcal{T}u-\mathcal{T}v\|))\leq F(\|u-v\|+\varphi(\|u-v\|)) \quad (8.6)$$

for all $u,v\in\mathcal{X}$, $\tau>0$, $F\in\mathfrak{F}$, $\tau\in\mathbb{R}^+$, and a continuous and strictly increasing mapping $\varphi:\mathbb{R}^+\to\mathbb{R}^+$. Then $\mathcal{T}$ has a unique fixed point.

Proof. Let $\mu:\mathfrak{M}_{\mathcal{X}}\to\mathbb{R}^+$ be a set quantity defined by the formula $\mu(\Omega)=diam\Omega$, where $diam\Omega=\sup\{\|u-v\|:u,v\in\Omega\}$ stands for the diameter of Ω. It is easily seen that μ is a MNC in a space $\mathcal{X}$ in the sense of Definition 8.1.1. Therefore from (8.6), we have

$$\sup_{u,v\in\Omega}\|\mathcal{T}u-\mathcal{T}v\|>0$$

which implies

$$\begin{aligned}
&\tau(\sup_{u,v\in\Omega}\|u-v\|+\varphi(\sup_{u,v\in\Omega}\|u-v\|))+F(\sup_{u,v\in\Omega}\|\mathcal{T}u-\mathcal{T}v\|\\
&\quad+\varphi(\sup_{u,v\in\Omega}\|\mathcal{T}u-\mathcal{T}v\|))\\
&=\tau(\sup_{u,v\in\Omega}\|u-v\|+\varphi(\sup_{u,v\in\Omega}\|u-v\|))\\
&\quad+\sup_{u,v\in\Omega}F(\|\mathcal{T}u-\mathcal{T}v\|+\varphi(\|\mathcal{T}u-\mathcal{T}v\|))\\
&\leq\sup_{u,v\in\Omega}F(\|u-v\|+\varphi(\|u-v\|))\\
&\leq F(\sup_{u,v\in\Omega}\|u-v\|+\varphi(\sup_{u,v\in\Omega}\|u-v\|))
\end{aligned}$$

that is, $diam(\mathcal{T}(\Omega))>0$ implies

$$\begin{aligned}
\tau(diam(\Omega)+\varphi(diam(\Omega))+F(diam(\mathcal{T}(\Omega))+\varphi(diam(\mathcal{T}(\Omega))))\\
\leq F(diam(\Omega)+\varphi(diam(\Omega)).
\end{aligned}$$

Thus following Proposition 8.2.6, $\mathcal{T}$ has an unique fixed point. □

In the following, we derive $n-$tuple Darbo type fixed point result in product spaces.

Here, we recall a useful theorem concerning the construction of a MNC on a finite product space.

Theorem 8.2.8. *[5] Suppose $\chi_1,\chi_2,...,\chi_n$ are MNCs in Banach spaces $\mathcal{X}_1,\mathcal{X}_2,...,\mathcal{X}_n$, respectively. Moreover assume that the function $F:\mathbb{R}_+^n\to\mathbb{R}_+$ is convex and $F(\vartheta_1,\vartheta_2,...,\vartheta_n)=0$ if and only if $\vartheta_i=0$ for $i=1,2,...,n$. Then*

$$\chi(\Omega)=F(\chi_1(\Omega_1),\chi_2(\Omega_2),...,\chi_n(\Omega_n))$$

define a MNC in $\Pi_{i=1}^n\mathcal{X}_i$ where Ω_i denotes the natural projection of Ω into $\mathcal{X}_i$ for $i=1,2,...,n$.

Now, as results from Theorem 8.2.8, we present the following example.

Example 8.2.9. *Let $\chi_i(1\leq i\leq n)$ be MNCs in Banach spaces $\mathcal{X}_1,\mathcal{X}_2,\ldots,\mathcal{X}_n$ respectively, considering $F_1(\vartheta_1,\ldots,\vartheta_n)=k\max\{\vartheta_1,\vartheta_2,...,\vartheta_n\}$ and $F_2(\vartheta_1,\ldots,\vartheta_n)=k(\sum_{i=1}^n\vartheta_i)$, $k\in\mathbb{R}_+$ for any $(\vartheta_1,\ldots,\vartheta_n)\in\mathbb{R}_+^n$, then all the conditions of Theorem 8.2.8 are satisfied. Therefore, $\widehat{\chi}_1:=k\max\{\chi(\Omega_1),\chi(\Omega_2),...,\chi(\Omega_n)\}$ and $\widehat{\chi}_2:=k(\chi(\sum_{i=1}^n\Omega_i))$ define MNCs in the space $\Pi_{i=1}^n\mathcal{X}_i$ where Ω_i, $i=1,2,\ldots,n$ denote the natural projections of Ω into $\mathcal{X}_i$.*

Theorem 8.2.10. *Let $\Omega_i \in \Lambda_i$ $(i = 1, 2, \ldots, n)$, and $\mathcal{T}_i : \Pi_{i=1}^n \Omega_i \longrightarrow \Omega_i$ be a continuous. Assume that there exist $F \in \mathbb{F}$, $\tau \in \mathbb{R}^+$ and $\varphi : \mathbb{R}^+ \to \mathbb{R}^+$ is continuous, increasing such that*

$$\tau(\max_i \chi(\Omega_i) + \varphi(\max_i \chi(\Omega_i))) + F(\chi(\mathcal{T}_i(\Pi_{i=1}^n \Omega_i)) + \varphi(\chi(\mathcal{T}_i(\Pi_{i=1}^n \Omega_i)))) \\ \leq F(\max_i \chi(\Omega_i) + \varphi(\max_i \chi(\Omega_i))), \quad (8.7)$$

for any subset Ω_i of Ω_i where χ_i is an arbitrary MNC on $\mathcal{X}_i$ $(i = 1, 2, \ldots, n)$. Then there exist $(\vartheta_1^*, \vartheta_2^*, \ldots, \vartheta_n^*) \in \Pi_{i=1}^n \Omega_i$ such that for all $1 \leq i \leq n$

$$\mathcal{T}_i(\vartheta_1^*, \vartheta_2^*, \ldots, \vartheta_n^*) = \vartheta_i^*. \quad (8.8)$$

Proof. Consider the MNC $\widehat{\chi}$, defined by

$$\widehat{\chi}(\Omega) = \max\{\chi(\Omega_1), \chi(\Omega_2), ..., \chi(\Omega_n)\}.$$

for any bounded subset $\Omega \subset \Pi_{i=1}^n \mathcal{X}_i$, where Ω_i $(i = 1, 2, \ldots, n)$ denote the natural projections of Ω into $\mathcal{X}_i$. Also, we define a mapping $\widehat{\mathcal{T}} : \Pi_{i=1}^n \Omega_i \longrightarrow \Pi_{i=1}^n \Omega_i$ as follows

$$\widehat{\mathcal{T}}(\vartheta_1, \vartheta_2, \ldots, \vartheta_n) = (\mathcal{T}_1(\vartheta_1, \vartheta_2, \ldots, \vartheta_n), \mathcal{T}_2(\vartheta_1, \vartheta_2, \ldots, \vartheta_n), \\ \ldots, \mathcal{T}_n(\vartheta_1, \vartheta_2, \ldots, \vartheta_n)).$$

It is obvious that $\widehat{\mathcal{T}}$ is continuous. Now we claim that $\widehat{\mathcal{T}}$ satisfies all the conditions of Theorem 8.2.5. To prove this, let Ω be any nonempty and bounded subset of $\Omega_1 \times \Omega_2 \times \ldots \times \Omega_n$. Then by (2°), (8.7), we obtain

$$\begin{aligned}
&\tau(\widehat{\chi}(\Omega) + \varphi(\widehat{\chi}(\Omega))) + F(\widehat{\chi}(\widehat{\mathcal{T}}(\Omega)) + \varphi(\widehat{\chi}(\mathcal{T}\Omega))) \\
&\leq \tau(\widehat{\chi}(\Omega) + \varphi(\widehat{\chi}(\Omega))) + F(\widehat{\chi}(\mathcal{T}_1(\Pi_{i=1}^n \Omega_i) \times \ldots \times \mathcal{T}_n(\Pi_{i=1}^n \Omega_i)) \\
&\quad + \varphi(\widehat{\chi}(\mathcal{T}_1(\Pi_{i=1}^n \Omega_i) \times \ldots \times \mathcal{T}_n(\Pi_{i=1}^n \Omega_i)))) \\
&= \tau(\max_{1 \leq i \leq n} \chi(\Omega_i) + \varphi(\max_{1 \leq i \leq n} \chi(\Omega_i))) \\
&\quad + F(\max_{1 \leq k \leq n} \chi(\mathcal{T}_k(\Pi_{i=1}^n \Omega_i)) + \varphi(\max_{1 \leq k \leq n} \chi(\mathcal{T}_k(\Pi_{i=1}^n \Omega_i)))) \\
&\leq \max_{1 \leq k \leq n} [\tau(\max_{1 \leq i \leq n} \chi(\Omega_i) + \varphi(\max_{1 \leq i \leq n} \chi(\Omega_i))) \\
&\quad + F(\chi(\mathcal{T}_k(\Pi_{i=1}^n \Omega_i)) + \varphi(\chi(\mathcal{T}_k(\Pi_{i=1}^n \Omega_i))))] \\
&\leq \max_{1 \leq k \leq n} F(\max_{1 \leq i \leq n} \chi(\Omega_i) + \varphi(\max_{1 \leq i \leq n} \chi(\Omega_i))) \\
&\leq F(\widehat{\chi}(\Omega) + \varphi(\widehat{\chi}(\Omega))).
\end{aligned}$$

Hence, from Theorem 8.2.5, $\widehat{\mathcal{T}}$ has at least one fixed point, i.e., there exist $(\vartheta_1^*, \vartheta_2^*, \ldots, \vartheta_n^*) \in \Pi_{i=1}^n \Omega_i$ such that

$$\begin{aligned}
(\vartheta_1^*, \vartheta_2^*, \ldots, \vartheta_n^*) &= \widehat{\mathcal{T}}(\vartheta_1^*, \vartheta_2^*, \ldots, \vartheta_n^*) \\
&= (\mathcal{T}_1(\vartheta_1^*, \vartheta_2^*, \ldots, \vartheta_n^*), \mathcal{T}_2(\vartheta_1^*, \vartheta_2^*, \ldots, \vartheta_n^*), \\
&\quad \ldots, \mathcal{T}_n(\vartheta_1^*, \vartheta_2^*, \ldots, \vartheta_n^*))
\end{aligned}$$

which gives (8.8) and the proof is complete. □

Theorem 8.2.11. *Let $\Omega_i \in \Lambda_i$ $(i = 1, 2, \ldots, n)$, and $\mathcal{T}_i : \Pi_{i=1}^n \Omega_i \longrightarrow \Omega_i$ be a continuous. Assume that there exist $F \in \mathbb{F}$, $\tau \in \mathbb{R}^+$ and $\varphi : \mathbb{R}^+ \to \mathbb{R}^+$ is continuous, increasing such that*

$$\tau\left(\sum_i \chi(\Omega_i) + \varphi\left(\sum_i \chi(\Omega_i)\right)\right) + F\left(\sum_i \chi(\mathcal{T}_i(\Pi_{i=1}^n \Omega_i))\right.$$
$$\left. + \varphi\left(\sum_i \chi(\mathcal{T}_i(\Pi_{i=1}^n \Omega_i))\right)\right) \leq F\left(\sum_i \chi(\Omega_i) + \varphi\left(\sum_i \chi(\Omega_i)\right)\right), \quad (8.9)$$

for any subset Ω_i of Ω_i where χ_i is an arbitrary MNC on $\mathcal{X}_i$ $(i = 1, 2, \ldots, n)$. Then there exist $(\vartheta_1^*, \vartheta_2^*, \ldots, \vartheta_n^*) \in \Pi_{i=1}^n \Omega_i$ such that for all $1 \leq i \leq n$

$$\mathcal{T}_i(\vartheta_1^*, \vartheta_2^*, \ldots, \vartheta_n^*) = \vartheta_i^*. \quad (8.10)$$

Proof. Consider the MNC $\widehat{\chi}$, defined by

$$\widehat{\chi}(\Omega) = \sum_i \chi(\Omega_i)$$

for any bounded subset $\Omega \subset \Pi_{i=1}^n \mathcal{X}_i$, where Ω_i $(i = 1, 2, \ldots, n)$ denote the natural projections of Ω into $\mathcal{X}_i$. Also, we define a mapping $\widehat{\mathcal{T}} : \Pi_{i=1}^n \Omega_i \longrightarrow \Pi_{i=1}^n \Omega_i$ as follows

$$\widehat{\mathcal{T}}(\vartheta_1, \vartheta_2, \ldots, \vartheta_n) = (\mathcal{T}_1(\vartheta_1, \vartheta_2, \ldots, \vartheta_n), \mathcal{T}_2(\vartheta_1, \vartheta_2, \ldots, \vartheta_n), \ldots, \mathcal{T}_n(\vartheta_1, \vartheta_2, \ldots, \vartheta_n)).$$

It is obvious that $\widehat{\mathcal{T}}$ is continuous. Now we claim that $\widehat{\mathcal{T}}$ satisfies all the conditions of Theorem 8.2.5. To prove this, let Ω be any nonempty and bounded subset of $\Omega_1 \times \Omega_2 \times \ldots \times \Omega_n$. Then by (2°), (8.9), we obtain

$$\begin{aligned}
&\tau(\widehat{\chi}(\Omega) + \varphi(\widehat{\chi}(\Omega))) + F(\widehat{\chi}(\widehat{\mathcal{T}}(\Omega)) + \varphi(\widehat{\chi}(\mathcal{T}\Omega)))\\
&\leq \tau(\widehat{\chi}(\Omega) + \varphi(\widehat{\chi}(\Omega))) + F(\widehat{\chi}(\mathcal{T}_1(\Pi_{i=1}^n \Omega_i) \times \ldots \times \mathcal{T}_n(\Pi_{i=1}^n \Omega_i))\\
&\quad + \varphi(\widehat{\chi}(\mathcal{T}_1(\Pi_{i=1}^n \Omega_i) \times \ldots \times \mathcal{T}_n(\Pi_{i=1}^n \Omega_i))))\\
&= \tau\left(\sum_{1 \leq i \leq n} \chi(\Omega_i) + \varphi(\sum_{1 \leq i \leq n} \chi(\Omega_i))\right)\\
&\quad + F\left(\sum_{1 \leq k \leq n} \chi(\mathcal{T}_k(\Pi_{i=1}^n \Omega_i)) + \varphi\left(\sum_{1 \leq k \leq n} \chi(\mathcal{T}_k(\Pi_{i=1}^n \Omega_i))\right)\right)\\
&\leq F\left(\sum_i \chi(\Omega_i) + \varphi(\sum_i \chi(\Omega_i))\right)\\
&\leq F(\widehat{\chi}(\Omega) + \varphi(\widehat{\chi}(\Omega))).
\end{aligned}$$

Hence, from Theorem 8.2.5, $\widehat{\mathcal{T}}$ has at least one fixed point, i.e., there exist $(\vartheta_1^*, \vartheta_2^*, \ldots, \vartheta_n^*) \in \Pi_{i=1}^n \Omega_i$ such that

$$\begin{aligned}(\vartheta_1^*, \vartheta_2^*, \ldots, \vartheta_n^*) &= \widehat{\mathcal{T}}(\vartheta_1^*, \vartheta_2^*, \ldots, \vartheta_n^*) \\ &= (\mathcal{T}_1(\vartheta_1^*, \vartheta_2^*, \ldots, \vartheta_n^*), \mathcal{T}_2(\vartheta_1^*, \vartheta_2^*, \ldots, \vartheta_n^*), \\ &\quad \ldots, \mathcal{T}_n(\vartheta_1^*, \vartheta_2^*, \ldots, \vartheta_n^*))\end{aligned}$$

which gives (9.2) and the proof is complete. □

Following result is a generalization of results in [3, 22].

Corollary 8.2.12. *Let $\Omega_i \in \Lambda_i$ $(i = 1, 2, \ldots, n)$, and let $\mathcal{T}_i : \Pi_{i=1}^n \Omega_i \longrightarrow \Omega_i$ $(i = 1, 2, \ldots, n)$ be a continuous operator such that*

$$\chi(\mathcal{T}_i(\Pi_{i=1}^n \Omega_i)) \leq k[\max_i \chi(\Omega_i) + \varphi(\max_i \chi(\Pi_{i=1}^n \Omega_i))]$$

for any subset Ω_i of Ω_i, where χ_i is an arbitrary MNC on $\mathcal{X}_i$ and $k \in [0, 1)$. Then there exist $(\vartheta_1^*, \vartheta_2^*, \ldots, \vartheta_n^*) \in \Pi_{i=1}^n \Omega_i$ such that for all $1 \leq i \leq n$

$$\mathcal{T}_i(\vartheta_1^*, \vartheta_2^*, \ldots, \vartheta_n^*) = \vartheta_i^*.$$

8.3 Solvability of Volterra integral equation in Banach algebra

In this chapter, we shall work in the space $\mathfrak{E} = C(I)$ which consists of the set of real continuous on I where $I = [0, 1]$. The space $\mathfrak{E}$ is equipped with the norm

$$\| x \| = \sup \{|x(t)| : t \in I\}, \; x \in \mathfrak{E}.$$

The space $\mathfrak{E}$ has the Banach algebra structure.

Let $\mathfrak{X}$ be a fixed nonempty and bounded subset of the space $\mathfrak{E} = C(I)$ and for $x \in \mathfrak{X}$ and $\epsilon > 0$, denote by $\omega(x, \epsilon)$ the modulus of the continuity of x i.e.

$$\omega(x, \epsilon) = \sup \{|x(t) - x(s)| : t, s \in I, |t - s| \leq \epsilon\}.$$

Further, we define

$$\omega(\mathfrak{X}, \epsilon) = \sup \{\omega(x, \epsilon) : x \in \mathfrak{X}\}.$$

$$\omega_0(\mathfrak{X}) = \lim_{\epsilon \to 0} \omega(\mathfrak{X}, \epsilon).$$

In this part, we study the following Volterra integral equation,

$$z(t) = F(t, z(t)) \int_0^t U(t, s, z(s)) ds, \; t \in [0, 1] = I. \tag{8.11}$$

We consider the following assumptions

(A) $F : I \times \mathbb{R} \to \mathbb{R}$ is continuous and there exists nonnegative constants P, Q such that

$$|F(t,0)| \leq P$$

and

$$|F(t,z) - F(t,\bar{z})| \leq Q\,|z - \bar{z}|\,,\ t \in I,\ z, \bar{z} \in \mathbb{R}.$$

(B) $U : I \times I \times \mathbb{R} \to \mathbb{R}$ is continuous function and there exists nonnegative constants A, B such that

$$|U(t,s,z)| \leq A + B\,|z|\,,\ t, s \in I,\ z \in \mathbb{R}$$

and

$$K = \sup\left\{|U(t,s,z)| : t, s \in I,\ z \in C(I)\right\}.$$

(C) There exists a solution $r_0 > 0$ for the inequality

$$QBr^2\,(QA + PB)\,r + PA \leq r.$$

Also, $QK < 1$.

Theorem 8.3.1. *If the assumptions from (A)–(C) are satisfied then the equation (8.11) has at least one solution in $C(I)$.*

Proof. Let us define the operator T on $C(I)$ in the following way,

$$(Tz)(t) = F(t, z(t)) \int_0^t U(t,s,z(s))ds,\ t \in I.$$

Observe that by virtue of our assumptions the function Tz is continuous on I for any $x \in C(I)$.

Let us fix $z \in C(I)$, then applying for $t \in I$, we have

$$\begin{aligned}
|(Tz)(t)| &= \left|F(t,z(t)) \int_0^t U(t,s,z(s))ds\right| \\
&\leq \left\{|F(t,z(t)) - F(t,0)| + |F(t,0)|\right\} \int_0^t |U(t,s,z(s))|\,ds \\
&\leq \left\{Q\,|z(t)| + P\right\} \left\{A \int_0^t ds + B \int_0^t |z(s)|\,ds\right\} \\
&\leq \left\{Q \parallel z(t) \parallel + P\right\} \left\{A + B \parallel z \parallel\right\}.
\end{aligned}$$

Therefore,

$$\parallel Tz \parallel \leq QB \parallel z \parallel^2 + (QA + PB) \parallel z \parallel + PA$$

for $z \in C(I)$.

Let $B_{r_0} = \left\{z \in C(I) : \parallel z \parallel \leq r_0\right\}.$

For $\| z \| \leq r_0$ we have,

$$\| Tz \| \leq QBr_0^2 + (QA + PB)r_0 + PA \leq r_0$$

i.e. T maps B_{r_0} onto itself.

Now, we shall show that T is continuous on B_{r_0}. To do this take $\epsilon > 0$ and take arbitrary $z, \bar{z} \in B_{r_0}$ such that $\| z - \bar{z} \| \leq \epsilon$. Then for $t \in I$, we get,

$$\begin{aligned}
&|(Tz)(t) - (T\bar{z})(t)| \\
&= \left| F(t, z(t)) \int_0^t U(t, s, z(s))ds - F(t, \bar{z}(t)) \int_0^t U(t, s, \bar{z}(s))ds \right| \\
&\leq \left| F(t, z(t)) \left\{ \int_0^t U(t, s, z(s))ds - \int_0^t U(t, s, \bar{z}(s))ds \right\} \right| \\
&\quad + \left| \{F(t, z(t)) - F(t, \bar{z}(t))\} \int_0^t U(t, s, \bar{z}(s))ds \right| \\
&\leq |F(t, z(t))| \int_0^t |U(t, s, z(s)) - U(t, s, \bar{z}(s))| \, ds \\
&\quad + |F(t, z(t)) - F(t, \bar{z}(t))| \int_0^t |U(t, s, \bar{z}(s))| \, ds \\
&\leq (Q \| z \| + P)\omega(U, \epsilon) + P \| z - \bar{z} \| (A + B \| z \|)
\end{aligned}$$

Using the fact that U is uniformly continuous on the bounded set $I \times I \times [-r_0, r_0]$ we have $\omega(U, \epsilon) \to 0$ as $\epsilon \to 0$. The above estimate shows that T is continuous on B_{r_0}.

Let us consider a nonempty subset Z of B_{r_0} and $z \in Z$ then for a fix $\epsilon > 0$ and $t_1, t_2 \in I$ such that $t_1 \leq t_2$ and $t_2 - t_1 \leq \epsilon$ we get,

$$\begin{aligned}
&|(Tz)(t_2) - (Tz)(t_1)| \\
&= \left| F(t_2, z(t_2)) \int_0^{t_2} U(t_2, s, z(s))ds - F(t_1, z(t_1)) \int_0^{t_1} U(t_1, s, z(s))ds \right| \\
&\leq \left| F(t_2, z(t_2)) \left\{ \int_0^{t_2} U(t_2, s, z(s))ds - \int_0^{t_1} U(t_1, s, z(s))ds \right\} \right. \\
&\quad \left. + (F(t_2, z(t_2)) - F(t_1, z(t_1))) \int_0^{t_1} U(t_1, s, z(s))ds \right| \\
&\leq \{Q \| z \| + P\} \left| \int_0^{t_1} (U(t_2, s, z(s)) - U(t_1, s, z(s))) \, ds \right. \\
&\quad \left. + \int_{t_1}^{t_2} U(t_2, s, z(s))ds \right| + \Big\{ |F(t_2, z(t_2)) - F(t_2, z(t_1))| \\
&\quad + |F(t_2, z(t_1)) - F(t_1, z(t_1))| \int_0^{t_1} |U(t_1, s, z(s))| \, ds \Big\}
\end{aligned}$$

Define the following notations,

$$\omega_F(\epsilon) = \sup \{ |F(t, x) - F(\bar{t}, x)| : t, \bar{t} \in I, \ |t - \bar{t}| \leq \epsilon, \ x \in [-r_0, r_0] \}$$

and

$$\omega_U(\epsilon) = \sup\{|U(t,s,x) - U(\bar{t},s,x)| : t,\bar{t} \in I,\ |t-\bar{t}| \leq \epsilon,\ x \in [-r_0, r_0]\}.$$

We obtain

$$\begin{aligned}|(Tz)(t_2) - (Tz)(t_1)| &\leq (Q \parallel z \parallel + P)\,(\omega_U(\epsilon) + K\epsilon) \\ &\quad + (Q\,|z(t_2) - z(t_1)| + \omega_F(\epsilon)\,K,\end{aligned}$$

i.e.

$$\omega(Tz,\epsilon) \leq (Q \parallel z \parallel + P)\,(\omega_U(\epsilon) + K\epsilon) + (Q\omega(z,\epsilon) + \omega_F(\epsilon)\,K.$$

As F is uniformly continuous on $I \times [-r_0, r_0]$ and U is uniformly continuous $I \times I \times [-r_0, r_0]$ so $\omega_F(\epsilon) \to 0$ and $\omega_U(\epsilon) \to 0$ as $\epsilon \to 0$. Consequently,

$$\omega_0(TZ) \leq QK\omega_0(Z).$$

Thus by assumption (C) and using example 9.1 on Theorem 8.2.5 with $\phi = 0$, $\tau > 0$, we have T has at least one fixed point in $Z \subseteq B_{r_0}$. Hence equation (8.11) has at least one solution in $C(I)$. This completes the proof. □

Example 8.3.2. *Consider the following equation*

$$z(t) = \cos(tz(t)) \int_0^t \frac{sz(s)}{(2+t^4)(1+z^2(s))}ds,\ t \in I. \tag{8.12}$$

Here

$$F(t, z(t)) = \cos(tz(t))$$

and

$$U(t,s,z(s)) = \frac{sz(s)}{(2+t^4)(1+z^2(s))}.$$

It can be observed that both F and U are continuous functions.

Now,

$$\begin{aligned}&|F(t,z) - F(t,\bar{z})| \\ &= |\cos(tz(t)) - \cos(t\bar{z}(t))| \\ &\leq |z - \bar{z}|\end{aligned}$$

and

$$|F(t,0)| = 1.$$

Therefore, $P = 1$, $Q = 1$.

Also,

$$|U(t,s,z)| \leq \frac{1}{2}\,|z|$$

which gives $A = 0$, $B = \frac{1}{2}$ and $K = \frac{1}{2}$. Therefor, $QK = \frac{1}{2} < 1$.

The inequality in assumption (C) takes the following form,

$$\frac{r^2}{2} + \frac{r}{2} \leq r$$

which gives

$$r^2 - r \leq 0$$

so for $r = 1$ *the above inequality is satisfied i.e. for* $r = 1$ *all the assumptions of Theorem 8.3.1 are satisfied. Thus applying the Theorem 8.3.1, we conclude that the integral equation (8.12) has at least one solution in* $C(I)$.

Bibliography

[1] A. Aghajani, R. Allahyari, M. Mursaleen, A generalization of Darbo's theorem with application to the solvability of systems of integral equations, J. Comput. Appl. Math. 260 (2014) 68–77.

[2] A. Aghajani, J. Banas, N. Sabzali, Some generalizations of Darbo fixed point theorem and applications, Bull. Belg. Math. Soc. Simon Stevin 20 (2013) 345–358.

[3] A. Aghajani, N. Sabzali, A coupled fixed point theorem for condensing operators with application to system of integral equations, J. Nonlinear Convex Anal. 15(5) (2014) 941–952.

[4] R. Arab, Some fixed point theorems in generalized Darbo fixed point theorem and the existence of solutions for system of integral equations, J. Korean Math. Soc. 52 (2015) 125–139.

[5] J. Banas, K. Goebel, Measures of noncompactness in Banach Spaces, LectureNotes in Pure and Applied Mathematics, p. 60. Dekker, New York (1980).

[6] G. Darbo, Punti uniti in transformazioni a codominio non compatto (Italian), Rend. Sem. Math. Univ. Padova 24 (1955) 84–92.

[7] J. G. Falset, K. Latrach, On Darbo-Sadovskii's fixed point theorems type for abstract measures of (weak) noncompactness, Bull. Belg. Math. Soc. Simon Stevin 22 (2015) 797–812.

[8] D. Guo, V. Lakshmikantham, X. Liu, Nonlinear Integral Equations in Abstract Spaces, vol. 373 of Mathematics and Its Applications, Kluwer Academic Publishers, Dordrecht, The Netherlands (1996).

[9] R. Akmerov, M. Kamenski, A. Potapov, A. Rodkina, B. Sadovskii, Measures of Noncompactness and Condensing Operators, Birkhauser-Verlag, Basel (1992).

[10] K. Kuratowski, Sur les espaces completes, Fund. Math. 15 (1930) 301–309.

[11] M. Mursaleen, S. A. Mohiuddine, Applications of measures of noncompactness to the infinite system of differential equations in l_p spaces, Nolinear Anal. 75 (2012) 2111–2115.

[12] M. Mursaleen, A. Alotaibi, Infinite system of differential equations in some spaces, Abstr. Appl. Anal. Volume 2012, Article ID 863483, 20 pages (2012).

[13] M. Mursaleen, S. M. H. Rizvi, Solvability of infinite system of second order differential equations in c_0 and ℓ_1 by Meir-Keeler condensing operator, Proc. Amer. Math. Soc. 144(10) (2016) 4279–4289.

[14] M. Mursaleen, B. Bilalov, S. M. H. Rizvi, Applications of measures of noncompactness to infinite system of fractional differential equations, Filomat 31(11) (2017) 3421–3432.

[15] H. K. Nashine, R. Arab, Existence of solutions to nonlinear functional-integral equations via the measure of noncompactness, J. Fixed Point Theory Appl. 20 (2018), Article 66.

[16] H. K. Nashine, R. Arab, R. P. Agarwal, A. S. Haghighi, Darbo type fixed and coupled fixed point results and its application to integral equation, Period Math. Hung. 77 (2018) 94–107.

[17] H. K. Nashine, R. Arab, R. P. Agarwal, M. De la Sen, Positive solutions of fractional integral equations by the technique of measure of noncompactness, J. Inequa. Appl. 2017 (2017), Article 225.

[18] H. K. Nashine, R. Arab, Some new type generalization of Darbo's fixed point and its applications to integral equations, Indian J. Math. 61(3) (2019) 395–420.

[19] H. K. Nashine, R. W. Ibrahim, Monotone solutions of iterative fractional equations found by modified Darbo-type fixed-point theorems, J. Fixed Point Theory Appl. 19(14) (2017) 3217–3229.

[20] H. K. Nashine, R. W. Ibrahim, R. P. Agarwal, Moments solution of fractional evolution equation found by new krasnoselskii type fixed point theorems, Fixed Point Theory 22(1) (2021) 263–278.

[21] H. K. Nashine, J. R. Roshan, Fixed point theorem via measure of noncompactness and application to Volterra integral equations in Banach algebras, J. Math. Ext. 13(4) (2019) 91–116.

[22] B. Rezpecki, On measure of noncompactness in topological vector spaces, Comment. Math. Univ. Carolinae 23 (1982) 105–116.

[23] B. N. Sadovskii, Limit-compact and condensing operators (Russian), Uspehi Mat. Nauk. 27(1)(163) (1972) 81–146.

[24] D. Wardowski, Fixed points of new type of contractive mappings in complete metric spaces, Fixed Point Theory Appl. 2012 (2012), Article 94.

Chapter 9

Solution of a pair of Nonlinear Matrix Equation using Fixed Point Theory

Hemant Kumar Nashine

Vellore Institute of Technology, hemant.nashine@vit.ac.in

Sourav Shil

Vellore Institute of Technology, souravcuh@gmail.com

9.1 Introduction and Preliminaries

Nonlinear matrix equations (NMEs, in short) play an important role in various fields of mathematics and allied sciences. These equations intervene in many practical problems of computational mathematics, biology, electricity, Markov chains, dynamic programming, stochastic filtering, statistics, solid mechanics, ladder networks, queuing theory, statistics, control and system theory, spectral factorization, and other, providing a rich variety of problems where matrix equations play a fundamental role. Among other ways, NMEs can be solved by using various types of fixed point theorems in metric spaces and partially ordered metric spaces. Fixed point theory is an efficient and reliable tool for computing solutions of various classes of Hermitian (semi-Hermitian) positive definite (semi-definite) NMEs. The investigation of solutions can further be interpreted numerically via Matlab.

Let $\mathcal{H}(n)$ stand for the set of all $n \times n$ Hermitian matrices over $\mathbb{C}$, $\mathcal{K}(n)$ $\big(\subset \mathcal{H}(n)\big)$ stand for the set of all $n \times n$ positive semi-definite matrices, $\mathcal{P}(n)$ $\big(\subset$

DOI: 10.1201/9781003178200-9

$\mathcal{K}(n)$) stand for the set of $n \times n$ positive definite matrices, $\mathcal{M}(n)$ stand for the set of all $n \times n$ matrices over $\mathbb{C}$.

Ran and Reuring [8] discussed the solutions of

$$\mathcal{X} + \mathcal{B}^* F(\mathcal{X})\mathcal{B} = \mathcal{Q} \tag{9.1}$$

in $\mathcal{K}(n)$, where $\mathcal{B} \in \mathcal{M}(n)$, $\mathcal{Q}$ is positive definite and F is a map from $\mathcal{K}(n)$ into $\mathcal{M}(n)$. Note that $\mathcal{X}$ is a solution of (9.1) if and only if it is a fixed point of the map $G(\mathcal{X}) = \mathcal{Q} - \mathcal{B}^* F(\mathcal{X})\mathcal{B}$.

In [9], Ran and Reurings used the notion of partial ordering and established an analogous result of the Banach fixed point theorem in metric spaces, and further used it to solve a class of NMEs of the form $\mathcal{X} = \mathcal{Q} + \sum_{i=1}^{m} \mathcal{B}_i^* F(\mathcal{X})\mathcal{B}_i$ for Ky Fan norm in $\mathcal{M}(n)$. In [10], Sawangsup and Sintunavarat studied an NME of the form $\mathcal{X} = \mathcal{Q} + \sum_{i=1}^{m} \mathcal{B}_i^* F(\mathcal{X})\mathcal{B}_i$ for the spectral norm of matrices using a generalized contraction condition in metric spaces endowed with a transitive binary relation, and tested numerically an approximate solution of it. In the papers [1, 4, 5], the authors discussed positive definite solution (PDS, in short) of a pair of NMEs. Recently, Garai and Dey [3] obtained some sufficient conditions for the existence and uniqueness of solution of a system of NMEs, using some common fixed point results in Banach spaces using a pair of altering distance functions.

In this study, we discuss a sufficient condition for the existence of a unique common PDS of a pair of NMEs of the form

$$\mathcal{X} = \mathcal{R}_1 + \sum_{i=1}^{m} \mathcal{B}_i^* \mathcal{U}(\mathcal{X})\mathcal{B}_i, \quad \mathcal{X} = \mathcal{R}_2 + \sum_{i=1}^{m} \mathcal{B}_i^* \mathcal{V}(\mathcal{X})\mathcal{B}_i,$$

where $\mathcal{R}_1$, $\mathcal{R}_2$ are two $n \times n$ Hermitian matrices, $\mathcal{B}_1, \mathcal{B}_2, \dots. \mathcal{B}_m$ are arbitrary $n \times n$ matrices, and $\mathcal{U}$, $\mathcal{V}$ are self-mappings from the set of all Hermitian matrices into itself, continuous with respect to the trace norm. Further, we apply this result to two different types of NMEs and discuss its convergence behaviour with respect to three different initializations with graphical representations using MATLAB. To establish sufficient condition for NMEs, we first design two common fixed point results for orbitally weakly contractive mappings on Banach spaces, and justify them by two illustrations. Thus, modifications of the results from [3] are obtained.

9.2 Result 1

Denote by $\mathbb{R}$ the set of real numbers, $\mathbb{R}_+ = [0, +\infty)$, by $\mathbb{N}$ the set of natural numbers, and $\mathbb{N}^* = \mathbb{N} \cup \{0\}$.

Definition 9.2.1. *[2] Let $\mathcal{S}, \mathcal{T}$ be a pair of self-mappings on a normed space $(\Xi, \|\cdot\|)$.*

1. *If for a point $\nu_0 \in \Xi$, there exists a sequence $\{\nu_n\}$ in Ξ such that $\nu_{2n+1} = \mathcal{S}\nu_{2n}$, $\nu_{2n+2} = \mathcal{T}\nu_{2n+1}$, $n \in \mathbb{N}^*$, then the set $\mathcal{O}(\nu_0; \mathcal{S}, \mathcal{T}) = \{\nu_n : n \in \mathbb{N}\}$ is called the orbit of $(\mathcal{S}, \mathcal{T})$ at ν_0.*

2. *The normed space Ξ is said to be $(\mathcal{S}, \mathcal{T})$-orbitally complete at ν_0 if every Cauchy sequence in $\mathcal{O}(\nu_0; \mathcal{S}, \mathcal{T})$ converges in Ξ (with respect to the norm $\|\cdot\|$).*

3. *The map $\mathcal{S}$ is said to be orbitally continuous at ν_0 if it is continuous on $\mathcal{O}(\nu_0; \mathcal{S}, \mathcal{T})$.*

The set of all fixed (common fixed) points of a self-mapping $\mathcal{S}$ (and self-mapping $\mathcal{T}$) on a set $\Xi \neq \emptyset$ will be denoted by $Fix(\mathcal{S})$ ($CFP(\mathcal{S}, \mathcal{T})$).

We will use the following sets of functions from [6,7]:

$\Psi_1 = \{\psi_1 \colon [0, +\infty) \to [0, +\infty) \mid \psi_1$ is nondecreasing and lower semicontinuous$\}$,

$\Psi_2 = \{\psi_2 \colon [0, +\infty) \to [0, +\infty) \mid \psi_2$ is upper semicontinuous$\}$,

$\Psi_3 = \{\psi_3 \colon [0, +\infty) \to [0, +\infty) \mid \psi_3$ is lower semicontinuous$\}$.

We now introduce the notion of a ψ_i-weakly contractive mapping on a normed space.

Definition 9.2.2. *Let $(\Xi, \|\cdot\|)$ be a normed space and $\mathcal{S}, \mathcal{T} \colon \Xi \to \Xi$ be given mappings. We say that $(\mathcal{S}, \mathcal{T})$ is a ψ_i-weakly contractive pair of mappings ($i \in \{1, 2, 3\}$) if $\psi_i \in \Psi_i$ ($i \in \{1, 2, 3\}$) are such that*

$$t > 0 \text{ and } (s = t \text{ or } s = 0) \text{ imply } \psi_1(t) - \psi_2(s) + \psi_3(s) > 0, \tag{9.2}$$

and, for all $u, v \in \Xi$,

$$\psi_1(\|\mathcal{S}u - \mathcal{T}v\|) \leq \psi_2(\Theta(u, v)) - \psi_3(\Theta(u, v)), \tag{9.3}$$

where

$$\Theta(u, v) = \max\{\|u - v\|, \|u - \mathcal{S}u\|, \|v - \mathcal{T}v\|\}.$$

If (9.3) is satisfied for $u, v \in \overline{\mathcal{O}(\nu_0; \mathcal{S}, \mathcal{T})}$ (for some $\nu_0 \in \Xi$), we say that $(\mathcal{S}, \mathcal{T})$ is an orbitally ψ_i-weakly contractive pair of mappings (at ν_0).

Theorem 9.2.3. *Let Ξ be a normed space, and let $\mathcal{S}, \mathcal{T} : \Xi \to \Xi$ be an orbitally ψ_i-weakly contractive pair of mappings for $i \in \{1, 2, 3\}$ (for some ν_0). We assume the following hypotheses:*

(i) Ξ is $(\mathcal{S}, \mathcal{T})$-orbitally complete at ν_0;

(ii) $\mathcal{S}$ or $\mathcal{T}$ is $(\mathcal{S}, \mathcal{T})$-orbitally continuous at ν_0.

Then $CFP(\mathcal{S}, \mathcal{T})$ is a singleton.

Proof. First of all, we will show that every fixed point of $\mathcal{S}$ is also a fixed point of $\mathcal{T}$ (and vice versa). Indeed, assume that $\vartheta = \mathcal{S}\vartheta$ and $\|\vartheta - \mathcal{T}\vartheta\| > 0$ for some $\vartheta \in \Xi$. Then from (9.3), for $u = v = \vartheta$, we have

$$\psi_1(\|\vartheta - \mathcal{T}\vartheta\|) \leq \psi_2(\|\vartheta - \mathcal{T}\vartheta\|) - \psi_3(\|\vartheta - \mathcal{T}\vartheta\|),$$

which is, by (9.2), a contradiction. Similarly, it can be shown that if $\mathcal{T}\vartheta = \vartheta$, then $\mathcal{S}\vartheta = \vartheta$.

For the given point ν_0, define a sequence $\{\nu_n\}$ in Ξ as follows:

$$\nu_{2n+1} = \mathcal{S}\nu_{2n} \text{ and } \nu_{2n+2} = \mathcal{T}\nu_{2n+1} \text{ for } n \in \{0, 1, \dots\}. \tag{9.4}$$

We can assume that $\nu_n \neq \nu_{n+1}$ for all $n \in \mathbb{N}^*$ (otherwise there is nothing to prove). We claim that

$$\lim_{n\to+\infty} \|\nu_n - \nu_{n+1}\| = 0. \tag{9.5}$$

Putting $u = \nu_{2n}$, $v = \nu_{2n+1}$, i.e., $\mathcal{S}u = \nu_{2n+1}$, $\mathcal{T}v = \nu_{2n+2}$ and

$$\begin{aligned}\Theta(u, v) &= \max\{\|\nu_{2n} - \nu_{2n+1}\|, \|\nu_{2n} - \nu_{2n+1}\|, \|\nu_{2n+1} - \nu_{2n+2}\|\} \\ &= \max\{\|\nu_{2n} - \nu_{2n+1}\|, \|\nu_{2n+1} - \nu_{2n+2}\|\}\end{aligned}$$

in (9.3), we get

$$\begin{aligned}\psi_1(\|\nu_{2n+1} - \nu_{2n+2}) &= \psi_1(\|\mathcal{S}\nu_{2n} - \mathcal{T}\nu_{2n+1}\|) \\ &\leq \psi_2(\max\{\|\nu_{2n} - \nu_{2n+1}\|, \|\nu_{2n+1} - \nu_{2n+2}\|\}) \\ &\quad - \psi_3(\max\{\|\nu_{2n} - \nu_{2n+1}\|, \|\nu_{2n+1} - \nu_{2n+2}\|\}).\end{aligned}$$

Denote $\varrho_n = \|\nu_n - \nu_{n+1}\|$ for $n \in \mathbb{N}^*$. It follows that

$$\psi_1(\varrho_{2n+1}) \leq \psi_2(\max\{\varrho_{2n}, \varrho_{2n+1}\}) - \psi_3(\max\{\varrho_{2n}, \varrho_{2n+1}\}). \tag{9.6}$$

We shall show that $\{\varrho_n\}$ is a nonincreasing sequence. Indeed, suppose that $\varrho_{2n} < \varrho_{2n+1}$ for some $n \in \mathbb{N}$. Then (9.6) implies that

$$\psi_1(\varrho_{2n+1}) \leq \psi_2(\varrho_{2n+1}) - \psi_3(\varrho_{2n+1}),$$

which is, by (9.2), a contradiction. Hence, $\varrho_{2n+1} \leq \varrho_{2n}$ for all $n \in \mathbb{N}^*$. Similarly, we have $\varrho_{2n} \leq \varrho_{2n-1}$ for all $n \in \mathbb{N}$. Therefore, for all $n \in \mathbb{N}$, $\varrho_n \leq \varrho_{n-1}$ and thus $\{\varrho_n\}$ is a nonincreasing sequence of positive real numbers. Hence there exists an $r \geq 0$ such that $\lim_{n\to+\infty} \varrho_n = \lim_{n\to+\infty} \|\nu_{n+1} - \nu_n\| = r$. This implies that

$$\lim_{n\to+\infty} \varrho_{2n} = \lim_{n\to+\infty} \varrho_{2n-1} = r.$$

If $r > 0$ then, by condition (9.3) and using the properties of functions ψ_i we get that

$$\begin{aligned}\psi_1(r) &\leq \liminf \psi_1(\varrho_{2n}) \leq \limsup \psi_1(\varrho_{2n}) \leq \limsup[\psi_2(\varrho_{2n-1}) - \psi_3(\varrho_{2n-1})] \\ &= \limsup \psi_2(\varrho_{2n-1}) - \liminf \psi_3(\varrho_{2n-1}) \leq \psi_2(r) - \psi_3(r),\end{aligned}$$

a contradiction to condition (9.2), with $t = s = r > 0$. Hence, $\lim_{n\to+\infty} \varrho_n = \lim_{n\to+\infty} \|\nu_n - \nu_{n+1}\| = 0$, which establishes (9.5).

Next, we will show that $\{\nu_n\}$ is a Cauchy sequence in $\mathcal{O}(\nu_0; \mathcal{S}, \mathcal{T})$. Following (9.5), it is enough to prove that $\{\nu_{2n}\}$ is a Cauchy sequence in the normed space $\mathcal{O}(\nu_0; \mathcal{S}, \mathcal{T})$. We proceed by negation and suppose that the sequence $\{\nu_{2n}\}$ is not Cauchy. Then we can find an $\eta > 0$ such that for each even integer $2k$ there exist even integers $2m_k > 2n_k > 2k$ such that

$$\|\nu_{2n_k} - \nu_{2m_k}\| \geq \eta \text{ for } k \in \{1, 2, \dots\}. \tag{9.7}$$

We may also assume

$$\|\nu_{2m_k-2} - \nu_{2n_k}\| < \eta \tag{9.8}$$

by choosing $2m_k$ to be the smallest number exceeding $2n_k$ for which (9.7) holds. Now (9.5), (9.7) and (9.8) imply

$$\begin{aligned} 0 < \eta &\leq \|\nu_{2n_k} - \nu_{2m_k}\| \\ &\leq \|\nu_{2n_k} - \nu_{2m_k-2}\| + \|\nu_{2m_k-2} - \nu_{2m_k-1}\| + \|\nu_{2m_k-1} - \nu_{2m_k}\| \\ &\leq \eta + \|\nu_{2m_k-2} - \nu_{2m_k-1}\| + \|\nu_{2m_k-1} - \nu_{2m_k}\|, \end{aligned}$$

and so

$$\lim_{k\to+\infty} \|\nu_{2n_k} - \nu_{2m_k}\| = \eta. \tag{9.9}$$

Also, by the triangular inequality,

$$\big|\|\nu_{2n_k} - \nu_{2m_k-1}\| - \|\nu_{2n_k} - \nu_{2m_k}\|\big| \leq \|\nu_{2m_k-1} - \nu_{2m_k}\|$$

and

$$\big|\|\nu_{2n_k+1} - \nu_{2m_k-1}\| - \|\nu_{2n_k} - \nu_{2m_k}\|\big| \leq \|\nu_{2m_k-1} - \nu_{2m_k}\| + \|\nu_{2n_k} - \nu_{2n_k+1}\|.$$

Therefore, we get

$$\lim_{k\to+\infty} \|\nu_{2n_k} - \nu_{2m_k-1}\| = \eta \tag{9.10}$$

and

$$\lim_{k\to+\infty} \|\nu_{2n_k+1} - \nu_{2m_k-1}\| = \eta. \tag{9.11}$$

Applying condition (9.3) to the pair $v = 2m_k + 1$, $u = 2n_k + 1$, we get

$$\begin{aligned} \psi_1(\|\nu_{2n_k+1} - \nu_{2m_k+1}\|) &\leq \psi_1(\|\nu_{2n_k+1} - \nu_{2m_k+1}\|) \\ &\leq \psi_2(\Theta(\nu_{2n_k}, \nu_{2m_k})) - \psi_3(\Theta(\nu_{2n_k}, \nu_{2m_k})), \end{aligned} \tag{9.12}$$

where

$$\Theta(\nu_{2n_k}, \nu_{2m_k}) = \max\left\{\|\nu_{2n_k} - \nu_{2m_k}\|, \|\nu_{2n_k} - \nu_{2n_k+1}\|, \|\nu_{2m_k} - \nu_{2m_k+1}\|\right\},$$

implying that

$$\lim_{k\to+\infty} \Theta(\nu_{2n_k}, \nu_{2m_k}) = \max\{0, \eta\} = \eta.$$

Taking the upper limit in (9.12), and using the properties of functions ψ_i, we get that

$$\begin{aligned}\psi_1(\eta) &\leq \liminf_{k\to+\infty} \psi_1(\|\nu_{2n_k+1} - \nu_{2m_k+1}\|) \leq \limsup_{k\to+\infty} \psi_1(\|\nu_{2n_k+1} - \nu_{2m_k+1}\|) \\ &\leq \limsup_{k\to+\infty} [\psi_2(\Theta(\nu_{2n_k}, \nu_{2m_k}))) - \psi_3(\Theta(\nu_{2n_k}, \nu_{2m_k}))] \\ &\leq \limsup_{k\to+\infty} \psi_2(\Theta(\nu_{2n_k}, \nu_{2m_k}))) - \liminf_{n\to+\infty} \psi_3(\Theta(\nu_{2n_k}, \nu_{2m_k}))) \\ &\leq \psi_2(\eta) - \psi_3(\eta),\end{aligned}$$

a contradiction to condition (9.2). Therefore, $\eta = 0$. Thus $\{\nu_{2n}\}$ is a Cauchy sequence in the normed space $\mathcal{O}(\nu_0; \mathcal{S}, \mathcal{T})$, and so is $\{\nu_n\}$. Since Ξ is $(\mathcal{T}, \mathcal{S})$-orbitally complete at ν_0, there exists a $\vartheta \in \overline{\mathcal{O}(\nu_0; \mathcal{S}, \mathcal{T})}$ with $\lim_{n\to+\infty} \nu_n = \vartheta$. By the orbital continuity of $\mathcal{S}$ or $\mathcal{T}$, and the remark from the beginning of this proof, $\vartheta = \mathcal{S}\vartheta = \mathcal{T}\vartheta$.

Finally, to check the uniqueness of the CFP ϑ, let $\mu \neq \vartheta$ be such that $\mu = \mathcal{S}\mu = \mathcal{T}\mu$. From (9.3), it follows

$$\begin{aligned}\psi_1(\|\vartheta - \mu\|) &= \psi_1(\|\mathcal{S}\vartheta - \mathcal{T}\mu\|) \\ &\leq \psi_2(\max\{\|\vartheta - \mu\|, \|\vartheta - \mathcal{S}\vartheta\|, \|\mu - \mathcal{T}\mu\|\}) \\ &\quad - \psi_3(\max\{\|\vartheta - \mu\|, \|\vartheta - \mathcal{S}\vartheta\|, \|\mu - \mathcal{T}\mu\|\}) \\ &= \psi_2(\|\vartheta - \mu\|) - \psi_3(\|\vartheta - \mu\|)\end{aligned}$$

a contradiction to condition (9.2), i.e., the $CFP(\mathcal{S}, \mathcal{T})$ is unique. □

Example 9.2.4. *Let us consider the normed linear space $(\Xi, \|\cdot\|)$, where $\Xi = \mathcal{P}[0,1]$, the set all polynomials with real coefficients defined on $[0,1]$ and $\|x\| = \max_{0\leq t\leq 1} |x(t)|$ for $x \in \Xi$. We now define two mappings $\mathcal{S}$, $\mathcal{T} : \Xi \to \Xi$ by*

$$\mathcal{S}x(t) = \frac{1}{4}x(t), \qquad \mathcal{T}x(t) = x\left(\frac{t}{2}\right)$$

for all $x \in \Xi$. We choose $\nu_0(t) = \frac{t}{2} \in \Xi$. Then the orbit of $(\mathcal{S}, \mathcal{T})$ at ν_0 is $\mathcal{O}(\nu_0; \mathcal{S}, \mathcal{T}) = \{\nu_n : n \in \mathbb{N}\}$, where

$$\nu_n(t) = \begin{cases} \dfrac{t}{2^{3m+1}}, & \text{if } n = 2m; \\ \dfrac{t}{2^{3m}}, & \text{if } n = 2m-1. \end{cases}$$

We choose $\psi_1(t) = t$, $\psi_2(t) = t$ and $\psi_3(t) = \frac{t}{100}$. Let $u, v \in \overline{\mathcal{O}(\nu_0; \mathcal{S}, \mathcal{T})}$ be arbitrary. Then the following cases arise.

Case I: *Let $u = \nu_{2m}$, $v = \nu_{2n}$ for some $n, m \in \mathbb{N}$. Then $\|\mathcal{S}u - \mathcal{T}v\| = \left|\frac{1}{2^{3n+2}} - \frac{1}{2^{3m+3}}\right|$, $\|v - \mathcal{T}v\| = \frac{1}{2^{3n+2}}$, $\|u - \mathcal{S}u| = \frac{3}{2^{3m+3}}$ and $\|u - v\| = \left|\frac{1}{2^{3n+1}} - \frac{1}{2^{3m+1}}\right|$. In this case two subcases arise.*

Subcase 1: Let $n = m$. Then $\|\mathcal{T}u - \mathcal{S}v\| = \frac{1}{2^{3n+3}}$ and $\Theta(u, v) = \max\{\|u - v\|, \|u - \mathcal{S}u\|, \|v - \mathcal{T}v\|\} = \frac{3}{2^{3n+3}}$. So (9.3) holds.

Subcase 2: Let $n \neq m$. WLOG, we assume that $n > m$. Then $\|\mathcal{T}u - \mathcal{S}v\| = \frac{1}{2^{3m+3}} - \frac{1}{2^{3n+2}}$ and $\Theta(u,v) = \max\{\|u-v\|, \|u-\mathcal{S}u\|, \|v-\mathcal{T}v\|\} = \frac{1}{2^{3m+1}} - \frac{1}{2^{3n+1}}$. Therefore, (9.3) is satisfied.

Case II: *Let $v = \nu_{2m-1}$, with subcases $u = \nu_{2n-1}$ or $u = \nu_{2n}$, for some $n, m \in \mathbb{N}$. It can be checked in a similar way as in Case I that (9.3) is satsisfied.*

Case III: *Let $u(t) = 0$ for all $t \in [0,1]$ and $v = \nu_n$ for some $n \in \mathbb{N}$. Then $\|\mathcal{S}u - \mathcal{T}v\| = v_n\left(\frac{1}{2}\right)$ and $\Theta(u,v) = \max\{\|u-v\|, \|u-\mathcal{S}u\|, \|v-\mathcal{T}v\|\} = v_n(1)$. So (9.3) holds in this case.*

Case IV: *Let $u(t) = 0$ and $v(t) = 0$ for all $t \in [0,1]$. Then it is obvious that (9.3) holds true.*

Thus, $\mathcal{S}, \mathcal{T}$ is an orbitally ψ_i-weakly contractive pair of mappings. Further, it is easy to verify that Ξ is $(\mathcal{S}, \mathcal{T})$-orbitally complete space at ν_0 and that $\mathcal{S}$ and $\mathcal{T}$ are $(\mathcal{S}, \mathcal{T})$-orbitally continuous at ν_0. So all the conditions of Theorem 11.2.3 are fulfilled. Hence, $CFP(\mathcal{S}, \mathcal{T})$ is a singleton. Note that $\vartheta \in \Xi$ defined by $\vartheta(t) = 0$ for all $t \in [0,1]$ is the unique CFP of $\mathcal{S}$ and $\mathcal{T}$.

9.3 Result 2

We introduce the notion of an (ζ, φ)-weakly contractive mapping in a normed space. Denote:

$\Upsilon := \{\zeta \mid \zeta : \mathbb{R}^+ \to \mathbb{R}^+$ is a continuous function, satisfying $\zeta(t_n) \to 0 \Rightarrow t_n \to 0\}$, and

$\Phi := \{\varphi \mid \varphi : \mathbb{R}_+ \to \mathbb{R}_+$ is a nondecreasing and lower semicontinuous function such that $\varphi(0) = 0$ and $\varphi(t) > 0$ for $t > 0\}$.

Definition 9.3.1. *Let $(\Xi, \|\cdot\|)$ be a normed space and $\mathcal{S}, \mathcal{T} \colon \Xi \to \Xi$ be given mappings. We say that $(\mathcal{S}, \mathcal{T})$ is a (ζ, φ)-weakly contractive pair of mappings if for all $\vartheta, \nu \in \Xi$,*

$$\|\mathcal{S}\vartheta - \mathcal{T}\nu\| \leq \Theta(\vartheta, \nu) - \varphi(\zeta(\Theta(\vartheta, \nu))), \tag{9.13}$$

where $\zeta \in \Upsilon$, and $\varphi \in \Phi$.

If (9.13) is satisfied for $\vartheta, \nu \in \overline{\mathcal{O}(\varpi_0; \mathcal{S}, \mathcal{T})}$ (for some $\varpi_0 \in \Xi$), we say that $(\mathcal{S}, \mathcal{T})$ is an orbitally (ζ, φ)-weakly contractive pair of mappings (at ϖ_0).

Theorem 9.3.2. *Let Ξ be a normed space, and let $\mathcal{S}, \mathcal{T} : \Xi \to \Xi$ be an orbitally (ζ, φ)-weakly contractive pair of mappings for some ϖ_0. We assume the following hypotheses:*

(i) Ξ is $(\mathcal{S}, \mathcal{T})$-orbitally complete space at ϖ_0;

(ii) $\mathcal{S}$ or $\mathcal{T}$ is $(\mathcal{S}, \mathcal{T})$-orbitally continuous at ϖ_0.

Then $CFP(\mathcal{S}, \mathcal{T})$ is a singleton.

Proof. Starting with the given element $\varpi_0 \in \Xi$, define a sequence $\{\varpi_m\}$ in Ξ as follows:

$$\varpi_{2m+1} = \mathcal{S}\varpi_{2m} \text{ and } \varpi_{2m+2} = \mathcal{T}\varpi_{2m+1} \text{ for } m \in \mathbb{N}^*.$$

WLOG, assume $\varpi_m \neq \varpi_{m+1}$ for each $m \in \mathbb{N}^*$. Indeed, if $\varpi_{m_0} = \varpi_{m_0+1}$ for some $m_0 \in \mathbb{N}^*$, then $\vartheta^* = \varpi_{m_0}$ is a common fixed point of $\mathcal{S}, \mathcal{T}$ and this completes the proof. More precisely, to see that $\{\vartheta^*\} = CFP(\mathcal{S}, \mathcal{T})$, we can discuss two cases. First, we assume that m_0 is even, that is, $m_0 = 2p$. In this case, we have $\varpi_{2p} = \varpi_{2p+1} = \mathcal{S}\varpi_{2p}$, that is, ϖ_{2p} is a fixed point of $\mathcal{S}$. Now we shall prove that $\varpi_{2p} = \varpi_{2p+1} = \mathcal{S}\varpi_{2p} = \mathcal{T}\varpi_{2p+1}$. Suppose on the contrary that $\|\mathcal{S}\varpi_{2p} - \mathcal{T}\varpi_{2p+1}\| > 0$. For $\vartheta = \varpi_{2p}$ and $\nu = \varpi_{2p+1}$ in (9.13),

$$\|\varpi_{2p+1}-\varpi_{2p+2}\| = \|\mathcal{S}\varpi_{2p}-\mathcal{T}\varpi_{2p+1}\| \leq \Theta(\varpi_{2p}, \varpi_{2p+1})-\varphi(\zeta(\Theta(\varpi_{2p}, \varpi_{2p+1}))),$$

where

$$\begin{aligned}\Theta(\varpi_{2p}, \varpi_{2p+1}) &= \max\{\|\varpi_{2p} - \varpi_{2p+1}\|, \|\varpi_{2p} - \mathcal{S}\varpi_{2p}\|, \|\varpi_{2p+1} - \mathcal{T}\varpi_{2p+1}\|\}\\ &= \max\{\|\varpi_{2p} - \varpi_{2p+1}\|, \|\varpi_{2p} - \varpi_{2p+1}\|, \|\varpi_{2p+1} - \varpi_{2p+2}\|\}\\ &= \|\varpi_{2p+1} - \varpi_{2p+2}\|.\end{aligned}$$

Therefore,

$$\|\varpi_{2p+1} - \varpi_{2p+2}\| \leq \|\varpi_{2p+1} - \varpi_{2p+2}\| - \varphi(\zeta(\|\varpi_{2p+1} - \varpi_{2p+2}\|)),$$

which implies that $\varphi(\zeta(\|\varpi_{2p+1}-\varpi_{2p+2}\|)) \leq 0$, and hence $\|\mathcal{S}\varpi_{2p}-\mathcal{T}\varpi_{2p+1}\| = 0$, $\varpi_{2p} = \varpi_{2p+1} = \mathcal{S}\varpi_{2p} = \mathcal{T}\varpi_{2p+1}$, that is, $\varpi_{2p} = \varpi_{2p+1} = \vartheta^*$ is a $CFP(\mathcal{S}, \mathcal{T})$. Similarly, it can be shown when m_0 is odd, say, $m_0 = 2p - 1$.

Thus, we assume $\varpi_m \neq \varpi_{m+1}$ for each $m \in \mathbb{N}^*$. Let $\varrho_m = \|\varpi_m - \varpi_{m+1}\|$ for $m \in \mathbb{N}^*$. Then $\varrho_m > 0$ for all $m \in \mathbb{N}^*$. We will prove that $\lim_{n\to+\infty} \varrho_m = 0$.

Case (I): Take $\vartheta = \varpi_{2m-1}$ and $\nu = \varpi_{2m}$ in (9.13), implying that

$$\varrho_{2m} = \|\mathcal{S}\varpi_{2m-1} - \mathcal{T}\varpi_{2m-1}\| \leq \Theta(\varpi_{2m-1}, \varpi_{2m}) - \varphi(\zeta(\Theta(\varpi_{2m-1}, \varpi_{2m}))),$$

where

$$\begin{aligned}&\Theta(\varpi_{2m-1}, \varpi_{2m})\\ &\quad= \max\{\|\varpi_{2m-1} - \varpi_{2m}\|, \|\varpi_{2m-1} - \mathcal{S}\varpi_{2m-1}\|, \|\varpi_{2m} - \mathcal{T}\varpi_{2m}\|\}\\ &\quad= \max\{\|\varpi_{2m-1} - \varpi_{2m}\|, \|\varpi_{2m-1} - \varpi_{2m}\|, \|\varpi_{2m} - \varpi_{2m+1}\|\}\\ &\quad= \max\{\|\varpi_{2m-1} - \varpi_{2m}\|, \|\varpi_{2m} - \varpi_{2m+1}\|\} = \max\{\varrho_{2m-1}, \varrho_{2m}\}.\end{aligned}$$

Therefore,

$$\varrho_{2m} \leq \max\{\varrho_{2m-1}, \varrho_{2m}\} - \varphi(\zeta(\max\{\varrho_{2m-1}, \varrho_{2m}\})). \tag{9.14}$$

We shall show that $\{\varrho_m\}$ is a nonincreasing sequence. Indeed, if $\varrho_{2m-1} < \varrho_{2m}$ for some $m \in \mathbb{N}$, then (9.14) implies that $\varrho_{2m} \leq \varrho_{2m} - \varphi(\zeta(\varrho_{2m}))$. It follows that $\varphi(\zeta(\varrho_{2m})) \leq 0$, and by the properties of ζ, φ, we get $\varrho_{2m} \to 0$, a contradiction. Hence, $\varrho_{2m} \leq \varrho_{2m-1}$ for some $n \in \mathbb{N}$.

Case (II): Take $\vartheta = \varpi_{2m}$ and $\nu = \varpi_{2m+1}$ in (9.13). By a similar line of arguments, we get $\varrho_{2m+1} \leq \varrho_{2m}$ for all $n \in \mathbb{N}^*$.

Therefore, for all $n \in \mathbb{N}$, $\varrho_m \leq \varrho_{m-1}$, and thus the sequence $\{\varrho_m\}$ is a nonincreasing sequence of positive real numbers. Hence, there exists an $s \geq 0$ such that

$$\lim_{m\to+\infty} \varrho_m = \lim_{m\to+\infty} \|\varpi_{m+1} - \varpi_m\| = s.$$

This implies that

$$\lim_{n\to+\infty} \varrho_{2m} = \lim_{n\to+\infty} \varrho_{2m-1} = s.$$

If $s > 0$ then, by the condition (9.13),

$$\begin{aligned} s &\leq \liminf \varrho_{2m} \leq \limsup \varrho_{2m} \leq \limsup[\varrho_{2m-1} - \varphi(\zeta(\varrho_{2m-1}))] \\ &= \limsup \varrho_{2m-1} - \liminf \varphi(\zeta(\varrho_{2m-1})) \\ &\leq s - \liminf \varphi(\zeta(\varrho_{2m-1})). \end{aligned}$$

By the nondecreasing property of φ,

$$\varphi(\zeta(s)) \leq \liminf_{m\to+\infty} \varphi(\zeta(\varrho_{2m-1})) = 0.$$

Using a property of φ, $\varphi(\zeta(\varrho_{2m-1})) \to 0$ as $m \to +\infty$. Since $\zeta \in \Upsilon$, $s = 0$, and hence

$$\lim_{m\to+\infty} \|\varpi_{m+1} - \varpi_m\| = 0.$$

Next, we will show that $\{\varpi_n\}$ is a Cauchy sequence in $\mathcal{O}(\varpi_0; \mathcal{S}, \mathcal{T})$. Following the line of discussion in the proof of Theorem 11.2.3, we get

$$\begin{aligned} &\lim_{k\to+\infty} \|\varpi_{2n_k} - \varpi_{2m_k}\| = \eta, \quad \lim_{k\to+\infty} \|\varpi_{2n_k} - \varpi_{2m_k-1}\| = \eta, \\ &\lim_{k\to+\infty} \|\varpi_{2n_k+1} - \varpi_{2m_k-1}\| = \eta, \quad \lim_{k\to+\infty} \Theta(\varpi_{2n_k}, \varpi_{2m_k}) = \eta. \end{aligned}$$

Taking the pair $\vartheta = 2m_k + 1$, $\nu = 2n_k + 1$ in (9.13) and taking the upper limit, we obtain

$$\begin{aligned} \eta &\leq \liminf_{k\to+\infty} \|\varpi_{2n_k+1} - \varpi_{2m_k+1}\| \leq \limsup_{k\to+\infty} \|\varpi_{2n_k+1} - \varpi_{2m_k+1}\| \\ &\leq \limsup_{k\to+\infty} [\Theta(\varpi_{2n_k}, \varpi_{2m_k})) - \varphi(\zeta(\Theta(\varpi_{2n_k}, \varpi_{2m_k})))] \\ &\leq \limsup_{k\to+\infty} \Theta(\varpi_{2n_k}, \varpi_{2m_k})) - \liminf_{n\to+\infty} \varphi(\zeta(\Theta(\varpi_{2n_k}, \varpi_{2m_k}))). \end{aligned}$$

By the nondecreasing property of φ,

$$\varphi(\zeta(\eta)) \leq \liminf_{m\to+\infty} \varphi(\zeta(\Theta(\varpi_{2n_k}, \varpi_{2m_k}))) = 0.$$

Using the properties of ζ, φ, we get $\eta = 0$, a contradiction.

Thus $\{\varpi_{2n}\}$ is a Cauchy sequence in the normed space $\mathcal{O}(\varpi_0;\mathcal{S},\mathcal{T})$, and so is $\{\varpi_n\}$. Since Ξ is $(\mathcal{T},\mathcal{S})$-orbitally complete at ϖ_0, there exists a $\vartheta^* \in \mathcal{O}(\varpi_0;\mathcal{S},\mathcal{T})$ with $\lim_{n\to+\infty}\varpi_n = \vartheta^*$. By the orbital continuity of $\mathcal{S}$ or $\mathcal{T}$, $\vartheta^* = \mathcal{S}\vartheta^* = \mathcal{T}\vartheta^*$.

Finally, to check the uniqueness of $\vartheta^* \in CFP(\mathcal{S},\mathcal{T})$, let $\mu^* \neq \vartheta^*$ be such that $\mu^* = \mathcal{S}\mu^* = \mathcal{T}\mu^*$. From (9.13),

$$\begin{aligned}\|\vartheta^* - \mu^*\| &= \|\mathcal{S}\vartheta^* - \mathcal{T}\mu^*\| \\ &\leq \max\{\|\vartheta^* - \mu^*\|, \|\vartheta^* - \mathcal{S}\vartheta^*\|, \|\mu^* - \mathcal{T}\mu^*\|\}) \\ &\quad - \varphi(\zeta(\max\{\|\vartheta^* - \mu^*\|, \|\vartheta^* - \mathcal{S}\vartheta^*\|, \|\mu^* - \mathcal{T}\mu^*\|\})) \\ &= \|\vartheta^* - \mu^*\|) - \varphi(\zeta(\|\vartheta^* - \mu^*\|)).\end{aligned}$$

Applying the properties of φ, ζ, $\vartheta^* = \mu^*$, i.e., $CFP(\mathcal{S},\mathcal{T})$ is a singleton. □

Example 9.3.3. *Consider the normed space $\Xi = c_0$ equipped with the $l_{+\infty}$ norm. Define two mappings $\mathcal{S}$, $\mathcal{T} : \Xi \to \Xi$ by*

$$\mathcal{S}(\{x_n\}) = \{z_n\}, \quad \mathcal{T}(\{x_n\}) = \{y_n\}$$

for all $\{x_n\} \in \Xi$, where $y_n = \frac{x_n}{3}$, $z_1 = \frac{x_1}{3^2}$ and $z_n = 0$ for all $n \geq 2$. We choose $\nu_0 \in \Xi$ as the sequence $\{1, 2, \cdots, 100, 0, 0, \cdots\}$. Then the orbit of $(\mathcal{S},\mathcal{T})$ at ν_0 is $\mathcal{O}(\nu_0;\mathcal{S},\mathcal{T}) = \{\nu_n : n \in \mathbb{N}\}$, where the first terms of $\{\nu_{2n}\}$ and $\{\nu_{2n-1}\}$ are respectively $\frac{1}{3^{3n}}$ and $\frac{1}{3^{3n-1}}$ and all other terms of $\{\nu_n\}$ are 0. Then Ξ is an $(\mathcal{S},\mathcal{T})$-orbitally complete space at ν_0 and $\mathcal{S}$ and $\mathcal{T}$ are $(\mathcal{S},\mathcal{T})$-orbitally continuous at ν_0. We choose $\varphi \in \Phi$ as $\varphi(t) = \frac{t}{1000}$ and $\zeta \in \Upsilon$ as $\zeta(t) = t^2$. Then it can be checked in a similar manner to that of Example 9.2.4 that

$$\|\mathcal{S}\vartheta - \mathcal{T}\nu\| \leq \Theta(\vartheta,\nu) - \varphi(\zeta(\Theta(\vartheta,\nu))),$$

$\vartheta, \nu \in \overline{\mathcal{O}(\nu_0;\mathcal{S},\mathcal{T})}$. So $\mathcal{S},\mathcal{T} : \Xi \to \Xi$ is an orbitally (ζ,φ)-weakly contractive pair of mappings. Thus we see that all the assumptions of Theorem 11.2.4 are satisfied. So, $CFP(\mathcal{S},\mathcal{T})$ is unique in Ξ, which is the sequence $\{0, 0, \ldots\}$.

9.3.1 Consequences

By taking different values of $\varphi \in \Phi$ and $\zeta \in \Upsilon$ in (9.13), we obtain particular forms of Theorems 11.2.4. We state some of them.

(R1) If $\zeta(t) = t$ $(t > 0)$ in (9.13), then $(\mathcal{S},\mathcal{T})$ is a φ-weakly contractive mapping if there exists $\varphi \in \Phi$ such that for all $\vartheta, \nu \in \Xi$,

$$\|\mathcal{S}\vartheta - \mathcal{T}\nu\| \leq \Theta(\vartheta,\nu) - \varphi(\Theta(\vartheta,\nu)).$$

(R2) If $\varphi(t) = (1-\lambda)t$ $(t > 0)$, $\lambda \in [0,1)$ in (9.13), then $(\mathcal{S},\mathcal{T})$ is a (ζ,λ)-contractive mapping if there exist $\zeta \in \Upsilon$ and $\varphi \in \Phi$ such that for all $\vartheta, \nu \in \Xi$,

$$\|\mathcal{S}\vartheta - \mathcal{T}\nu\| \leq \lambda\zeta(\Theta(\vartheta,\nu)). \tag{9.15}$$

(R3) If $\zeta(t) = t$ $(t > 0)$ in (9.15), then $(\mathcal{S}, \mathcal{T})$ is a λ-contractive mapping if there exists $\lambda \in [0,1)$ such that for all $\vartheta, \nu \in \Xi$,

$$\|\mathcal{S}\vartheta - \mathcal{T}\nu\| \leq \lambda\Theta(\vartheta, \nu).$$

9.4 Application

For a matrix $\mathcal{B} \in \mathcal{H}(n)$, we will denote by $s(\mathcal{B})$ any of its singular values and by $s^+(\mathcal{B})$ the sum of all of its singular values, that is, the trace norm $\|\mathcal{B}\| = s^+(\mathcal{B})$. For $\mathcal{C}, \mathcal{D} \in \mathcal{H}(n)$, $\mathcal{C} \succeq \mathcal{D}$ (resp. $\mathcal{C} \succ \mathcal{D}$) will mean that the matrix $\mathcal{C} - \mathcal{D}$ is positive semi-definite (resp. positive definite).

Theorem 9.4.1. *Consider the pair of NMEs:*

$$\begin{aligned} \mathcal{X} &= \mathcal{R}_1 + \sum_{j=1}^{m} \mathcal{B}_j^* \mathcal{U}(\mathcal{X}) \mathcal{B}_j, \\ \mathcal{X} &= \mathcal{R}_2 + \sum_{j=1}^{m} \mathcal{B}_j^* \mathcal{V}(\mathcal{X}) \mathcal{B}_j, \end{aligned} \tag{9.16}$$

where $\mathcal{R}_1, \mathcal{R}_2 \in \mathcal{P}(n)$, $\mathcal{B}_j \in \mathcal{M}(n)$, and $\mathcal{U}, \mathcal{V}$ are two functions, continuous in the trace norm of $\mathcal{P}(n)$. Let, for any $\mathcal{X} \in \mathcal{P}(n)$ with $\|\mathcal{X}\| \leq \alpha$, all singular values of $\mathcal{U}(\mathcal{X})$ and $\mathcal{V}(\mathcal{X})$ satisfy $s(\mathcal{U}(\mathcal{X})), s(\mathcal{V}(\mathcal{X})) \leq \beta$ for some α, $\beta \in \mathbb{R}$. Assume that

(i) $\|\mathcal{R}_1\|, \|\mathcal{R}_2\| \leq \alpha - n\beta\gamma$, *where* $\sum_{j=1}^{m} \|\mathcal{B}_j^*\| \, \|\mathcal{B}_j\| = \gamma$;

(ii) for any $\mathcal{X} \in \mathcal{P}(n)$ *with* $\|\mathcal{X}\| \leq \alpha$, $\sum_{j=1}^{m} \mathcal{B}_j^* \mathcal{U}(\mathcal{X}) \mathcal{B}_j \succeq O$ *or* $\sum_{j=1}^{m} \mathcal{B}_j^* \mathcal{V}(\mathcal{X}) \mathcal{B}_j \succeq O$;

(iii) for any $\mathcal{K} \in \mathcal{P}(n)$ *with* $\|\mathcal{K}\| \leq \alpha$, $\mathcal{K} \preceq \mathcal{R}_1 + \sum_{j=1}^{m} \mathcal{B}_j^* \mathcal{U}(\mathcal{K}) \mathcal{B}_j$ *and* $\mathcal{K} \preceq \mathcal{R}_2 + \sum_{j=1}^{m} \mathcal{B}_j^* \mathcal{V}(\mathcal{K}) \mathcal{B}_j$ *hold.*

(iv) for any $s(\mathcal{B}_1 - \mathcal{B}_2)$ *with* $\sum_{j=1}^{m} \mathcal{B}_j^* \mathcal{U}(\mathcal{X}) \mathcal{B}_j \neq \sum_{j=1}^{m} \mathcal{B}_j^* \mathcal{V}(\mathcal{Y}) \mathcal{B}_j$,

$$\begin{aligned} & s(\mathcal{B}_1 - \mathcal{B}_2) + \frac{\gamma}{n} s^+(\mathcal{U}(\mathcal{X}) - \mathcal{V}(\mathcal{Y})) \\ & \quad \leq \frac{1}{n+1} \max \left\{ \begin{array}{l} \left| s^+(\sum_{j=1}^{m} \mathcal{B}_j^* \mathcal{U}(\mathcal{X}) \mathcal{B}_j + \mathcal{R}_1 - \mathcal{X}) \right|, \\ \left| s^+(\sum_{j=1}^{m} \mathcal{B}_j^* \mathcal{V}(\mathcal{Y}) \mathcal{B}_j + \mathcal{R}_2 - \mathcal{Y}) \right|, \\ s^+(\mathcal{X} - \mathcal{Y}) \end{array} \right\}, \end{aligned}$$

for all $\mathcal{X}$, $\mathcal{Y} \in \mathcal{P}(n)$ *with* $\|\mathcal{X}\|, \|\mathcal{Y}\| \leq \alpha$ *and* $\mathcal{X} \preceq \mathcal{Y}$.

Then the system (9.16) has a unique common PDS $\widehat{\mathcal{X}}$ with $\|\widehat{\mathcal{X}}\| \leq \alpha$. Moreover, the solution can be derived as the limit of the iterative sequence $\{\mathcal{X}_n\}$, where

$$\mathcal{X}_{2k+1} = \mathcal{R}_1 + \sum_{j=1}^{m} \mathcal{B}_j^* \mathcal{U}(\mathcal{X}_{2k}) \mathcal{B}_j \text{ for } k \geq 0,$$

$$\mathcal{X}_{2k+2} = \mathcal{R}_2 + \sum_{j=1}^{m} \mathcal{B}_j^* \mathcal{V}(\mathcal{X}_{2k+1}) \mathcal{B}_j \text{ for } k \geq 0,$$

and $\mathcal{X}_0$ is an arbitrary element of $\mathcal{P}(n)$ satisfying $\|\mathcal{X}_0\| \leq \alpha$.

Proof. Consider $\Lambda = \{\mathcal{X} \in \mathcal{P}(n) : \|\mathcal{X}\| \leq a\}$, a closed subset of $\mathcal{P}(n)$. Assumption (ii) ensures that any solution of (9.16) in Λ must be positive definite. For any $\mathcal{X} \in \Lambda$,

$$\begin{aligned}\|\mathcal{R}_1 + \sum_{i=1}^{m} \mathcal{B}_i^* \mathcal{U}(\mathcal{X}) \mathcal{B}_j\| &\leq \|\mathcal{R}_1\| + \|\sum_{j=1}^{m} \mathcal{B}_j^* \mathcal{U}(\mathcal{X}) \mathcal{B}_j\| \\ &\leq \|\mathcal{R}_1\| + \sum_{j=1}^{m} \|\mathcal{B}_j^*\| \, \|\mathcal{B}_j\| \, \|\mathcal{U}(\mathcal{X})\| \\ &= \|\mathcal{R}_1\| + \gamma \|\mathcal{U}(\mathcal{X})\|. \end{aligned} \tag{9.17}$$

Also, since $s(\mathcal{U}(\mathcal{X})) \leq \beta$ for all singular values of $\mathcal{U}(\mathcal{X})$, adding over all $s(\mathcal{U}(\mathcal{X}))$, we get $\|\mathcal{U}(\mathcal{X})\| \leq n\beta$. Therefore, the equation (9.17) implies that

$$\|\mathcal{R}_1 + \sum_{j=1}^{m} \mathcal{B}_j^* \mathcal{U}(\mathcal{X}) \mathcal{B}_j\| \leq \|\mathcal{R}_1\| + n\beta\gamma = \alpha.$$

Similarly for any $\mathcal{X} \in \Lambda$, $\|\mathcal{R}_2 + \sum_{j=1}^{m} \mathcal{B}_j^* \mathcal{V}(\mathcal{X}) \mathcal{B}_j\| \leq \alpha$. Define $\mathcal{S}, \mathcal{T} : \Lambda \to \mathcal{P}(n)$ by

$$\mathcal{S}(\mathcal{X}) = \mathcal{R}_1 + \sum_{j=1}^{m} \mathcal{B}_j^* \mathcal{U}(\mathcal{X}) \mathcal{B}_j, \quad \mathcal{T}(\mathcal{X}) = \mathcal{R}_2 + \sum_{j=1}^{m} \mathcal{B}_j^* \mathcal{V}(\mathcal{X}) \mathcal{B}_j,$$

for all $\mathcal{X} \in \Lambda$. Now, finding common solution(s) of the equations (9.16) is equivalent to finding common fixed point(s) of $\mathcal{S}$ and $\mathcal{T}$. For any $\mathcal{X}$, $\mathcal{Y} \in \Lambda$,

$$\begin{aligned}\|\mathcal{S}(\mathcal{X}) - \mathcal{T}(\mathcal{Y})\| &= \|\mathcal{R}_1 + \sum_{j=1}^{m} \mathcal{B}_j^* \mathcal{U}(\mathcal{X}) \mathcal{B}_j - \mathcal{R}_2 - \sum_{j=1}^{m} \mathcal{B}_j^* \mathcal{V}(\mathcal{Y}) \mathcal{B}_j\| \\ &\leq \|\mathcal{R}_1 - \mathcal{R}_2\| + \|\sum_{j=1}^{m} \mathcal{B}_j^* \mathcal{U}(\mathcal{X}) \mathcal{B}_j - \sum_{j=1}^{m} \mathcal{B}_j^* \mathcal{V}(\mathcal{Y}) \mathcal{B}_j\| \\ &\leq \|\mathcal{R}_1 - \mathcal{R}_2\| + \sum_{j=1}^{m} \|\mathcal{B}_j^* \mathcal{U}(\mathcal{X}) \mathcal{B}_j - \mathcal{B}_j^* \mathcal{V}(\mathcal{Y}) \mathcal{B}_j\| \\ &\leq \|\mathcal{R}_1 - \mathcal{R}_2\| + \sum_{j=1}^{m} \|\mathcal{B}_j^*\| \, \|\mathcal{B}_j\| \|\mathcal{U}(\mathcal{X}) - \mathcal{V}(\mathcal{Y})\|, \end{aligned}$$

implying that

$$\|\mathcal{S}(\mathcal{X}) - \mathcal{T}(\mathcal{Y})\| \leq \|\mathcal{R}_1 - \mathcal{R}_2\| + \gamma(\|\mathcal{U}(\mathcal{X}) - \mathcal{V}(\mathcal{Y})\|), \tag{9.18}$$

for any $\mathcal{X},\ \mathcal{Y} \in \Lambda$,

For some fixed $\mathcal{X},\ \mathcal{Y} \in \Lambda$, if

$$\max\{|s^+(\sum_{j=1}^{m} \mathcal{B}_j^*\mathcal{U}(\mathcal{X})\mathcal{B}_j + \mathcal{R}_1 - \mathcal{X})|, |s^+(\sum_{j=1}^{m} \mathcal{B}_j^*\mathcal{V}(\mathcal{Y})\mathcal{B}_j + \mathcal{R}_2 - \mathcal{Y})|,$$

$$s^+(\mathcal{X} - \mathcal{Y})\} = |s^+(\sum_{j=1}^{m} \mathcal{B}_j^*\mathcal{U}(\mathcal{X})\mathcal{B}_j + \mathcal{R}_1 - \mathcal{X})|,$$

following the condition (iii), we get

$$\begin{aligned} s(\mathcal{R}_1 - \mathcal{R}_2) + \frac{\gamma}{n}(\|\mathcal{U}(\mathcal{X}) - \mathcal{V}(\mathcal{Y})\|) &\leq \frac{1}{n+1}|s^+(\sum_{j=1}^{m} \mathcal{B}_j^*\mathcal{U}(\mathcal{X})\mathcal{B}_j + \mathcal{R}_1 - \mathcal{X})| \\ &\leq \frac{1}{n+1}|s^+(\mathcal{S}(\mathcal{X}) - \mathcal{X})| \\ &= \frac{1}{n+1}\|\mathcal{S}(\mathcal{X}) - \mathcal{X}\|. \end{aligned}$$

The above relation holds for all singular values of $(\mathcal{R}_1 - \mathcal{R}_2)$, so adding over all these values we obtain

$$\|\mathcal{R}_1 - \mathcal{R}_2\| + \gamma(\|\mathcal{U}(\mathcal{X}) - \mathcal{V}(\mathcal{Y})\|) \leq \frac{n}{n+1}\|\mathcal{S}(\mathcal{X}) - \mathcal{X}\|.$$

Hence, from the equation (9.18), it follows that

$$\|\mathcal{S}(\mathcal{X}) - \mathcal{T}(\mathcal{Y})\| \leq \frac{n}{n+1}\|\mathcal{S}(\mathcal{X}) - \mathcal{X}\|.$$

For some fixed $\mathcal{X},\ \mathcal{Y} \in \Lambda$, if

$$\max\{|s^+(\sum_{j=1}^{m} \mathcal{B}_j^*\mathcal{U}(\mathcal{X})\mathcal{B}_j + \mathcal{R}_1 - \mathcal{X})|,\ |s^+(\sum_{\mathrm{i}=1}^{m} \mathcal{B}_j^*\mathcal{V}(\mathcal{Y})\mathcal{B}_j + \mathcal{R}_2 - s^+(\mathcal{Y})|,$$

$$s^+(\mathcal{X} - \mathcal{Y})\} = |s^+(\sum_{j=1}^{m} \mathcal{B}_j^*\mathcal{V}(\mathcal{Y})\mathcal{B}_j) - s^+(\mathcal{Y} - \mathcal{R}_2)|,$$

using the similar arguments, we get

$$\|\mathcal{S}(\mathcal{X}) - \mathcal{T}(\mathcal{Y})\| \leq \frac{n}{n+1}\|\mathcal{T}(\mathcal{Y}) - Y\|.$$

Finally, for some fixed $\mathcal{X},\ \mathcal{Y} \in \Lambda$, if

$$\max\{|s^+(\sum_{j=1}^{m} \mathcal{B}_j^*\mathcal{U}(\mathcal{X})\mathcal{B}_j + \mathcal{R}_1 - \mathcal{X})|, |s^+(\sum_{j=1}^{m} \mathcal{B}_j^*\mathcal{V}(\mathcal{Y})\mathcal{B}_j + \mathcal{R}_2 - \mathcal{Y})|,$$

$$s^+(\mathcal{X} - \mathcal{Y})\} = s^+(\mathcal{X} - \mathcal{Y}).$$

we get similarly

$$\|\mathcal{S}(\mathcal{X}) - \mathcal{T}(\mathcal{Y})\| \le \frac{n}{n+1}\|\mathcal{X} - \mathcal{Y}\|.$$

Thus, for any $\mathcal{X}, \mathcal{Y} \in \Lambda$, we have

$$\|\mathcal{S}(\mathcal{X}) - \mathcal{T}(\mathcal{Y})\| \le \frac{n}{n+1}\max\{\|\mathcal{S}(\mathcal{X}) - \mathcal{X}\|, \ \|\mathcal{T}(\mathcal{Y}) - \mathcal{Y}\|, \|\mathcal{X} - \mathcal{Y}\|\}.$$

Now, considering $\psi_1, \psi_2, \psi_3 \colon \mathbb{R}_+ \to \mathbb{R}_+$ described by:

$$\psi_1(r) = r + \frac{3}{2}, \ \ \psi_2(r) = \left(\frac{2n}{n+1}\right) r + \frac{5}{2} \text{ and } \psi_3(r) = \left(\frac{n}{n+1}\right) r + 1, \text{ for } r > 0,$$

we get that the condition (9.3) is fulfilled. Therefore, from Theorem 11.2.3, $CFP(\mathcal{S}, \mathcal{T})$ is a singleton. Hence, the system (9.16) has a unique common PDS $\widehat{\mathcal{X}}$ with $\|\widehat{\mathcal{X}}\| \le \alpha$. □

9.5 Numerical Experiment

Here, we demonstrate two examples in support of Theorem 9.4.1 using MATLAB.

Example 9.5.1. *Consider the system of NMEs*

$$\begin{aligned} \mathcal{X} &= \mathcal{R}_1 + \mathcal{B}_1^*\mathcal{U}(\mathcal{X})\mathcal{B}_1 + \mathcal{B}_2^*\mathcal{U}(\mathcal{X})\mathcal{B}_2, \\ \mathcal{X} &= \mathcal{R}_2 + \mathcal{B}_1^*\mathcal{V}(\mathcal{X})\mathcal{B}_1 + \mathcal{B}_2^*\mathcal{V}(\mathcal{X})\mathcal{B}_2, \end{aligned} \tag{9.19}$$

where $\mathcal{B}_1 = \begin{pmatrix} \frac{2}{3} & 0 & 0 \\ 0 & \frac{1}{3} & 0 \\ 0 & 0 & \frac{5}{6} \end{pmatrix}$, $\mathcal{B}_2 = \begin{pmatrix} \frac{5}{12} & 0 & 0 \\ 0 & \frac{3}{12} & 0 \\ 0 & 0 & \frac{4}{12} \end{pmatrix}$ *and*

$\mathcal{R}_1 = \mathcal{R}_2 = \begin{pmatrix} 9.98335 & -11.98913 & 0 \\ -11.98913 & 14.99301 & 0 \\ 0 & 0 & 16.9339 \end{pmatrix}$. *Also, we choose* $\mathcal{U}, \mathcal{V}$: $\mathcal{P}(n) \to \mathcal{P}(n)$, *defined by*

$$\mathcal{U}(\mathcal{X}) = \tilde{\mathcal{X}} \text{ and } \mathcal{V}(\mathcal{X}) = \frac{\mathcal{X}^2}{9 \times 10^3},$$

where if we choose $\mathcal{X} = \begin{pmatrix} x_1 & x_2 & x_3 \\ x_4 & x_5 & x_6 \\ x_7 & x_8 & x_9 \end{pmatrix}$, *then* $\tilde{\mathcal{X}} = \begin{pmatrix} \frac{x_1}{369} & \frac{x_2}{392} & \frac{x_3}{100} \\ \frac{x_4}{392} & \frac{x_5}{372} & \frac{x_6}{100} \\ \frac{x_7}{100} & \frac{x_8}{100} & \frac{x_9}{207} \end{pmatrix}$.

We choose $\alpha = 200$, $\beta = 10.5$ *and we have* $\gamma = \sum_{j=1}^{m} \|\mathcal{B}_j^*\| \, \|\mathcal{B}_j\| = 4.3611$, *and* $\|\mathcal{R}_1\| = \|\mathcal{R}_2\| = 41.9102 \le \alpha - n\beta\gamma$.

The conditions of Theorem 9.4.1 can be checked numerically, taking various special values for matrices involved. For example, they can be tested (and verified to be true) for:

$$\mathcal{X} = \begin{bmatrix} 4 & 2 & 1 \\ 2 & 4 & 1 \\ 1 & 1 & 4 \end{bmatrix}, \quad \mathcal{Y} = \begin{bmatrix} 7 & 3 & 1.5 \\ 3 & 7 & 1.5 \\ 1.5 & 1.5 & 7 \end{bmatrix},$$

$$\mathcal{K} = \begin{bmatrix} 0.002154285714286 & 0.000124897959184 & 0.000205714285714 \\ 0.000124897959184 & 0.002257142857143 & 0.000124897959184 \\ 0.000205714285714 & 0.000124897959184 & 0.002301224489796 \end{bmatrix}.$$

To see the convergence of the sequence $\{\mathcal{X}_n\}$ *defined in* (9.19), *we start with three different initial values:*

$$\mathcal{L}_0 = \begin{bmatrix} 1.03173 & 0.07557 & 0.0253 \\ 0.07557 & 1.19038 & 0.0637 \\ 0.0253 & 0.0637 & 1.023080 \end{bmatrix}, \quad \textit{with} \quad \|\mathcal{L}_0\| = 3.24519,$$

$$\mathcal{M}_0 = \begin{bmatrix} 1 & 0 & 0 \\ 0 & 1 & 0 \\ 0 & 0 & 1 \end{bmatrix}, \quad \textit{with} \quad \|\mathcal{M}_0\| = 3,$$

$$\mathcal{W}_0 = \begin{bmatrix} 10.86303 & 1.002555 & 1.167492 \\ 1.002555 & 11.78101 & 1.628121 \\ 1.167492 & 1.628121 & 11.69929 \end{bmatrix} \quad \textit{with}\ \|\mathcal{W}_0\| = 34.343329999999995.$$

After 10 *iterations, we have the following approximation of the unique PDS of the system* (9.19) *as*

$$\widehat{\mathcal{L}} \approx \mathcal{L}_{10}$$
$$= \begin{bmatrix} 10.000255068232548 & -12.000008946471102 & 0.000000000000000 \\ -12.000008946471102 & 15.000127654940146 & -0.000000000000000 \\ 0.000000000000000 & -0.000000000000000 & 16.959766980137900 \end{bmatrix}$$

with $\|\widehat{\mathcal{L}}\| = 41.960149703310599$ *and* $Error(10) = 0.041911600373504$,

$$\widehat{\mathcal{M}} \approx \mathcal{M}_{10}$$
$$= \begin{bmatrix} 10.000255068232548 & -12.000008946471102 & 0.000000000000000 \\ -12.000008946471102 & 15.000127654940146 & -0.000000000000000 \\ 0.000000000000000 & -0.000000000000000 & 16.959766980137900 \end{bmatrix}$$

with $\|\widehat{\mathcal{M}}\| = 41.960149703310599$ *and* $Error(10) = 0.041911600373504$,

$$\widehat{\mathcal{W}} \approx \mathcal{W}_{10}$$
$$= \begin{bmatrix} 10.000255068232548 & -12.000008946471102 & 0.000000000000000 \\ -12.000008946471102 & 15.000127654940146 & -0.000000000000000 \\ 0.000000000000000 & -0.000000000000000 & 16.959766980137900 \end{bmatrix}$$

with $\|\widehat{\mathcal{W}}\| = 41.960149703310599$ *and* $Error(10) = 0.041911600373504$.

The graphical representation of convergence of $\{\mathcal{X}_n\}$ *is shown in Fig. 9.1.*

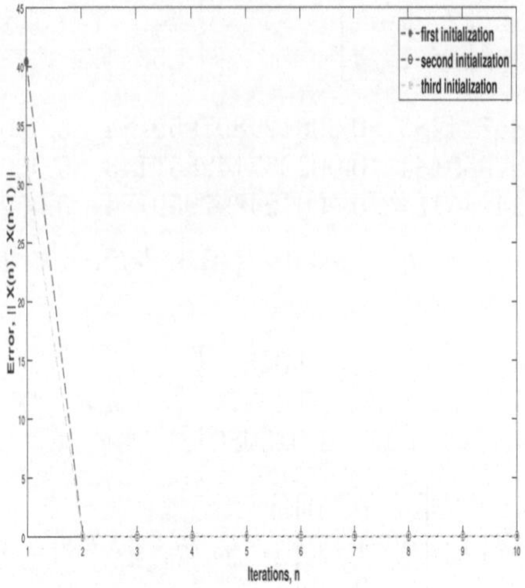

FIGURE 9.1: Convergence analysis

Example 9.5.2. *Consider the system of NMEs*

$$\begin{aligned} \mathcal{X} &= \mathcal{P}_1 + \mathcal{A}_1^*\mathcal{U}(\mathcal{X})\mathcal{A}_1 + \mathcal{A}_2^*\mathcal{U}(\mathcal{X})\mathcal{A}_2 \\ \mathcal{X} &= \mathcal{P}_2 + \mathcal{A}_1^*\mathcal{V}(\mathcal{X})\mathcal{A}_1 + \mathcal{A}_2^*\mathcal{V}(\mathcal{X})\mathcal{A}_2, \end{aligned} \tag{9.20}$$

where $\mathcal{A}_1 = \begin{pmatrix} 0.235 & 0.234 & -0.254 \\ 0.265 & 0.245 & 0.285 \\ -0.215 & 0.263 & 0.243 \end{pmatrix}$, $\mathcal{A}_2 = \begin{pmatrix} 0.265 & -0.243 & 0.214 \\ 0.234 & 0.217 & 0.292 \\ 0.286 & -0.275 & 0.236 \end{pmatrix}$

and $\mathcal{P}_1 = \begin{pmatrix} 14.6 & 10.5 & 6.6 \\ 10.5 & 10.5 & 4.9 \\ 6.6 & 4.9 & 5.4 \end{pmatrix}$,

$$\mathcal{P}_2 = \begin{pmatrix} 14.252377163452010 & 10.449039646116516 & 6.325091506209104 \\ 10.449039646116516 & 10.235404899450467 & 4.945154595038549 \\ 6.325091506209104 & 4.945154595038549 & 5.155446741281554 \end{pmatrix}.$$

Also we choose $\mathcal{U},\ \mathcal{V} : \mathcal{P}(n) \to \mathcal{P}(n)$, *defined by*

$$\mathcal{U}(\mathcal{X}) = \mathcal{X}^2/1000 \ \text{ and } \ \mathcal{V}(\mathcal{X}) = \frac{\mathcal{X}^3}{9 \times 10^3},$$

$\alpha = 200, \beta = 10.5$. *We have* $\gamma = \sum_{j=1}^{m} \|\mathcal{A}_j^*\|\ \|\mathcal{A}_j\| = 2.648420741616673$, *and* $\|\mathcal{P}_1\| = 30.499999999999996$, $\|\mathcal{P}_2\| = 29.643228804184027 \leq \alpha - n\beta\gamma$.

In what follows, we take various special values for matrices, to verify numerically the conditions of Theorem 9.4.1:

$$\mathcal{X} = \begin{bmatrix} 1.15 & 0.5998 & 0.9888 \\ 0.5998 & 0.5396 & 0.8192 \\ 0.9888 & 0.8192 & 1.3648 \end{bmatrix}, \quad \mathcal{Y} = \begin{bmatrix} 3.23 & 1.1996 & 1.9776 \\ 1.1996 & 2.0792 & 1.6384 \\ 1.9776 & 1.6384 & 3.7296 \end{bmatrix},$$

$$\mathcal{K} = \begin{bmatrix} 0.002154285714286 & 0.000124897959184 & 0.000205714285714 \\ 0.000124897959184 & 0.002257142857143 & 0.000124897959184 \\ 0.000205714285714 & 0.000124897959184 & 0.002301224489796 \end{bmatrix}.$$

To see the convergence of the sequence $\{\mathcal{X}_n\}$ *defined as in* (9.4)*, we start with three different initial values:*

$$\mathcal{X}_0 = \begin{bmatrix} 37.157200000000003 & 34.745800000000003 & 36.310000000000002 \\ 34.745800000000003 & 34.745800000000003 & 35.085799999999999 \\ 36.310000000000002 & 35.085799999999999 & 37.568399999999997 \end{bmatrix}$$

with $\|\mathcal{X}_0\| = 109.4714$,

$$\mathcal{U}_0 = \begin{bmatrix} 18.078600000000002 & 17.372900000000001 & 18.155000000000001 \\ 17.372900000000001 & 16.872900000000001 & 17.542899999999999 \\ 18.155000000000001 & 17.542899999999999 & 18.284199999999998 \end{bmatrix}$$

with $\|\mathcal{U}_0\| = 53.235699999999987$*, and*

$$\mathcal{W}_0 = \begin{bmatrix} 10.86303 & 1.002555 & 1.167492 \\ 1.002555 & 11.78101 & 1.628121 \\ 1.167492 & 1.628121 & 11.69929 \end{bmatrix}$$

with $\|\mathcal{W}_0\| = 34.343329999999995$.

After 10 *iterations, we have the following approximation of the unique PDS of the system* (9.20) *as*

$$\begin{aligned} \widehat{\mathcal{X}} &\approx \mathcal{X}_{10} \\ &= \begin{bmatrix} 14.777867902822258 & 10.525342476943340 & 6.739770133820008 \\ 10.525342476943340 & 10.636102533038679 & 4.877172415845318 \\ 6.739770133820008 & 4.877172415845318 & 5.525218947956358 \end{bmatrix} \end{aligned}$$

with $\|\widehat{\mathcal{X}}\| = 30.9391893838173009$ *and* $Error(10) = 3.202 \times 10^{-15}$,

$$\begin{aligned} \widehat{\mathcal{U}} &\approx \mathcal{U}_{10} \\ &= \begin{bmatrix} 14.777867902822258 & 10.525342476943340 & 6.739770133820008 \\ 10.525342476943340 & 10.636102533038679 & 4.877172415845318 \\ 6.739770133820008 & 4.877172415845318 & 5.525218947956358 \end{bmatrix} \end{aligned}$$

with $\|\widehat{\mathcal{U}}\| = 30.9391893838173009$ *and* $Error(10) = 3.202 \times 10^{-15}$, *and*

$$\widehat{\mathcal{W}} \approx \mathcal{W}_{10}$$
$$= \begin{bmatrix} 14.777867902822258 & 10.525342476943340 & 6.739770133820008 \\ 10.525342476943340 & 10.636102533038679 & 4.877172415845318 \\ 6.739770133820008 & 4.877172415845318 & 5.525218947956358 \end{bmatrix}$$

with $\|\widehat{\mathcal{W}}\| = 30.9391893838173009$ *and* $Error(10) = 3.202 \times 10^{-15}$.

The graphical representation of convergence of these sequences with three initial values $\{\mathcal{X}_0\}$, $\{\mathcal{U}_0\}$, $\{\mathcal{W}_0\}$ *are shown below (Fig. 9.2–Fig. 9.8).*

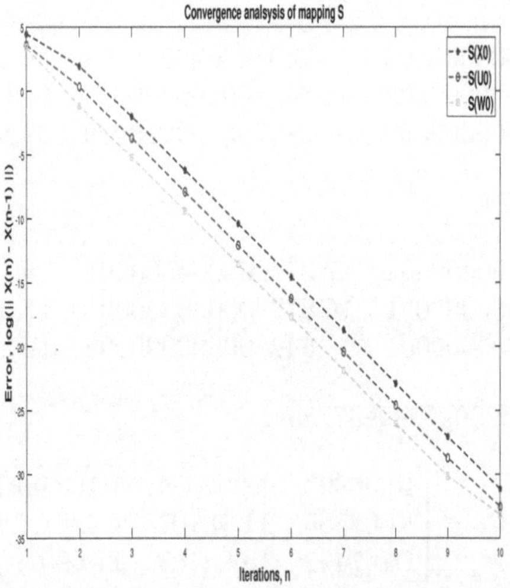

FIGURE 9.2: Convergence analysis of mapping S

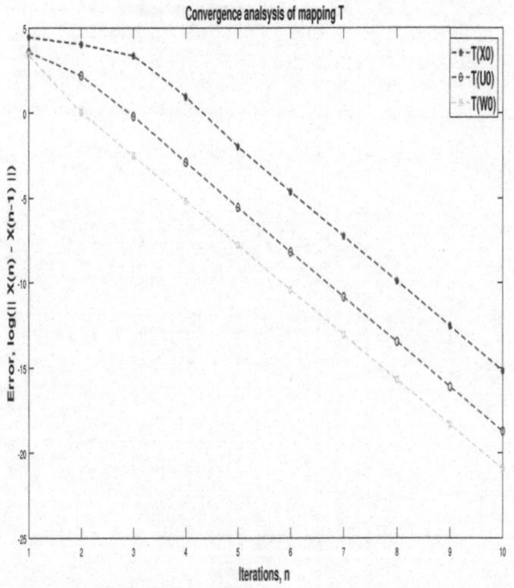

FIGURE 9.3: Convergence analysis of mapping T

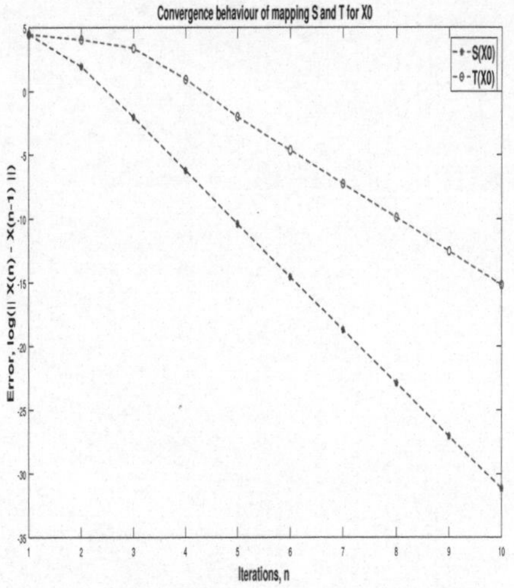

FIGURE 9.4: Convergence analysis of mapping S and T for X_0

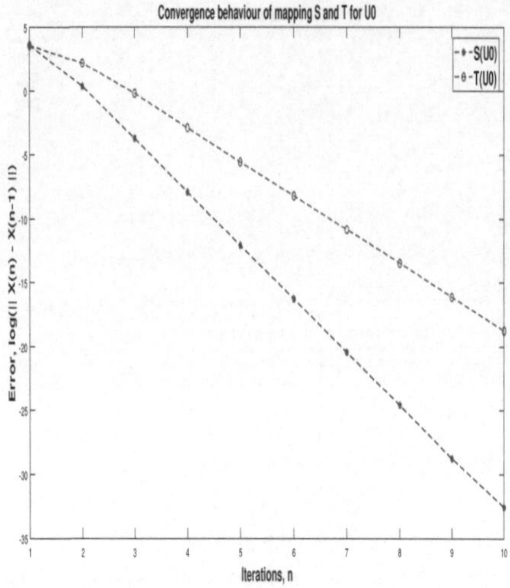

FIGURE 9.5: Convergence analysis of mapping S and T for U_0

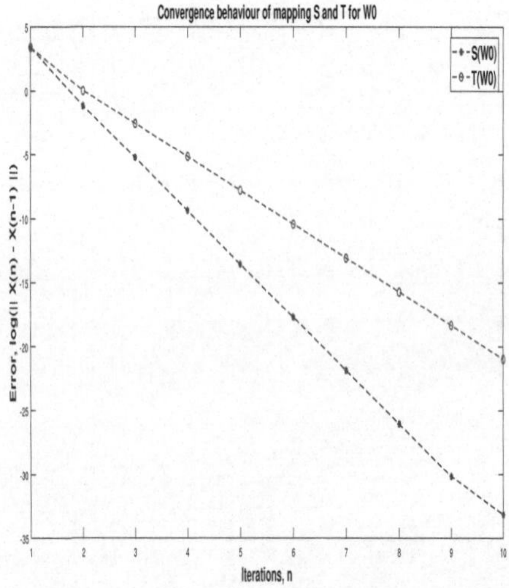

FIGURE 9.6: Convergence analysis of mapping S and T for W_0

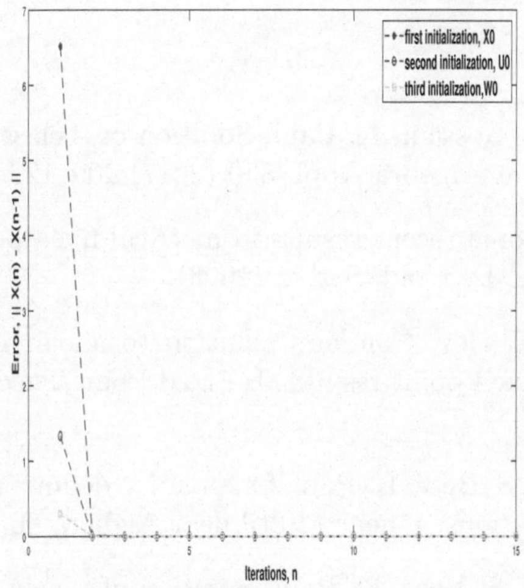

FIGURE 9.7: Common convergence behaviour

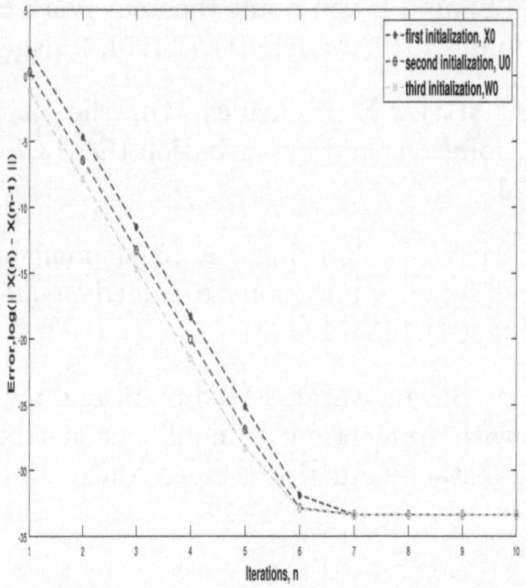

FIGURE 9.8: Common convergence behaviour (log plot)

Bibliography

[1] S. Bose, Sk. M. Hossein, K. Paul, Solution of a class ofnonlinear matrix equations, Linear Algebra Appl. 530 (2017) 109–126.

[2] C. Ljubomir, Some recent results in metrical fixed point theory, University of Belgrade, Beograd, Serbia (2003).

[3] H. Garai, L. K. Dey, Common solution to a pair of nonlinear matrix equations via fixed point results, J. Fixed Point Theory Appl. 21 (2019), Article 61.

[4] Sk. M. Hossein, S. Bose, K. Paul, On positive definite solution of a nonlinear matrix equations, Linear Multilinear Algbera 66(5) (2018) 881–893.

[5] Sk. M. Hossein, S. Bose, K. Paul, Solution of a pair of nonlinear matrix equations, Fixed Point Theory 19(1) (2018) 265–274.

[6] Z. Kadelburg, S. Radenović, Fixed points under ψ-α-β conditions in ordered partial metric spaces, Int. J. Anal. Appl. 5 (2014) 91–101.

[7] E. Karapinar, P. Salimi, Fixed point theorems via auxiliary functions, J. Appl. Math. Volume 2012, Article ID 792174, 9 pages (2012).

[8] A. C. M. Ran, M. C. B. Reurings, On the matrix equation $X + A^*F(X)A = Q$: solutions and perturbation theory, Linear Algebra Appl. 346 (2002) 15–26.

[9] A. C. M. Ran, M. C. B. Reurings, A fixed point theorem in partially ordered sets and some applications to matrix equations, Proc. Amer. Math. Soc. 132 (2004) 1435–1443.

[10] K. Sawangsup, W. Sintunavarat, Fixed point and multidimensional fixed point theorems with applications to nonlinear matrix equations in terms of weak altering distance functions, Open Math. 15 (2017) 111–125.

Chapter 10

Sequence Spaces and Matrix Transformations

Ekrem Savaş

Uşak University, ekremsavas@yahoo.com

10.1 Introduction

Let s be the set of all sequences real or complex and let ℓ_∞, c and c_0 be the Banach spaces of bounded, convergent and null sequences $x = (x_k)$ with the usual norm $\|x\| = \sup_k |x_k|$.

Definition 10.1.1. *[1] A linear functional* L *on* ℓ_∞ *is said to be a Banach Limit if it satisfies the properties:*

(i) $\mathrm{L}(x) \geq 0$ *if* $x \geq 0$ *(i.e* $x_k \geq 0$ *for all* k*),*

(ii) $\mathrm{L}(e) = 1$*, where* $e = (1, 1, 1, ...)$*,*

(iii) $\mathrm{L}(Dx) = \mathrm{L}(x)$*,*

where D *is the shift operator defined by* $(Dx_n) = (x_{n+1})$*. Let* B *be the set of all Banach limits on* ℓ_∞*.*

Definition 10.1.2. *[3] A bounded sequence* x *is said to be almost convergent to be a number* ℓ *if* $\mathrm{L}(x) = \ell$ *for all* $\mathrm{L} \in \mathrm{B}$*. Equivalently* x *is almost convergent*

DOI: 10.1201/9781003178200-10

to be ℓ if

$$lim_{n\to\infty} \frac{1}{n}\sum_{k=1}^{n} x_{k+m} = \ell$$

uniformly in m.

Note that $\hat{c}$ will denote the set of all almost convergent sequences.

Definition 10.1.3. *[11] A sequence $x = (x_n) \in \ell_\infty$ is said to be strongly almost convergent to be a number ℓ if*

$$\lim_{n\to\infty} \frac{1}{n}\sum_{k=1}^{n} |x_{k+m} - \ell| = 0$$

uniformly in m.

By $[\hat{c}]$ we denote the space of all strongly almost convergent sequences. It has been observed that $c \subset [\hat{c}] \subset \hat{c} \subset \ell_\infty$ and the inclusions are strict (see [13]).

Almost convergent sequences were studied by Duran [2], Lorentz [3], King [6], Nanda [18, 19], Savas [29, 31, 38, 40] and others.

Let σ be a mapping of the set of positive integers into itself. A continuous linear functional φ on ℓ_∞ is said to be an invariant mean or a σ-mean if and only if

(i) $\varphi(x) \geq 0$ when the sequence $x = (x_n)$ has $x_n \geq 0$ for all n,

(ii) $\varphi(e) = 1$ where $e = (1, 1, ...)$

(iii) $\varphi(x_{\sigma(n)}) = \varphi(x)$ for all $x \in \ell_\infty$.

In case σ is the translation mapping $n \to n+1$, a $\sigma-$mean is often called a Banach limit and V_σ, the set of bounded sequences all of whose invariant means are equal, is the set of almost convergent sequences which is defined in [3]. If $x = (x_n)$, set $Tx = (Tx_n) = (x_{\sigma(n)})$. It can be shown [52] that

$$V_\sigma = \{x \in \ell_\infty : \lim_{m\to\infty} t_{mn}(x) = L \text{ uniformly in } n,\ L = \sigma - \lim x\}, \tag{10.1}$$

where

$$t_{mn}(x) = (x_n + Tx_n + ... + T^n x_n)/(m+1)$$

and

$$t_{-1,n}(x) = 0$$

The special case of (12.1) in which $\sigma(n) = n+1$ was given by Lorentz.
By $V_{0\sigma}$, we denote the set of sequences all of whose σ-convergent to 0. Using the concept of σ-mean P. Schaefer [52] defined σ-conservative, σ-regular and σ-coercive matrices and obtained the conditions to characterize them which generalized almost conservative, almost regular (see, [6]) and almost coercive matrices (see, [2]), respectively. The concept of strong almost convergence was introduced by Maddox [13] who later on generalized this concept analogous to Stieglitz's extension of almost convergence.

Recently Mursaleen [17] defined strongly σ-convergent sequences, replacing the Banach limits by σ-means, in the following manner:

A bounded sequence $x = (x_k)$ is said to be strongly σ-convergent to a number L if and only if

$$\lim_p \frac{1}{p+1} \sum_{i=0}^{p} |x_{\sigma} i_{(n)} - L| = 0$$

uniformly in n. By $[V_\sigma]$ we denote the set of all strongly σ-convergent sequences.

Put

$$\phi_{mn}(x) = t_{mn}(x) - t_{m-1,n}(x)$$

assuming that $t_{mn}(x) = 0$ for $m = -1$. A straightforward calculation shows that

$$\phi_{mn}(x) = \begin{cases} \frac{1}{m(m+1)} \sum_{j=1}^{m} j(T^j x_n - T^{j-1} x_n), & (m \geq 1) \\ x_n & (m = 0). \end{cases}$$

A sequence $x = (x_k)$ is said to be absolutely σ-convergent if $\sum\limits_{m=1}^{\infty} |\phi_{mn}(x)|$ converges uniformly in n.

By ℓ^σ, we denote the set of all absolutely σ-convergent sequences, (see, [30]). Some more results concerning this convergence can be find in [17,24,25,34–36].

10.2 On Strong σ-Convergence

Savaş [30] proved the following interesting lemma and theorem.

Lemma 10.2.1. $[\sigma] - \lim x = L$, *if and only if,*

(i) $\sigma - \lim x = L$ *and*

(ii) $\frac{1}{p} \sum_{i=1}^{p} |x_{\sigma^i(n)} - t_{i-1,n}(x)| = o(1)$ *as* $p \to \infty$*, uniformly in* n.

Proof. Suppose that $[\sigma] - \lim x - L$. It follows from the inequality

$$|\frac{1}{p} \sum_{i=1}^{p} x_{\sigma^i(n)} - L| \leq \frac{1}{p} \sum_{i=1}^{p} |x_{\sigma^i(n)} - L|$$

that $\sigma - \lim x = L$. This proves (i)

Again

$$\frac{1}{p} \sum_{i=1}^{p} |x_{\sigma^i(n)} - t_{i-1,n}(x)| \leq \frac{1}{p} \sum_{i=1}^{p} |x_{\sigma^i(n)} - L| + \frac{1}{p} \sum_{i=1}^{p} |t_{i-1,n}(x) - L|$$

$= \sum_1 + \sum_2$ (say).

Now, by hypothesis, $\sum_1 = 0(1)$ as $p \to \infty$, uniformly in n. Also, (i) gives us

$$|t_{i-1,n}(x) - L| \to 0 \quad as \quad i \to \infty, \text{ uniformly in } n,$$

and hence $\sum_2 = o(1)$ as $p \to \infty$ uniformly in n. This proves (ii).

Since

$$\left(\frac{1}{p}\sum_{i=1}^{p}|x_{\sigma^i(n)} - L|\right) \leq \left(\frac{1}{p}\sum_{i=1}^{p}|x_{\sigma^i(n)} - t_{i-1,n}(x)|\right) + \left(\frac{1}{p}\sum_{i=1}^{p}|t_{i-1,n}(x) - L|\right),$$

the converse part follows immediately. □

Theorem 10.2.2. *$l^\sigma \subset [V_\sigma]$, and if $x \in [V_\sigma]$, then $[\sigma] - \lim x = \sigma - \lim x = L$.*

Proof. Suppose that $x \in l^\sigma$, hence, if we write

$$U_{k,n}(x) = \sum_{m=k}^{\infty}|t_{mn}(x) - t_{m-1,n}(x)|,$$

then $U_{k,n}(x)$ is finite for each $k \geq 1$ and $U_{k,n}(x) \to 0$ as $k \to \infty$, uniformly in n. Since every absolutely σ-convergent sequence is σ-convergent, we have $\sigma - \lim x = L$ (say). Hence, to prove the theorem, it is enough to show, (by use of the lemma 10.2.1) that

$$\frac{1}{p}\sum_{i=1}^{p}|x_{\sigma^i(n)} - t_{i-1,n}(x)| \to 0 \quad as \quad p \to \infty,$$

uniformly in n. Since for $i \geq 1$

$$x_{\sigma} i_{(n)} - t_{i-1,n}(x) = i(t_{i,n}(x) - t_{i-1,n}(x)),$$

we have

$$\begin{aligned}
\frac{1}{p}\sum_{i=1}^{p}|x_{\sigma^i(n)} - t_{i-1,n}(x)| &= \frac{1}{p}\sum_{i=1}^{p} i|t_{i,n}(x) - t_{i-1,n}(x)| \\
&= \frac{1}{p}\sum_{i=1}^{p} i(U_{i,n}(x) - U_{i-1,n}(x)) \\
&= \frac{U_{1,n}(x)}{p} - U_{p+1,n}(x) + \frac{1}{p}\sum_{i=1}^{p-1} U_{i+1,n}(x) \\
&= o(1)
\end{aligned}$$

as $p \to \infty$, uniformly in n. This completes the proof. □

Let $A = (a_{nk})$ be an infinite matrix of real or complex numbers. We write $Ax = (A_n(x))$, if $A_n(x) = \sum_k a_{nk}x_k$ converges for each n. Let X and Y be two nonempty subsets of the space s of all sequences. If $x = (x_k) \in X$ implies that $Ax = (A_n(x)) \in Y$, we say that A defines a matrix transformation from X into Y and we denote it by $A : X \to Y$. By (X, Y) we mean the class of matrices A such that $A : X \to Y$.

Now let $\mathbb{A} = A^i = (a^i_{n,k})$ be the generalized three parametric real matrix and $s = (s_k)$ be a sequence with complex terms. If the series

$$A^i_n(s) = \sum_{k=1}^{\infty} a^i_{n,k} s_k \tag{10.2}$$

is convergent for $n, i = 1, 2, ...,$ then the sequence $A^i(s) = (A^i_n(s))$ is said to be the $\mathbb{A}$-transform of the sequence $s = (s_k)$. Hence, the summability method $\mathbb{A}$ is sequence method.

The matrix $A^i = (a^i_{n,k})$ is called regular if the $\mathbb{A}$-transform of s is convergent to the limit of s for each $s \in c$.

A sequence s is said to be invariant $\mathbb{A}$-summable if

$$A^i(s) = \left(\sum_{k=1}^{\infty} a^i_{n,k} s_k \right) \in V_\sigma.$$

The matrix $\mathbb{A} = A^i = (a^i_{n,k})$ is called σ-conservative if and only if $A^i(s) = \left(\sum\limits_{k=1}^{\infty} a^i_{n,k} s_k \right) \in V_\sigma$ for all $s \in c$. When $\mathbb{A}$ is σ-conservative, we write this as $\mathbb{A} \in (c, V_\sigma)$.

The matrix $\mathbb{A} = A^i = (a^i_{n,k})$ is called σ-regular if and only if it is σ-conservative and $\sigma - \lim A^i(s) = \lim s$ for all $s \in c$.

For the purpose of characterizing the matrix mappings $A^i : [V_\sigma] \to c$ it is necessary to introduce the concept of sets which are uniform invariant zero density.

A set E of positive integers is said to be uniform invariant density of X if and only if the number of elements of E which lie in the interval $[\sigma^0(n), \sigma^{p-1}(n)]$ is $o(p)$ as $p \to \infty$, uniformly in $n \geq 0$.
We have

Theorem 10.2.3. *Let $\mathbb{A} = A^i = (a^i_{nk})$ be the generalized three parametric complex matrix and $x = (x_k) \in [V_\sigma]$. Then $\mathbb{A} \in ([V_\sigma], c)$ if and only if*

(i) $\mathbb{A}$ is regular

(ii) $\sum_{k \in E_\sigma} |\Delta a^i_{nk}| \to 0$ as $n \to \infty$,
for each set E_σ of uniform invariant zero density and where

$$\Delta a^i_{nk} = a^i_{n,k} - a^i_{n,k+1}.$$

Proof. Sufficiency. Since $\mathbb{A}$ is regular there without loss of generality in supposing that x is strongly σ-convergent to 0. It is then sufficient to prove that $\mathbb{A}$−transform sequence of $x \in [V_\sigma]$ converges to 0. If $\varepsilon > 0$, then there is a natural number p such that

$$\frac{1}{p}\sum_{j=0}^{p-1}|x_{\sigma^j(k)}|< \varepsilon, \tag{10.3}$$

for all $k \geq 0$. Let

$$E_\sigma = \{x : |x_k|\geq \varepsilon/p\},$$

then E_σ is uniformly of invariant zero density. By using Lorentz's method, it can be verified that

$$\sum_{k=0}^{\infty} a^i_{n,k}(x_{\sigma^0(k)} + x_{\sigma^1(k)} + ... + x_{\sigma^{p-1}(k)}/p) \tag{10.4}$$

$$= \frac{1}{p}\sum_{k=0}^{p-1} a^i_{n,k} \sum_{j=0}^{p-(k+1)} x_{\sigma^j(k)} + \frac{1}{p}\sum_{j=1}^{\infty}\sum_{r=j}^{p+j-1} a_{nr}x_{\sigma^{p-r+j-1}(r)}.$$

Since $x \in \ell_\infty$ and $\mathbb{A}$ is regular, we have

$$\sum_{k=0}^{p-1} a^i_{n,k} \sum_{j=0}^{p-(k+1)} x_{\sigma^j(k)} = o(1) \quad as \quad n \to \infty, \quad for \quad each \quad k = 0, ..., p-1$$

We note that

$$\frac{1}{p}\sum_{j=1}^{\infty}\sum_{r=j}^{p+j-1} a^i_{n,r}x_{\sigma^{p-r+j-1}(r)} = \frac{1}{p}\sum_{j=p}^{\infty}\sum_{r=j-p+1}^{j} a^i_{n,r}x_{\sigma^{j-r}(r)} \tag{10.5}$$

$$= \frac{1}{p}\sum_{j=p}^{\infty}\sum_{r=j-p+1}^{j} (a^i_{n,r} - a^i_{n,j})x_{\sigma^{j-r}(r)} + \frac{1}{p}\sum_{j=p}^{\infty}\sum_{r=j-p+1}^{j} x_{\sigma^{j-r}(r)}.$$

Therefore, we get from (10.4) and (10.5) that

$$\frac{1}{p}\sum_{j=1}^{\infty} a^i_{n,j} \sum_{r=j-p+1}^{j} x_{\sigma^{j-r}(r)} = \frac{1}{p}\sum_{k=0}^{\infty} a^i_{n,r} \sum_{j=0}^{p-1} x_{\sigma^j(k)} \tag{10.6}$$

$$+\frac{1}{p}\sum_{j=1}^{p-1} a^i_{n,r} \sum_{r=j-p+1}^{j} x_{\sigma^{j-r}(r)} - \frac{1}{p}\sum_{j=p}^{\infty}\sum_{r=j-p+1}^{j} (a^i_{n,r} - a^i_{n,j})x_{\sigma^{j-r}(r)} + o(1).$$

Again for all i as $a^i_{n,j} = o(1)$ as $n \to \infty$ for fixed j, we obtain

$$\frac{1}{p}\sum_{j=1}^{p-1} a^i_{n,j} \sum_{r=j-p+1}^{j} x_{\sigma^{j-r}(r)} = o(1) \quad as \quad n \to \infty, j = 1, ..., p-1.$$

Now since $\|\mathbb{A}\| = \sup_n \sum_{k=0}^{\infty} |a^i_{n,k}| < \infty$, it follows from (10.2) that

$$\frac{1}{p}|\sum_{k=0}^{\infty} a^i_{n,j}(x_{\sigma^0(k)} + x_{\sigma^1(k)} + ... + x_{\sigma^{p-1}(k)})| \leq \varepsilon. \|A\|.$$

Furthermore,

$$\frac{1}{p}\sum_{j=p}^{\infty} \sum_{r=j-p+1}^{j} (a^i_{n,r} - a^i_{n,j})x_{\sigma^{j-r}(r)} = \frac{1}{p}\sum_{j=p}^{\infty} \sum_{r=j-p+1}^{p} x_{\sigma^{j-r}(r)} \sum_{\theta=r}^{j} \Delta a^i_{n,\theta}.$$

Hence for $j - p + 1 \leq \mu \leq j$,

$$\frac{1}{p}|\sum_{j=p}^{\infty} x_{\sigma^{j-\mu}(\mu)} \sum_{\theta=\mu}^{j} \Delta a^i_{n,\theta}| \leq \|x\| \sum_{k \in E_\sigma} |\Delta a^i_{n,j}| + \varepsilon p^{-1} \sum_{j=1}^{\infty} |\Delta a^i_{n,j}|.$$

Therefore, combining the above facts, we get from (10.6) that

$$|\sum_{j=1}^{\infty} x^{j-\mu}_{\sigma}(\mu) a^i_{n,j}| \leq o(1) + \varepsilon \|A\| p + p \|x\| \sum_{k \in E_\sigma} |\Delta a^i_{n,j}| + \varepsilon \sum_{j=1}^{\infty} |\Delta a^i_{n,j}|.$$

Since ε is arbitrary, by hypothesis

$$\sum_{j=1}^{\infty} x^{j-\mu}_{\sigma}(\mu) a^i_{n,j} \to 0 \text{ as } n \to \infty \text{ for each } j - p + 1 \leq \mu \leq j.$$

This completes the proof of sufficiency.

Necessity. Since $x_k \to s$ implies $x_k \to s[V_\sigma]$, we have that $\mathbb{A}$ is regular. Now as in Maddox [13] let E_σ be a set of uniform invariant zero density. For $x \in \ell_\infty$, define $y_k = x_k$, if $k \in E_\sigma$ and $y_k = 0$, otherwise. Also define

$$u_1 = y_1 \quad and \quad u_k = y_k - y_{k-1}, \quad for \quad k > 1.$$

Now, for any p and n let $e_{\sigma^0(s)}, e_{\sigma^1(s)}, ..., e_{\sigma^t(s)}$ be the elements of E_σ lying in the interval $[\sigma^0(n), \sigma^{p-1}(n)]$. Since E_k is uniform invariant zero density. We have

$$\frac{|u_{\sigma^0(k)}| + u_{\sigma^1(k)}| + ... + |u_{\sigma^{p-1}(k)}|}{p} \leq \frac{2\|x\|}{p}(t+2) \to 0$$

as $p \to \infty$ uniformly in k. Hence, $u_k \to 0[V_\sigma]$ so by hypothesis

$$\sum_k a^i_{n,k} u_k \to 0 \quad as \quad n \to \infty.$$

But

$$\sum_k a^i_{n,k} u_k = \sum_k (\Delta a_{nk}) y_k = \sum_{k \in E_\sigma} (\Delta a^i_{n,k}) x_k.$$

Hence for $x \in \ell_\infty$

$$\sum_{k \in E_\sigma} (\Delta a^i_{n,k}) x_k \to 0 \quad as \quad n \to \infty$$

and therefore by Schur's theorem

$$\sum_{k \in E_\sigma} |\Delta a^i_{n,k}| \to \infty \quad as \quad n \to \infty.$$

This completes the proof. □

10.3 σ-Regular Dual Summability Methods

In this section, we define $\sigma-$regularity for a series method and determined necessary and sufficient conditions for a series method to be $\sigma-$regular and also investigated $\sigma-$regularity of dual summability methods. Throughout $\sum\limits_k$ denotes summation over k from $k = 1$ to $k = \infty$. It is easy to prove the following theorem, so omitted.

Theorem 10.3.1. *Let $\mathbb{A} = A^i = (a^i_{nk})$ be the generalized three parametric real matrix. The matrix $\mathbb{A}$ is σ-regular if and only if*

(i) $\|\mathbb{A}\| = \sup_{n,i}(\sum_k |a^i_{n,k}|) < +\infty$,

(ii) $a_{(k)} = (a^i_{n,k})_{n=1}^{\infty} \in V_\sigma$ *with σ-limit = zero for each k, uniformly in i*

(iii) $a = (\sum\limits_k a^i_{n,k})_{n=1}^{\infty} \in V_\sigma$ *with σ-limit = 1.*

The following result may easily be obtained:

Corollary 10.3.2. *The matrix $A = (a^i_{n,k})$ transforms c_0 into $V_{0\sigma}$ if and only if*

(i) $\|A\| = \sup_n(\sum_k |a^i_{n,k}|) < +\infty$,

(ii) $a_{(k)} \in V_\sigma$ *with σ-limit = zero for each k.*

10.3.1 Dual summability methods

Let $\mathbb{A} = (a^i_{n,k})$ be a sequence method given by (10.2). Suppose that, for each n, i, the series

$$\sum_k a^i_{n,k}$$

is convergent; this is a much weaker assumption than the regularity of $\mathbb{A}$. Then we shall define the matrix $\mathbb{B} = (b^i_{n,k})$ as follows:

$$b^i_{n,k} = \sum_{j=k}^{\infty} a^i_{n,j}.$$

We also suppose that the sequence (s_k) is formed by taking the partial sums of the series $\sum_k x_k$.

Let $\mathbb{B}$ denote the summability method given by the series to sequence transformation(series method)

$$r_n = \sum_{k=1}^{\infty} b^i_{n,k} x_k (n = 1, 2, ...). \tag{10.7}$$

The methods $\mathbb{A}$ and $\mathbb{B}$ are called dual summability methods (see, [8]). It is well-known that $\mathbb{A}$ is regular if and only if $\mathbb{B}$ is regular (see, [1]).

10.3.2 σ-Regular summability methods

In [42], Savas defined σ-regular dual summability methods. We now extend this definition by using the sequence of matrices.

Definition 10.3.3. *Let $\sum_k x_k$ be an infinite series such that its sequence of partial sums is convergent to the value a. If the method $\mathbb{B}$ invariant sums every series of this type to the same value a, then the series method $\mathbb{B}$ is said to be σ-regular.*
Now, let

$$\gamma = \{x = (x_k) : s_n = \sum_{k=1}^{n} x_k, (s_n) \in c\}.$$

If $\sigma - \lim \mathbb{B}(x) = \lim s$, for all $x \in \gamma$, then we write $\mathbb{B} \in (\gamma, V_\sigma)_{reg}$. Therefore, $\mathbb{B}$ is σ-regular if and only if $\mathbb{B} \in (\gamma, V_\sigma)_{reg}$.

Theorem 10.3.4. *The matrix $\mathbb{B} = (b^i_{n,k})$ is σ-regular if and only if*

(i) $\|B\| = \sup_{n,i} \left(\sum_k |\Delta b^i_{n,k}| \right) < +\infty,$

(ii) $b_k = \left(b^i_{n,k}\right)_{n=1}^{\infty} \in V_\sigma$ *with σ-limit =1 for each k, uniformly in i, where* $\Delta b^i_{n,k} = b^i_{n,k} - b^i_{n,k+1}$.

Proof. Necessity. Let s_n be nth partial sum of the series $\sum_k x_k$ and $\lim s = a$. Then $\mathbb{B}(x) = \sum_k b^i_{n,k} x_k$ exists for each n, i and for all $x \in \gamma$ and also $\sigma - \lim \mathbb{B}(x) = a$. If we put $x = e_k$, then the necessity of (ii) is easily obtained, since $e_k \in \gamma$ for such k, where e_k is the sequence whose kth component is one and the otherwise zero. We show that (i) is necessary.

By Abel's partial summation, we get

$$\sum_{k=1}^{m} b^i_{n,k} x_k = \sum_{k=1}^{m-1} \Delta b^i_{n,k}(s_k - a) + ab^i_{n,1} + (s_m - a)b^i_{n,m} \tag{10.8}$$

for each n, i, where $(x_k) \in \gamma$ and (s_m) is the mth partial sum of the series $\sum_k x_k$ and $\lim s = a$. On the other hand it is well-known that, if $\sum_k b^i_{n,k} x_k$ converges for each n, i, whenever $x = (x_k) \in \gamma$, then

$$\sup_k |b^i_{n,k}| < \infty, (n, i = 1, 2, ...). \tag{10.9}$$

Hence, letting $m \to \infty$ in (10.8), we get from (10.9) that

$$\mathbb{B}(x) = \sum_k b^i_{n,k} x_k = \sum_k \Delta b^i_{n,k}(s_k - a) + ab^i_{n,1}.$$

Then we must have

$$(\sum_k \Delta b^i_{n,k}(s_k - a)) \in V_{0\sigma}.$$

Since $\sigma - \lim \mathbb{B}(x) = a$ and $\sigma - \lim b^i_{n,1} = 1$. On the other hand, obviously, $(s_k - a) \in c_0$. Therefore, we get

$$C = (\Delta b^i_{n,k}) \in (c_0, V_{0\sigma}).$$

Now Corollary 10.3.2 yields the condition (i).

Sufficiency. Let us suppose that the conditions (i) and (ii) hold. Let $x = (x_k) \in \gamma$ and $s_m = \sum_{k=1}^{m} x_k$ and $\lim s = a$. From (i), the series $\sum_k (b^i_{n,k} - b^i_{n,k+1})$ is convergent and $\lim_m b^i_{n,m}$ exist for each n, i. So the statement (10.8) holds for every $x \in \gamma$ with $\lim s = a$. By (i) and (ii) it is easily seen that

$$C = (\Delta b^i_{n,k}) \in (c_0, V_{0\sigma}).$$

Since the conditions of Corollary 10.3.2 are satisfied. Therefore, the statement (10.8 implies that $\sigma - \lim B(x) = a$. This yields that B is $\sigma-$regular and the proof is completed. □

Theorem 10.3.5. *Let $\mathbb{A}$ and $\mathbb{B}$ be two dual summability methods. Then $\mathbb{A}$ is $\sigma-$regular if and only if $\mathbb{B}$ is $\sigma-$regular.*

Proof. We write

$$b^i_{n,k} = \sum_{i=k}^{\infty} b^i_{n,j}.$$

Then $\Delta b^i_{n,k} = b^i_{n,k} - b^i_{n,k+1} = a^i_{n,k}$ for all n, k, i. Hence $\sup_{n,i} \sum_k |a^i_{n,k}| < \infty$ if and only if $\sup_{n,i} \sum_k |\Delta b^i_{n,k}| < \infty$. Furthermore, $\sigma - \lim \sum_k a^i_{n,k} = 1$ and $\sigma - \lim a^i_{n,k} = 0$, for each k, if and only if $\sigma - \lim b^i_{n,k} = 1$, for each k, uniformly in i. Now the proof follows from Theorem 10.3.1 and Theorem 10.3.4 and the proof is completed. □

In [22], Orhan defined the strong almost regularity of a series method and determined the necessary and sufficient conditions for a series method $\mathbb{B}$ to be strong almost regular. Strong $\sigma-$regularity of a series method and investigate strong $\sigma-$regularity of dual summability methods were defined in [39]. We now generalize strong $\sigma-$regularity of a series method by using matrices sequences as follows: Now let $\mathbb{B} = B^i = (b^i_{nk})$ be a series method. We write the following:

Definition 10.3.6. *Let $\sum_k x_k$ be an infinite series with the partial sums (s_n) and let (s_n) be strongly σ-convergent to s. If the method B sums every series of this type to s, then the series method B is said to be strongly σ-regular.*

Let

$$\gamma_{[V_\sigma]} = \{x = (x_k) : s_n = x_1 + ...x_n, \quad \text{and} \quad (s_n) \in [V_\sigma]\}.$$

Hence the inclusion $\gamma \subset \gamma_{[V_\sigma]}$ is strict where γ denotes the class of all convergent series. We may show that $\gamma_{[V_\sigma]}$ is linearly isomorphic to $[V_\sigma]$ with the usual linear operations in sequence spaces. Therefore, $\mathbb{A}$ is strongly σ-regular means that $\mathbb{A} \in (\gamma_{[V_\sigma]}, c)_{reg}$. The following theorem gives the necessary and sufficient conditions for a series method to be strongly σ-regular.

Theorem 10.3.7. *A series method $\mathbb{B} = (b^i_{nk})$ is strongly σ-regular if and only if,*

(i) *B is regular,*

(ii) $\lim_k b^i_{nk} = 0$ *for every n, uniformly in i,*

(iii) $\sum_{k \in E_\sigma} |\Delta^2(b^i_{nk})| \to 0$ *as $n \to \infty$, uniformly in i,*
for each set E_σ of uniform invariant zero density, where $\Delta^2 b^i_{nk} = \Delta(\Delta b^i_{nk}) = \Delta(b^i_{nk} - b^i_{nk+1})$.

Proof. Suppose first that $\mathbb{B}$ is strongly σ-regular. Let $\sum_k x_k$ be an infinite series with the sequence of partial sums (s_n) and let $s_n \to s$. Then we have $s_n \to s[V_\sigma]$. Hence we must have that $\mathbb{B}$ is regular, which proves (i).

To prove (ii), consider the sequence (z_k) defined by $(z_k) = (-1)^k$ for $k \in E_\sigma$ and $z_k = 0$ for $k \notin E_\sigma$ where E_σ is the set of uniform invariant zero density. It can easily be shown that (z_k) fails to converge but $z_k \to 0[V_\sigma]$. We can find a series $\sum_k x_k$ such that its sequence of partial sums is (z_k). But $\mathbb{B}$ have to sum this series to zero. Using [14, Theorem 16], we see that (ii) holds.

By the Abel's partial summation we have, for all i,

$$\sum_{k=1}^{m} b^i_{nk} x_k = \sum_{k=1}^{m-1} \Delta b^i_{nk} s_k + s_m b^i_{nm} \tag{10.10}$$

using (ii) and letting $m \to \infty$ in (10.10). We get that

$$\sum_{k=1}^{\infty} b^i_{nk} x_k = \sum_{k=1}^{\infty} \Delta b^i_{nk} s_k \tag{10.11}$$

which shows that the sequence method $\mathbb{C} = (\Delta b^i_{nk})$ is strongly σ-regular. It follows from Theorem 10.2.3 that

$$\sum_{k \in E_\sigma} |\Delta(\Delta b^i_{nk})| \to 0 \quad as \quad n \to \infty, \text{ uniformly in } i,$$

for each set E_σ of uniform invariant zero density. This proves (iii).

To prove the converse, take an infinite series $\sum_k x_k$ for which $s_n \to s[V_\sigma]$. Conditions (i)–(iii) give us (10.8).

$$\lim_n \sum_k \Delta b^i_{nk} s_k = s$$

which yields that $\lim_n \sum_k b^i_{nk} x_k = s$ uniformly in i. Hence the proof is completed. □

Theorem 10.3.8. *Let $\mathbb{A}$ and $\mathbb{B}$ be two dual summability methods. Then $\mathbb{A}$ is strongly σ-regular if and only if $\mathbb{B}$ is strongly σ-regular.*

Proof. This follows from Theorem 10.2.3 and the present Theorem 10.3.7. □

10.4 Some New Sequence Spaces

Let $\mathbb{A} = A^i = (a^i_{n,k})$ be the generalized three parametric real matrix and let $p = \{p_k\}$ be a sequence of positive real numbers. We write $A^i_n(x) = \sum_k |x_k|^{p_k}$ if the series converges for each n, i and $t_{mni}(x) = \sum_k a^i_{mk} |x_{\sigma^k(n)}|^{p_k}$, if the series converges for each m, n and i. Here and afterwards summation without limits runs from 1 to ∞. We define

$$[\mathbb{A}_\sigma, p]_0 = \{x : t_{mni}(x) \to 0 \text{ as } m \to \infty, \text{ uniformly in } (n,i)\},$$

$$[\mathbb{A}_\sigma, p] = \{x : t_{mni}(x - Le) \to 0 \text{ as } m \to \infty \text{ for some } L, \text{ uniformly in } (n,i)\},$$

and

$$[\mathbb{A}_\sigma, p]_\infty = \{x : \sup_{m,n,i} t_{mni}(x) < \infty\},$$

which we call the spaces of strongly σ-summable to zero, strongly σ-summable and strongly σ-bounded sequences respectively. If $\sigma(n) = n+1$, we have

$$[\mathbb{A}_\sigma, p]_0 = [\hat{\mathbb{A}}, p]_0, [\mathbb{A}_\sigma, p] = [\hat{\mathbb{A}}, p] \quad \textit{and} \quad [\mathbb{A}_\sigma, p]_\infty = [\hat{\mathbb{A}}, p]_\infty.$$

If $[V_\sigma]$ denotes the set of all strongly σ-convergent sequences, then (see, Mursaleen [17])

$$[V_\sigma] = \left\{ x : \lim_{m\to\infty} \frac{1}{m} \sum_{k=1}^{m} |x_{\sigma^k(n)} - L| = 0 \text{ uniformly in } n \right\}.$$

Note that if $\mathbb{A} = (C,1)$ for all i and $p_k = 1$ for all k, then $[\mathbb{A}_\sigma, p] = [V_\sigma]$. It is evident that $[\mathbb{A}_\sigma, p]_0 \subset [\mathbb{A}_\sigma, p]$ and $[\mathbb{A}_\sigma, p]_0 \subset [\mathbb{A}_\sigma, p]_\infty$. But we have been able to prove "$[\mathbb{A}_\sigma, p] \subset [\mathbb{A}_\sigma, p]_\infty$" only under additional restriction on $\mathbb{A}$.

Proposition 10.4.1. $[\mathbb{A}_\sigma, p] \subset [\mathbb{A}_\sigma, p]_\infty$ *if*

$$\|\mathbb{A}\| = sup_{m,n,i} \sum_k a^i_{mk} < \infty. \tag{10.12}$$

Proof. If $H = \sup p_k$ and $K = \max(1, 2^{H-1})$, we have [see, Maddox [14], p. 346].

$$|a_k + b_k|^{p_k} \leqq K(|a_k|^{pk} + |b_k|^{pk}) \tag{10.13}$$

and for $\lambda \in \mathbb{C}$,

$$|\lambda|^{p_k} \leqq \max(1, |\lambda|^H). \tag{10.14}$$

Suppose that $x_k \to L[\mathbb{A}_\sigma, p]$ and (10.12) holds. Now by the inequality (10.13), we have

$$\begin{aligned} t_{mni}(x) &= t_{mni}(x - Le + Le) \\ &\leqq Kt_{mni}(x - Le) + K\sum_k a^i_{mk} |L|^{p_k} \\ &\leqq Kt_{mni}(x - Le) + K(\sup |L|^{p_k}) \sum_k a^i_{mk}. \end{aligned}$$

Therefore, $x \in [A_{(\lambda,\sigma)}, p]_\infty$ and this completes the proof.

□

Theorem 10.4.2. *If $p \in \ell_\infty$, then $[\mathbb{A}_\sigma, p]_0, [\mathbb{A}_\sigma, p]$ and $[\mathbb{A}_\sigma, p]_\infty$ are linear spaces over the complex field $\mathbb{C}$.*

Proof. We consider only $[\mathbb{A}_\sigma, p]$. If $H = \sup p_k$ and $K = \max(1, 2^{H-1})$. Now the linearity follows from (10.13) and (10.14). □

Definition 10.4.3. *If x is strongly σ-summable to L we write $x_k \to L[\mathbb{A}_\sigma, p]$. A pair $(\mathbb{A}, p)$ will be called strongly σ-regular if $x_k \to L$ implies that $x_k \to L[\mathbb{A}_\sigma, p]$.*

We now characterize the class of strongly σ-regular. It is known that $\mathbb{A} \in (c_0, c_0)$ if and only if $\|\mathbb{A}\| = \sup_{m,i} \sum_k |a^i_{mk}| < \infty$ and $a^i_{mk} \to 0$ $(m \to \infty, k$ fixed, uniformly in i fixed).

Theorem 10.4.4. *Suppose that $\mathbb{A} \in (c_0, c_0)$ and $p = \{p_k\}$ converges to a positive limit. Then $x_k \to L, x_k \to L[\mathbb{A}_\sigma, p]$ uniquely if and only if,*

$$\sum_k a^i_{mk} \to 0 \quad (m \to \infty), \text{ uniformly in } i. \tag{10.15}$$

Proof. Necessity. Suppose that $\mathbb{A} \in (c_0, c_0)$ and (p_k) is bounded. Let $x_k \to L$ imply that $x_k \to L[\mathbb{A}_\sigma, p]$ uniquely. We have $e \to L[\mathbb{A}_\sigma, p]$. Therefore condition (10.14) must hold, for otherwise $e \to 0[\mathbb{A}_\sigma, p]$ which is contradicts the uniqueness of L.

Sufficiency. Suppose that condition (10.15) holds, $\mathbb{A} \in (c_0, c_0)$ and $p_k \to r > 0$. Further assume that $x_k \to L$ implies $x_k \to L[\mathbb{A}_\sigma, p]$ and $x_k \to L'[\mathbb{A}_\sigma, p]$ where $|L - L'| = a > 0$. Then we get

$$\lim_{m\to\infty} \sum_k a^i_{mk} u_{nk} = 0 \quad (\text{uniformly in } (n, i)), \tag{10.16}$$

where

$$u_{nk} = |x_{\sigma^k}(n) - L|^{p_k} + |x_{\sigma^k(n)} - L'|^{p_k}.$$

By the assumption $u_{nk} \to a^r$. Since $\mathbb{A} \in (c_0, c_0), u_{nk} \to a^r$ implies that

$$\sum_k a^i_{mk} |u_{nk} - a^r| \to 0 \quad as \quad m \to \infty \quad (\text{uniformly in } (n, i)). \tag{10.17}$$

But we have

$$a^r \sum_k a^i_{mk} \le \sum_k a^i_{mk} u_{nk} + \sum_k a^i_{mk} |u_{nk} - a^r|. \tag{10.18}$$

By (10.16), (10.17) and (10.18) it follows that

$$\lim_{m\to\infty} \sum_k a^i_{mk} = 0. \tag{10.19}$$

Since this contradicts (10.15), we must have $L = L'$ and this completes the proof. □

Theorem 10.4.5. *Let* $0 < \theta = \inf p_k \leq p_k \leq H < \infty$. *Then* $(\mathbb{A}, p)$ *is strongly* $\sigma-$*regular if and only if* $\mathbb{A} \in (c_0, c_0)$.

Proof. Necessity. Suppose that $(\mathbb{A}, p)$ is strongly σ-regular. Therefore,

$$|x_k - L|^{1/p_k} \to 0 \to \sum_k a^i_{mk} |x_{\sigma^k(n)} - L| \to 0 \quad as \quad m \to \infty$$

uniformly in (n, i). Again since $1/p_k \geq 1/H > 0$, therefore we have

$$x_k \to L \Rightarrow |x_k - L|^{1/p_k} \to 0.$$

Thus

$$x_k \to L \Rightarrow \sum_k a^i_{mk} (x_{\sigma^k(n)} - L) \to 0 \quad as \quad m \to \infty$$

uniformly in (n, i), and therefore $\mathbb{A} \in (c_0, c_0)$.

Sufficiency. Since $p_k \geq \theta > 0$ we have

$$x_k \to L \Rightarrow |x_k - L|^{p_k} \to 0$$

Again we have $\mathbb{A} \in (c_0, c_0)$. Therefore, $x_k \to L[\mathbb{A}_\sigma, p]$ and this completes the proof. □

We now study locally boundedness and $r-$convexity for $[\mathbb{A}_\sigma, p]_0, [\mathbb{A}_\sigma, p]_\infty$ and $[\mathbb{A}_\sigma, p]$. We start with some definitions.

For $0 < r \leq 1$, a nonvoid subset U of a linear space is said to be absolutely $r-$convex if $x, y \in U$ and $|\lambda|^r + |\mu|^r \leq 1$ together imply that $\lambda x + \mu y \in U$. A linear topological space X is said to be r-convex if every neighbourhood of $0 \in X$ contains an absolutely $r-$convex neighbourhood of $0 \in X$. The r-convexity for $r > 1$ is of little interest, since X is r-convex for $r > 1$ if and only if X is the only neighbourhood of $0 \in X$, (see, Maddox and Roles [15]). A subset B of X is said to be bounded if for each neighbourhood U of X is said to be bounded if for each neighbourhood U of $0 \in X$ there exists an integer $N > 1$ such that $B \subset NU$. X is called locally bounded if there is a bounded neighbourhood of zero.

Theorem 10.4.6. *Let* $0 < p_k \leq 1$. *Then* $[\mathbb{A}_\sigma, p]_0$ *and* $[\mathbb{A}_\sigma, p]_\infty$ *are locally bounded if* $\inf p_k > 0$. *If* (10.12) *holds then* $[\mathbb{A}_\sigma, p]$ *has the same property.*

Proof. We consider only $[\mathbb{A}_\sigma, p]_\infty$. Let $\inf p_k = \theta > 0$. If $x \in [\mathbb{A}_\sigma, p]_\infty$ then there exists a constant $K > 0$ such that

$$\sum_k a_{mk}(i) |x_{\sigma^k(n)}|^{p_k} \leq K \quad (\forall m, n, i).$$

For this K and given $\delta > 0$, choose an integer $N > 1$ such that

$$N^\theta \geq K/\delta.$$

Since $N > 1$ and $p_k \geq \theta$, we have

$$\frac{1}{N^{p_k}} \leq \frac{1}{N^\theta} \quad (\forall k).$$

Then for all m, n and i, we get

$$\begin{aligned} \sum_k a_{mk}(i)|\frac{x_{\sigma^k(n)}}{N}|^{p_k} &\leq \frac{1}{N^\theta}\sum_k a_{mk}(i)|x_{\sigma^k(n)}|^{p_k} \\ &\leq \frac{K}{N^\theta} \leq \delta. \end{aligned}$$

Therefore by taking supremum over m, n and i we have

$$\{x : h(x) \leq K\} \subseteq N\{x : h(x) \leq \delta\}.$$

For every $\delta > 0$ there exists $N > 1$ for which the above inclusion holds and so

$$\{x : h(x) \leq K\}$$

is bounded. This completes the proof. □

Remark 10.4.7. *It is known that every locally bounded linear topological space is r-convex for some r such that $0 < r \leq 1$. But the following theorem gives exact conditions for r-convexity.*

Theorem 10.4.8. *Let $0 < p_k \leq 1$. Then $[\mathbb{A}_\sigma, p]_0$ and $[\mathbb{A}_\sigma, p]_\infty$ are r-convex for all r, where $0 < r < \liminf p_k$. Moreover, if $p_k = p \leq 1, \forall k$, then they are p-convex, $[\mathbb{A}_\sigma, p]$ has the same properties if condition (10.12) holds.*

Proof. We shall prove the theorem only for $[\mathbb{A}_\sigma, p]_\infty$. Let $x \in [\mathbb{A}_\sigma, p]_\infty$ and $r \in (0, \liminf p_k)$. Then there exists k_0 such that $r \leq p_k (\forall k > k_0)$. Now define,

$$g(x) = \sup_{m,n,i} [\sum_{k=1}^{k_0} a_{mk}(i)|x_{\sigma^k(n)}|^r + \sum_{k=k_{0+1}}^{\infty} a_{mk}(i)|x_{\sigma^k(n)}|^{p_k}].$$

Since $r \leq p_k \leq 1 \quad (\forall k > k_0)$, g is subsadditive. Further for $0 < |\lambda| \leq 1$,

$$|\lambda|^{p_k} \leq |\lambda|^r \quad (\forall k > k_0).$$

Therefore, for such λ we have

$$g(\lambda x) \leq |\lambda|^r g(x).$$

If $0 < \delta < 1$,

$$U = \{x : g(x) \leq \delta\}$$

is an absolutely r-convex set, for $|\lambda|^r + |\mu|^r \leq 1$ and $x, y \in U$ imply that

$$g(\lambda x + \mu y) \leq g(\lambda x) + g(\mu y) \leq |\lambda|^r g(x) + |\mu|^r g(y)$$

$$(|\lambda|^r + |\mu|^r)\delta \le \delta.$$

If $p_k = p, \forall k$, then for $0 < \delta < 1$,

$$V = \{x : g(x) \le \delta\}$$

is an absolutely p-convex set. This can be obtained by a similar analysis and therefore we omit details. This completes the proof. □

Now, let $\mathbb{A} = A^i = (a^i_{n,k})$ be the generalized three parametric real matrix and let $\{p_k\}$ be a bounded sequence of positive real numbers. We write $T_{mni}(x) = \sum_k a^i_{mk} x_{\sigma^k(n)}$ if the series converges for each m and n, i. We define

$$(\mathbb{A}_\sigma, p)_0 = \{x : |T_{mni}(x)|^{p_m} \to 0 \text{ as } m \to \infty, \text{ uniformly in } (n, i)\},$$

$$(\mathbb{A}_\sigma, p) = \{x : |T_{mni}(x - Le)|^{p_m} \to 0 \text{ as } m \to \infty \text{ for some } L \text{ uniformly in } (n, i)\},$$

$$(\mathbb{A}_\sigma, p)_\infty = \{x : sup_{m,n,i} |T_{mni}(x)|^{p_m} < \infty\}.$$

If $\sigma(n) = n + 1$, the spaces $(\mathbb{A}_\sigma, p)_0, (\mathbb{A}_\sigma, p)$ and $(\mathbb{A}_\sigma, p)_\infty$ reduce to $(\hat{\mathbb{A}}, p)_0, (\hat{\mathbb{A}}, p)$ and $(\hat{\mathbb{A}}, p)_\infty$, respectively. Note that if $\mathbb{A} = (C, 1)$ for all i, then $(\mathbb{A}_\sigma, p)_0 = V_{0\sigma}(p), (\mathbb{A}_\sigma, p) = V_\sigma(p)$ and $(\mathbb{A}_\sigma, p)_\infty = m_\sigma(p)$. The spaces $V_{0\sigma}(p), V_\sigma(p)$ and $m_\sigma(p)$ were introduced and discussed in [30, 33] respectively. We may, without loss of generality, assume that $0 < p_m \le 1$, for if $0 < p_m < \infty$ and $\sup_m p_m < \infty$, then $0 < \frac{p_m}{\sup p_m} \le 1$.

The following theorem gives inclusion relations among the spaces $(\mathbb{A}_\sigma, p)_0, (\mathbb{A}_\sigma, p)$ and $(\mathbb{A}_\sigma, p)_\infty$.

Theorem 10.4.9. *$(\mathbb{A}_\sigma, p)_0 \subset (\mathbb{A}_\sigma, p), (\mathbb{A}_\sigma, p)_0 \subset (\mathbb{A}_\sigma, p)_\infty$ and $(\mathbb{A}_\sigma, p) \subset (\mathbb{A}_\sigma, p)_\infty$ if*

$$\sup_{m,i} \left| \sum_k a^i_{mk} \right|^{p_m} < \infty. \tag{10.20}$$

Proof. Clearly $(\mathbb{A}_\sigma, p)_0 \subset (\mathbb{A}_\sigma, p)$ and $(\mathbb{A}_\sigma, p)_0 \subset (\mathbb{A}_\sigma, p)_\infty$. Let $x \in (\mathbb{A}_\sigma, p)$. Then there in some L such that $|T_{mni}(x - Le)|^{p_m} \to 0$. Now by the inequalities (10.13) and (10.14), we have

$$\begin{aligned} |T_{mni}(x)|^{p_n} &= |T_{mni}(x - Le) + T_{mni}(Le)|^{p_m} \\ &\le |T_{mni}(x - Le)|^{p_m} + |T_{mni}(Le)|^{p_m} \\ &= |T_{mni}(x - Le)|^{p_m} + |L \sum_k a^i_{mk}|^{p_m} \\ &\le |T_{mni}(x - Le)|^{p_m} + \max(1, |L|) |\sum_k a^i_{mk}|^{p_n}. \end{aligned} \tag{10.21}$$

Now suppose that (10.20) holds. Then it follows from (10.21) that $x \in (\mathbb{A}_\sigma, p)_\infty$ and thus $(\mathbb{A}_\sigma, p) \subset (\mathbb{A}_\sigma, p)_\infty$. Now if $(\mathbb{A}_\sigma, p) \subset (\mathbb{A}_\sigma, p)_\infty$ then $e \in (\mathbb{A}_\sigma, p)_\infty$ i.e. $Ae \in \ell_\infty(p)$ and this is (10.20). This completes the proof. □

It is easy to verify that $(\mathbb{A}_\sigma, p)_0$ is a linear topological space paranormed by h defined by

$$h(x) = \sup_{m,n,i} |T_{mni}(x)|^{p_m}$$

$(\mathbb{A}_\sigma, p)_\infty$ is paranormed by h if $\inf p_m > 0$. If further (10.20) holds, then $(\mathbb{A}_\sigma, p)$ also has the same paranorm. Moreover $(\mathbb{A}_\sigma, p)_0$ and $(\mathbb{A}_\sigma, p)_\infty$ are complete in their paranorm topologies and $(\mathbb{A}_\sigma, p)$ is complete if (10.21) holds;

$$\left|\sum_k a^i_{mk}\right|^{p_m} \to 0 \quad as \quad m \to \infty, \text{ uniformly in } i. \tag{10.22}$$

Theorem 10.4.10.

(i) $(\mathbb{A}_\sigma, q)_0 \subset (\mathbb{A}_\sigma, p)_0$ *if*

$$\liminf \frac{p_m}{q_m} > 0 \tag{10.23}$$

(ii) If $0 < p_m \leq q_m \leq 1$, *then* $(\mathbb{A}_\sigma, q)_\infty$ *is a closed sub-space of* $(\mathbb{A}_\sigma, p)_\infty$.

Proof. (i) Suppose that (10.23) holds and that $x \in (\mathbb{A}_\sigma, q)_0$. Then there is $\alpha > 0$ such that $p_m > \alpha q_m$ for large m. Hence for large m.

$$|T_{mni}(x)|^{p_m} \leq (|T_{mni}(x)|^{q_m})^\alpha$$

since $|T_{mni}(x)| \leq 1$ for such m, n, i. Hence $x \in (\mathbb{A}_\sigma, p)_0$. The converse of (10.23) is false. For example, if $\mathbb{A}_\sigma = 0$ then $(\mathbb{A}_\sigma, q)_0 = (\mathbb{A}_\sigma, p)_0 = s$ whatever the values of p and q.

(ii) Let $x \in (\mathbb{A}_\sigma, q)_\infty$. Then there is a constant $B > 1$ such that

$$|T_{mni}(x)|^{q_m} \leq B \quad for \quad all \quad (m, n, i).$$

This implies that

$$|T_{mni}(x)|^{p_m} \leq B \quad for \quad all \quad (m, n, i).$$

Thus $x \in (\mathbb{A}_\sigma, p)_\infty$. To show that $(\mathbb{A}_\sigma, q)_\infty$ is closed. Suppose that $x^j \in (\mathbb{A}_\sigma, q)_\infty$, $x^j \to x$ and $x \in (\mathbb{A}_\sigma, p)_\infty$. Then for every ε, $0 < \varepsilon < 1$, there is N such that for all m, n, i,

$$|T_{mni}(x^j - x)|^{p_m} < \varepsilon \quad for \quad j > N.$$

Now

$$|T_{mni}(x^j - x)|^{q_m} < |T_{mni}(x^j - x)|^{p_m} < \varepsilon$$

for $j > N$. Therefore $x \in (\mathbb{A}, q)_\infty$ and this completes the proof. □

10.5 Some New Sequence Spaces Defined by Modulus

In this section, we extend and examine some properties of three sequence spaces defined using a modulus function and sequence of matrices. If $A = (a_{mk})$ is a nonnegative matrix, $\sum_k$ denotes the summation $k = 1$ to ∞ and then Savaş [32] defined the following:

$$w_\sigma(A_\sigma) = \left\{ x \in w : \lim_m \sum_k a_{mk} |x_{\sigma^k(n)}| = 0 \text{ uniformly in } n \right\}$$

$$w(A_\sigma) = \left\{ x \in w : \lim_m \sum_k a_{mk} |x_{\sigma^k(n)} - L| = 0 \text{ for some } L \text{ uniformly in } n \right\}$$

and

$$w_\infty(A_\sigma) = \left\{ x \in w : \sup_{m,n} \sum_k a_{mk} |x_{\sigma^k(n)}| < \infty \right\}.$$

The collection $w(A_\sigma)$ is commonly referred to as the collection of the strongly invariant A−summable sequences. Nuray and Savaş [21] generalized the above sequence spaces by using modulus function.

We introduce some new sequence spaces by using a modulus function f and examine some properties of these sequence spaces.

Definition 10.5.1. *A function $f : [0, \infty) \to [0, \infty)$ is called a modulus [23] if*

(a) $f(x) = 0$ if and only if $x = 0$,

(b) $f(x + y) \leq f(x) + f(y)$,

(c) f is increasing and

(d) f is continuous from the right at 0.

Since $|f(x) - f(y)| \leq f(|x - y|)$, it follows from condition *(iv)* that f is continuous on $[0, \infty)$. Furthermore, from condition (ii), we have $f(nx) \leq nf(x)$ for all $n \in \mathbb{N}$, from condition *(ii)* and thus

$$f(x) = f\left(nx\frac{1}{n}\right) \leq nf\left(\frac{x}{n}\right)$$

hence

$$\frac{1}{n} f(x) \leq f\left(\frac{x}{n}\right) \qquad \text{for all } n \in \mathbb{N}.$$

From the definition, we can easily see that the space $L(f)$ is closely related to the space ℓ_1, if we consider $f(x) = x$ for all real numbers $x \geq 0$. Several authors study these types of spaces. For example, Maddox [12] introduced and examined some properties of the sequence spaces $w_0(f)$, $w(f)$ and $w_\infty(f)$, defined by using a modulus f, which generalized the well-known spaces w_0, w and w_∞ of strongly summable sequences (see [11]). Similarly, Savaş [49] generalized the concept of strong almost convergence by using a modulus f and examined some further properties of the corresponding new sequence spaces. On strong almost A-summability with respect to a modulus was given by Savaş [27].

Definition 10.5.2. *Let f be a modulus and $\mathbb{A} = A^i = (a^i_{n,k})$ be the generalized three parametric nonnegative regular matrix. We define*

$$w_\sigma(\mathbb{A}_\sigma, f) = \left\{ x \in w : \lim_m \sum_k a^i_{m,k} f(|x_{\sigma^k(n)}|) = 0 \textit{ uniformly in } (n,i) \right\}$$

$$w(\mathbb{A}_\sigma, f) = \left\{ x \in w : \lim_m \sum_k a^i_{m,k} f(|x_{\sigma^k(n)} - L|) = 0 \right.$$
$$\left. \textit{for some } L \textit{ uniformly in } (n,i) \right\}$$

and

$$w_\infty(\mathbb{A}_\sigma, f) = \left\{ x \in w : \sup_{m,n,i} \sum_k a^i_{m,k} f(|x_{\sigma^k(n)}|) < \infty \right\}.$$

If $x \in w(\mathbb{A}_\sigma, f)$, we say that x is strongly invariant $\mathbb{A}$-summable to L with respect to the modulus f.

When $\sigma(n) = n + 1$, we have the following sequences spaces:

$$w_0(\hat{\mathbb{A}}, f) = \left\{ x \in w : \lim_m \sum_k a^i_{m,k} f(|x_{k+n}|) = 0 \text{ uniformly in } (n,i) \right\}$$

$$w(\hat{\mathbb{A}}, f) = \left\{ x \in w : \lim_m \sum_k a^i_{m,k} f(|x_{k+n} - L|) = 0 \right.$$
$$\left. \text{for some L uniformly in } (n,i) \right\}$$

and

$$w_\infty(\hat{\mathbb{A}}, f) = \left\{ x \in w : \sup_{m,n,i} \sum_k a^i_{m,k} f(|x_{k+n}|) < \infty \right\}.$$

If $x \in w(\hat{\mathbb{A}}, f)$, we say that x is strongly almost $\mathbb{A}$-summable to L with respect to the modulus f. If we take $\mathbb{A} = (a^i_{nk})$ is a de la Valee poussin mean, i.e., for all i

$$a^i_{nk} := \begin{cases} \frac{1}{\lambda_m}, & \text{if } k \in I_m = [m - \lambda_m + 1, m], \\ 0 & \text{otherwise,} \end{cases}$$

where (λ_m) is a non-decreasing sequence of positive numbers tending to ∞ and $\lambda_{m+1} \leq \lambda_m + 1$, $\lambda_1 = 1$, then we have the following sequence spaces which were defined and studied by Savaş and Kiliçman [25], for all i

$$w(V_\sigma, \lambda, f) = \left\{ x : \lim_m \frac{1}{\lambda_m} \sum_{k \in I_m} f(|x_{\sigma^k(n)} - L|) = 0 \right.$$
$$\left. \text{for some } L \text{ uniformly in } n \right\},$$

$$w(V_\sigma, \lambda, f)_0 = \left\{ x : \lim_m \frac{1}{\lambda_m} \sum_{k \in I_m} f(|x_{\sigma^k(n)}|) = 0 \text{ uniformly in } n \right\},$$

$$w(V_\sigma, \lambda, f)_\infty = \left\{ x : \sup_{m,n} \frac{1}{\lambda_m} \sum_{k \in I_m} f(|x_{\sigma^k(n)}|) < \infty, \right\}.$$

Let f be any modulus. Then

$$w(V_\sigma, \lambda, f)_\infty = \ell_\infty^\sigma(f) = \{x \in s : (f(|x_{\sigma^k(n)}|)) \in \ell_\infty\}.$$

Let $x \in w(V_\sigma, \lambda, f)_\infty$. Then there is a constant $M > 0$ such that

$$\frac{1}{\lambda_1} f(|x_{\sigma^k(n)}|) \leq sup_{m,n} \frac{1}{\lambda_m} \sum_{k \in I_n} f\left(|x_{\sigma^k(n)}|\right) \leq M,$$

for all n, and so $(f(|x_{\sigma^k(m)}|)) \in \ell_\infty$.

Let $x \in \ell_\infty^\sigma(f)$. Then there is a constant $M > 0$ such that $(f(|x_{\sigma^k(n)}|)) \leq M$ for all k and n, and so

$$\frac{1}{\lambda_m} \sum_{k \in I_m} f\left(|x_{\sigma^k(n)}|\right) \leq M \frac{1}{\lambda_m} \sum_{k \in I_n} 1 \leq M,$$

for all k and n. Thus $x \in w(V_\sigma, \lambda, f)_\infty$.

If we consider $\sigma(n) = n + 1$, then the above sequence spaces reduce to the following sequence spaces which are studied by Malkowsky and Savaş [16]:

$$w(\hat{V}, \lambda, f) = \left\{ x : \lim_m \frac{1}{\lambda_m} \sum_{k \in I_m} f(|x_{k+n} - L|) = 0 \right.$$
$$\left. \text{for some L} \quad \text{uniformly in } n \right\},$$

$$w(\hat{V}, \lambda, f)_0 = \left\{ x : \lim_m \frac{1}{\lambda_n} \sum_{k \in I_m} f(|x_{k+n}|) = 0 \text{ uniformly in } n \right\}$$

and

$$w(\hat{V}, \lambda, f)_\infty = \left\{ x : \sup_{m,n} \frac{1}{\lambda_m} \sum_{k \in I_m} f(|x_{k+n}|) < \infty \right\}.$$

If $\mathbb{A} = A^i = (a^i_{nk})$ is considered as for all i

$$a_{nk} := \begin{cases} \frac{1}{h_r}, & \text{if} \quad k_{r-1} < k \leq k_r, \\ 0, & \text{otherwise} \end{cases}$$

then, we have the following sequence spaces:

$$N(\theta, \sigma, f) = \left\{ x : \lim_r \frac{1}{h_r} \sum_{k \in I_r} f(|x_{\sigma^k(n)} - L|) = 0 \right. \\ \left. \text{for some } L \text{ uniformly in } n \right\}$$

and

$$N(\theta, \sigma, f)_0 = \left\{ x : \lim_r \frac{1}{h_r} \sum_{k \in I_r} f(|x_{\sigma^k(n)}|) = 0 \text{ uniformly in } n \right\}.$$

Note that if we put $f(x) = x$, then $N(\theta, \sigma, f) = N(\theta, \sigma)$ and $N(\theta, \sigma, f)_0 = N(\theta, \sigma)_0$, (see, [37]), and also if we take $f(x) = x$ and $\sigma(n) = n + 1$, then $N(\theta, \sigma, f) = \hat{N}(\theta)$ and $N(\theta, \sigma, f)_0 = \hat{N}(\theta)_0$, (see, [3]), where by a lacunary $\theta = (k_r)$; $r = 0, 1, 2, ...$ where $k_0 = 0$, we shall mean an increasing sequence of non-negative integers with $k_r - k_{r-1} \to \infty$ as $r \to \infty$. The intervals determined by θ will be denoted by $I_r = (k_{r-1}, k_r]$ and $h_r = k_r - k_{r-1}$. The ratio $\frac{k_r}{k_{r-1}}$ will be denoted by q_r. The space of lacunary strongly convergent sequences N_θ was defined by Freedman et al. [5] as follows:

$$N_\theta = \left\{ x = (x_k) : \lim_r \frac{1}{h_r} \sum_{k \in I_r} |x_k - L|) = 0 \text{ for some } L \right\}.$$

More results on this convergence can be seen from [6, 15, 50, 51].

When $\mathbb{A} = (a^i_{m,k}) = (C, 1)$ Cesàro matrix, for all i, we have the following sequence spaces which are generalization of the sequence spaces $w(V_\sigma, f)_0, w(V_\sigma, f)$ and $w(V_\sigma, f)_\infty$ that were defined as following:

$$w(V_\sigma, f)_0 = \left\{ x \in w : \lim_m \frac{1}{m} \sum_{k=1}^{m} f(|x_{\sigma^k(n)}|) = 0 \text{ uniformly in } n \right\},$$

$$[w(V_\sigma, f) = \left\{ x \in w : \lim_m \frac{1}{m} \sum_{k=1}^{m} f(|x_{\sigma^k(n)} - L|) = 0 \right. \\ \left. \text{for some } L \text{ uniformly in } n \right\}$$

and

$$w(V_\sigma, \lambda, f)_\infty = \left\{ x \in w : \sup_{m,n} \frac{1}{m} \sum_{k=1}^{m} f(|x_{\sigma^k(n)}|) < \infty \right\}.$$

If $x \in [V_\sigma(f)]$, we say that x is strongly $\sigma-$convergent to L with respect to the modulus f.

We now establish a number of useful theorems.

Theorem 10.5.3. *$w_0(A_\sigma, f), w(A_\sigma, f)$ and $w_\infty(A_\sigma, f)$ are linear spaces over the complex field C.*

Proof. We consider only $w_0(\mathbb{A}_\sigma, f)$. Others can be treated similarly. Let $x, y \in w_0(\mathbb{A}_\sigma, f)$. For $\lambda, \mu \in C$, there exits M_λ and N_μ integers such that $|\lambda| \leq M_\lambda$ and $|\mu| \leq N_\mu$. From Definition 10.5.1(b), we write

$$\sum_k a^i_{m,k} f(|\lambda x_{\sigma^k(n)} + \mu y_{\sigma^k(n)}|) \leq M_\lambda \sum_k a^i_{m,k} f(|x_{\sigma^k(n)}|) + N_\mu \sum_k a^i_{m,k} f(|y^k_\sigma(n)|).$$

For $m \to \infty$, since $x, y \in w_0(\mathbb{A}_\sigma, f)$, we have $\lambda x + \mu y \in w_0(\mathbb{A}_\sigma, f)$. Thus $w_0(\mathbb{A}_\sigma, f)$ is linear space over $\mathbb{C}$. □

Theorem 10.5.4. *Let $\mathbb{A}$ be a nonnegative regular matrix and f be a modulus, then $w_\infty(\mathbb{A}_\sigma, f) \supset w(\mathbb{A}_\sigma, f)$.*

Proof. Let $x \in w(\mathbb{A}_\sigma, f)$. Now we write,

$$\sum_k a^i_{m,k} f(|x_{\sigma^k(n)}|) = \sum_k a^i_{m,k} f(|x_{\sigma^k(n)} - L + L|) \leq \sum_k a^i_{m,k} f(|x_{\sigma^k(n)} - L|) + f(|L|) \sum_k a^i_{m,k}.$$

There exists an integer M_L such that $|L| \leq M_L$. Hence we have

$$\sum_k a^i_{m,k} f(|x_{\sigma^k(n)}|) \leq \sum_k a^i_{m,k} f(|x_{\sigma^k(n)} - L|) + M_L f(1) \sum_k a^i_{m,k}.$$

Since $\mathbb{A}$ is regular and $x \in w(\mathbb{A}_\sigma, f)$, we get $x \in w_\infty(\mathbb{A}_\sigma, f)$ and this completes the proof. □

Theorem 10.5.5. *$w_0(\mathbb{A}_\sigma, f)$ and $w(\mathbb{A}_\sigma, f)$ are complete linear topological spaces paranormed by g defined by*

$$g(x) = \sup_{m,n,i} \sum_k a^i_{m,k} f(|x_{\sigma^k(n)}|).$$

Proof. From Theorem 5.4, for each $x \in w(\mathbb{A}_\sigma, f), g(x)$ exists. Clearly $g(\theta) = 0$ and $g(x) = g(-x)$ and $g(x+y) \leq g(x) + g(y)$. We now show that the scalar multiplication is continuous.

$$g(\lambda, x) = \sup_{m,n,i} \sum_k a^i_{m,k} f(|\lambda, x_{\sigma^k(n)}|) \leq (1 + [|\lambda|]) g(x),$$

where $[|\lambda|]$ denotes the integer part of λ. Hence $\lambda \to 0, x \to 0$ imply $g(\lambda x) \to 0$ and also $x \to 0, \lambda$ fixed imply $g(\lambda x \to 0)$. We now show that $\lambda \to 0,\ x$ fixed imply $g(\lambda x) \to 0$.

Let $x \in w(\mathbb{A}_\sigma, f)$. Then, as $m \to \infty$, we have

$$s_{mni} = \sum_k a^i_{m,k} f(|x_{\sigma^k(n)} - L|) \to 0 \text{ uniformly in } (n,i).$$

For $|\lambda| < 1$, we have

$$\begin{aligned}
\sum_k a^i_{m,k} f(|\lambda x_{\sigma^k(n)}|) &= \sum_k a^i_{m,k} f(|\lambda x_{\sigma^k(n)} - \lambda L + \lambda L|) \\
&\leq \sum_k a^i_{m,k} f(|\lambda x_{\sigma^k(n)} - \lambda L|) + \sum_k a^i_{m,k} f(|\lambda L|) \\
&\leq \sum_{k>M} a^i_{m,k} f(|\lambda x_{\sigma^k(n)} - \lambda L|) \\
&\quad + \sum_{k \leq M} a^i_{m,k} f(|\lambda x_{\sigma^k(n)} - \lambda L|) + \sum_k a^i_{m,k} f(|\lambda L|) \\
&\leq \sum_{k>M} a^i_{m,k} f(|\lambda x_{\sigma^k(n)} - \lambda L|) \\
&\quad + \sum_{k \leq M} a^i_{m,k} f(|\lambda x_{\sigma^k(n)} - \lambda L|) + \sum_k a^i_{m,k} f(|\lambda L|).
\end{aligned}$$

Let $\varepsilon > 0$ and choose M such that for each n, m, i and $k > M$ implies $s_{nm} < \varepsilon/2$. For each M, by continuity of f, as $\lambda \to 0$,

$$\sum_{k<M} a^i_{m,k} f(|\lambda x_{\sigma^k(n)} - \lambda L|) + \sum_{k \leq M} a^i_{m,k} f(|\lambda L|) \to 0.$$

Then choose $\delta < 1$ such that $|\lambda| < \delta$ implies

$$\sum_{k<M} a^i_{m,k} f(|\lambda x_{\sigma^k(n)} - \lambda L|) + \sum_{k \leq M} a^i_{m,k} f(|\lambda L|) < \varepsilon/2.$$

Hence, we have

$$\sum_k a^i_{m,k} f(|\lambda x_{\sigma^k(n)}|) < \frac{\varepsilon}{2} + \frac{\varepsilon}{2} = \varepsilon$$

and $g(\lambda x) \to 0 (\lambda \to 0)$ Thus $w(\mathbb{A}_\sigma, f)$ is paranormed linear topological space by g.

Now, we show that $w(\mathbb{A}_\sigma, f)$ is complete with respect to its paranorm topologies. Let $(x^{(s)})$ be a Cauchy sequence in $w(\mathbb{A}_\sigma, f)$. Then, we write

$$g(x^{(s)} - x^{(t)}) \to 0, \quad s, t \to \infty$$

i.e., as $s, t \to \infty$, for all n, i and m, we write.

$$\sum_k a^i_{m,k} f(|x^{(s)}_{\sigma^k(n)} - x^{(t)}_{\sigma^k(n)}|) \to 0 \tag{10.24}$$

Hence, for each fixed n, i and k, as $s, t \to \infty$, we have $f(|x^{(s)}_{\sigma^k(n)} - x^{(t)}_{\sigma^k(n)}|) \to 0$ and for each fixed n, i and k, $(x^{(s)}_{\sigma^k(n)})_s$ be a Cauchy sequence in $\mathbb{C}$. Since $\mathbb{C}$ is complete, as $s \to \infty$, $(x^{(s)}_{\sigma^k(n)})_s \to (x_{\sigma^k(n)})$ say. Now from (5.1), we have for $\varepsilon > 0$, there exists a natural number $\mathbb{N}$ such that

$$\sum_k a^i_{m,k} f(|x^{(s)}_{\sigma^k(n)} - x^{(t)}_{\sigma^k(n)}|) < \varepsilon \tag{10.25}$$

for all n, m, i and $s, t > N$. Since for any fixed natural number M, we have from (10.25),

$$\sum_{k \le M} a^i_{m,k} f(|x^{(s)}_{\sigma^k(n)} - x^{(t)}_{\sigma^k(n)}|) < \varepsilon \tag{10.26}$$

for all n, m, i and $s, t > N$. By taking $t \to \infty$ in the above expression we obtain

$$\sum_{k \le M} a^i_{m,k} f(|x^{(s)}_{\sigma^k(n)} - x_{\sigma^k(n)}|) < \varepsilon$$

for all n, m, i and $s > N$. Since M is arbitrary, by taking $M \to \infty$ we obtain

$$\sum_k a^i_{m,k} f(|x^{(s)}_{\sigma^k(n)} - x^{(t)}_{\sigma^k(n)}|) < \varepsilon$$

for all n, m, i and $s > N$ that is, $g(x^{(s)} - x) \to 0$ and $s \to \infty$ and thus $x_s \to x$ as $s \to \infty$.

Also, for each s, there exists $L^{(s)}$ with

$$\sum_k a^i_{m,k} f(|x^{(s)}_{\sigma^k(n)} - L^{(s)}|) \to 0 \quad (m \to \infty) \tag{10.27}$$

uniformly in (n, i). From regularity of $\mathbb{A}$. Definition 10.5.1(b) and (10.27), we have $f(|L^{(s)} - L^{(t)}|) \to 0$ $(s, t \to \infty)$ and $(L^{(s)})$ is a Cauchy sequence in $\mathbb{C}$, so $(L^{(s)})$ converges to L , say. Consequently we get

$$\sum_k a^i_{m,k} f(|x_{\sigma^k(n)} - L|) \to 0 (m \to \infty)$$

uniformly in (n, i). So that $x \in w(\mathbb{A}_\sigma, f)$ and the space is complete. □

It is easy to prove the following theorem so we omit the detail.

Theorem 10.5.6. *Let* $\mathbb{A}$ *be a nonnegative regular matrix and* f *be a modulus, then* $w_0(\mathbb{A}_\sigma, f) \supset w_0(\mathbb{A}_\sigma), w(\mathbb{A}_\sigma, f) \supset w(\mathbb{A}_\sigma)$ *and* $w_\infty(\mathbb{A}_\sigma, f) \supset w_\infty(\mathbb{A}_\sigma)$.

Theorem 10.5.7. *Let* $\mathbb{A}$ *be a nonnegative regular matrix and* f *be a modulus. If* $\beta = \lim_t (f(t)/t) > 0$ *then* $w(\mathbb{A}_\infty, f) = w(\mathbb{A}_\sigma)$.

Proof. In Theorem 10.5.6, it was shown that $w(\mathbb{A}_\sigma, f) \supset w(\mathbb{A}_\sigma)$. We must show that $w(\mathbb{A}_\sigma) \supset w_0(\mathbb{A}_\sigma, f)$. For any modulus function, the existence of positive limit given with β was given by Maddox [12]. Now $\beta > 0$ and let $x \in w(\mathbb{A}_\sigma, f)$. Since $\beta > 0$, for every $t > 0$, we write $f(t) \geq \beta t$. From this inequality, it is easy to see that $x \in w(A_\sigma)$. This completes the proof. □

10.6 Matrix Transformations

In this section, we characterize some matrix transformations connecting $\widehat{BV}_\theta$. We may remark here that the concept $\widehat{BV}$ of almost bounded variation have been recently introduced and investigated by Nanda and Nayak [20] as follow:

$$\widehat{BV} = \left\{ x : \sum_m |t_{mn}(x)| \text{ converges uniformly in } n \right\}$$

where

$$t_{mn}(x) = \frac{1}{m(m+1)} \sum_{v=1}^{m} v\left(x_{n+v} - x_{n+v-1}\right).$$

Das and Mishra [3] observed almost lacunary convergence:

$$M_\theta = \left\{ x : \text{ there exists } L \text{ such that } \lim_{r\to\infty} \frac{1}{h_r} \sum_{k\in I_r} (x_{k+i} - L) = 0 \text{ uniformly in } i \geq 0 \right\}.$$

It is natural question to expect that lacunary almost convergence must be related to some concept $\widehat{BV_\theta}$ in the same vein as almost convergence is related to the concept of $\widehat{BV}$. A sequence in $\widehat{BV_\theta}$ will mean a sequence of lacunary almost bounded variation.

Savaş and Karakaya [6] defined the following sequence spaces by using almost lacunary convergence,

$$\widehat{BV}_\theta = \left\{ x : \sum_r |\varphi_{rn}(x)| \text{ converges uniformly in } n \right\}$$

and

$$\widehat{\widehat{BV}}_\theta = \left\{ x : \sup_n \sum_r |\varphi_{rn}(x)| < \infty \right\},$$

where

$$\varphi_{rn}(x) = \varphi_{rn} = \frac{1}{h_r(h_r+1)} \sum_{u=1}^{h_r} u \left(x_{k_{r-1}+u+1+n} - x_{k_{r-1}+u+n} \right).$$

Furthermore, the following theorem was proved in [6]:

Theorem 10.6.1. *$\widehat{BV}_\theta \subset \widehat{\widehat{BV}}_\theta$ for every θ.*

But we do not know whether $\widehat{\widehat{BV}}_\theta \subset \widehat{BV}_\theta$, that is $\widehat{\widehat{BV}}_\theta = \widehat{BV}_\theta$. It is still open problem.

We write

$$\begin{aligned}\varphi_{rn}(Ax) &= \frac{1}{h_r(h_r+1)} \sum_{u=1}^{h_r} u \left(A_{k_{r-1}+u+1+n} - A_{k_{r-1}+u+n} \right)(x) \\ &= \sum_k b(n,k,r) x_k,\end{aligned}$$

where

$$b(n,k,r) = \frac{1}{h_r(h_r+1)} \sum_{u=1}^{h_r} u \left(a_{k_{r-1}+u+1+n,k} - a_{k_{r-1}+u+n,k} \right).$$

Some papers concerning matrix transformations were considered in [26,41,43–48]. Finally, we conclude this paper by stating the following theorem.

Theorem 10.6.2. *$A \in (\hat{c}, \widehat{BV}_\theta)$ if and only if*
(i) there exists a constant $K > 0$ such that

$$\sum_r \sum_k |b(n,k,r)| \leq K \quad (\forall n).$$

Proof. Necessity. Suppose that

$$A \in (\hat{c}, \widehat{BV}_\theta),$$

that is, $\sum_m |\varphi_{rn}(Ax)|$ converges uniformly in n, for every $a \in c \subset \hat{c}$. Write

$$q_n(a) = \sum_m |\varphi_{mn(Ax)}|.$$

Note that $\{q_n\}$ is a sequence of continuous seminorms on c such that $\sup_n q_n(x) < \infty$. Therefore by Banach-Steinhaus Theorem (see, Maddox [11], p.114) there exists a constant $K > 0$ such that

$$q_n(a) \leq K\,\|a\| \quad (\forall n, \forall a \in c). \tag{10.28}$$

Let us define $a^r = \{a_k^{(r)}\} \in \hat{c}$ by

$$a_k^{(r)} = \begin{cases} b(n,k,m), & 0 \leq k \leq r, \\ 0, & otherwise. \end{cases}$$

Then, it follows from (10.28) that

$$\sum_m \sum_{k=0}^{r} |b(n,k,m)| \leq K$$

for each n, r and $a \in \hat{c}$. Therefore holds.

Sufficiency. Suppose that the condition (i) holds and that $x \in \hat{c}$. Then give $\epsilon > 0$ there exists a p_0 so that for all $p \geq p_o$ and all k,

$$(x_k + x_{k+1} + ... + x_{k+p})/(p+1) = s + \alpha_{kp}, \tag{10.29}$$

where $|\alpha_{kp}| < K$, K being that of condition (i). Following Lorentz's proof of Theorem 7, we multiply by $b(n,k,r)$ and add obtaining

$$\sum_k b(n,k,r)(x_k + x_{k+1} + ... + x_{k+p})/(p+1) = s \sum_k b(n,k,r) + \sum_k b(n,k,r)\alpha_{kp}.$$

Then, by using the same method of the proof of Theorem 1 in Schaefer [52], we have

$$\begin{aligned} \sum_k b(n,k,r)s + \sum_k b(n,k,r)\alpha_{kp} &= \sum_k b(n,k,r)x_k \\ &+ \sum_{k=0}^{p-1} x_k \left[\frac{1}{p+1}\sum_{j=0}^{k} b(n,j,r) - b(n,k,r)\right] \\ + \sum_{k=p}^{\infty} x_k[(b(n,k-p,r) &+ ... + b(n,k,r))/(p+1) - b(n,k,r)]. \end{aligned} \tag{10.30}$$

Let us set

$$A_{npr} = \sum_{k=0}^{p-1} x_k \left[\frac{1}{p+1}\sum_{j=0}^{k} b(n,j,r) - b(n,k,r)\right]$$

$$B_{npm} = \sum_{k=p}^{\infty} x_k[(b(n,k-p,r) + ... + b(n,k,r))/(p+1) - b(n,k,r)].$$

Therefore, we get from (10.30) that

$$\sum_r \left|\sum_k b(n,k,r)x_k\right| \le \sum_r \left|\sum_k b(n,k,r)s\right| + \sum_r \left|\sum_k b(n,k,r)\alpha_{kp}\right| + \sum_r |B_{npr}| + \sum_r |A_{npr}|.$$

For all $p \ge p_0$,

$$\sum_r |\sum_k b(n,k,r)\alpha_{kp}| \le \sum_r \sum_k |b(n,k,r)||\alpha_{kp}| < \varepsilon.$$

As in Lorentz's proof, we get

$$\begin{aligned}\sum_r |B_{npr}| &\le \sum_r p\frac{\|x\|}{2}\sum_k |b(n,k,r)-b(n,k+1,r)| \\ &\le p\frac{\|x\|}{2}\sum_r\sum_k |b(n,k,r)-b(n,k+1,r)|.\end{aligned}$$

From the condition (i), there exists a constant M such that

$$\sum_r \sum_k |b(n,k,r)-b(n,k+1,r)| \le M, \quad \forall n.$$

So that

$$\sum_r |B_{npr}| < \infty \text{ uniformly in } n.$$

We see that

$$\begin{aligned}\sum_r |A_{npr}| &= \sum_r |\sum_{k=0}^{p-1}\left[\frac{1}{p+1}\sum_{j=0}^{k} b(n,j,r) - b(n,k,r)\right]x_k| \\ &\le \|x\|\left(\sum_{k=0}^{p-1}\frac{1}{p+1}\sum_r\sum_{j=0}^{k}|b(n,j,r)| \right.\\ &\quad \left. + \sum_m\sum_{k=0}^{p-1}\frac{1}{p+1}\sum_r\sum_{k=0}^{p-1}|b(n,k,r)|\right) \\ &\le \|x\|\left(\frac{p}{p+1}+1\right)K \\ &< \infty \text{ uniformly in } n.\end{aligned}$$

Moreover,

$$\sum_r |\sum_k b(n,k,r)s| \le \sum_r\sum_k |b(n,k,x)||s| < \infty \text{ uniformly in } n. \tag{10.31}$$

Hence, combining the above facts, we have

$$\sum_r |\varphi_{rn}(Ax)| = \sum_r |\sum_k b(n,k,r)x_k| < \infty \quad \text{uniformly in } n.$$

Now $Ax \in \widehat{BV}_\theta$ and this completes the proof. □

Bibliography

[1] S. Banach, Théorie des Opérations lineaires, Chelsea Publ. Co., New York (1955).

[2] R. C. Cook, Infinite matrices and Sequence spaces, Macmillan London (1950).

[3] G. Das, S. K. Mishra, Banach limits and lacunary strong almost convergence, J. Orissa Math. Soc. 2(2) (1983) 61–70.

[4] J. P. Duran, Infinite matrices and almost convergence, Math. Z. 128 (1972) 75–83.

[5] A. R. Freedman, J. J. Sember, M. Rapnael, Some Cesàro type summability spaces, Proc. London Math. Soc. 37 (1978) 508–520.

[6] V. Karakaya, E. Savaş, On almost θ-bounded variation of lacunary sequences, Comput. Math. Appl. 61(6) (2011) 1502–1506.

[7] J. P. King, Almost summable sequences, Proc. Amer. Math. Soc. 17 (1966) 1219–1225.

[8] B. Kuttner, On dual summability methods, Proc. Camb. Phil. Soc. 71 (1972) 67–73.

[9] J. Li, Lacunary statistical convergence and inclusion properties between lacunary methods, Int. J. Math. Math. Sc. 23(3) (2000) 175–180.

[10] G. G. Lorentz, A contribution to the theory of divergent sequences, Acta. Math. 80 (1948) 167–190.

[11] I. J. Maddox, Spaces of strongly summable sequences, Quart. J. Math. Oxford 18(2) (1967) 345–355.

[12] I. J. Maddox, Sequence spaces defined by a modulus, Math. Proc. Cambridge Philos. Soc. 100 (1986) 161–166.

[13] I. J. Maddox, A new type of convergence, Math. Proc. Cambridge Philos. Soc. 83 (1987) 61–64.

[14] I. J. Maddox, Elements of Functional Analysis, Cambridge University Press (1970).

[15] I. J. Maddox, J. W. Roles, Absolute convexity in certain topological linear spaces, Math. Proc. Cambridge Philos. Soc. 66 (1969) 541–545.

[16] E. Malkowsky, E. Savaş, Some λ-sequence spaces defined by a modulus, Arch. Math. (Brno) 36 (2000) 219–228.

[17] Mursaleen, Matrix transformations between some new sequence spaces, Houston J. Math. 9(4) (1983) 505–513.

[18] S. Nanda, Some sequence spaces and almost convergence, J. Austral. Math. Soc. 22 (1987) 446–455.

[19] S. Nanda, Strongly almost summable and strongly almost convergent sequences, Acta Math. Hung. 49 (1–2) 71–76.

[20] S. Nanda, K. C. Nayak, Some new sequence spaces, Indian J. Pure Appl. Math. 9(8) (1978) 836–846.

[21] F. Nuray, E. Savaş, Some new sequence spaces defined by a modulus funcion, Indian J. Pure. Appl. Math. 24(11) (1993) 657–663.

[22] C. Orhan, Strongly almost regular dual summability methods, Pure Appl. Math. Sci. 24(1) (1986) 9–15.

[23] W. H. Ruckle, FK Spaces in which the sequence of coordinate vectors in bounded, Cand. J. Math. 25 (1973) 973–978.

[24] S. K. Saraswat, S. K. Gupta, Spaces of strongly σ-summable sequences, Bull. Cal. Math. Soc. 75 (1983) 179–184.

[25] E. Savaş, A. Kiliçman, A Note on Some Strongly Sequence Spaces, Abstr. Appl. Anal. Volume 2011 (2011), Article ID 598393, 8 pages.

[26] E. Savaş, V. Karakaya, Some new sequence spaces defined by lacunary sequences, Math. Slovaca 57 (2007) 393–399.

[27] E. Savaş, On strong almost A-summability with respect to a modulus and statistical convergence, Indian. J. Pure Appl. Math. 23(3) (1992) 217–223.

[28] E. Savaş, Infinite matrices and summable series, J. Orissa Math. Soc. 4(1) (1985) 5–14.

[29] E. Savaş, Infinite matrices and generalized almost convergence, Doga Türk. J. Math. 5(3) (1987) 1–10.

[30] E. Savaş, On strong $\sigma-$convergence, J. Orissa Math. Soc. 6(1) (1987) 45–53.

[31] E. Savaş, Matrix transformations and absolute almost convergence, Bull. Ins. Math. Academia Sinica 15(3) (1987) 345–355.

[32] E. Savaş, Strongly $\sigma-$summable and strongly $\sigma-$convergence sequences, Bull. Call. Math. Soc. 81 (1989) 173–178.

[33] E. Savaş, Strongly $\sigma-$convergent sequences, Bull. Cal. Math. Soc. 81 (1989) 295–300.

[34] E. Savaş, Some sequence spaces involving invariant means, Indian J. Math. 31 (1989) 36–46.

[35] E. Savaş, Invariant means and generalization of a theorem of S. Mishra, Doga Türk. J. Math. 14 (1989) 8–14.

[36] E. Savaş, Invariant coregular and conull matrices of operators, Hacettepe Bull. Math. Sci. Eng. 19 (1990) 1–7.

[37] E. Savaş, On lacunary strong $\sigma-$convergence, Indian J. Pure Appl. Math. 21(4) (1990) 359–365.

[38] E. Savaş, Almost convergence and almost summability, Tamkang J. Math. 21(4) (1990) 27–35.

[39] E. Savaş, C. Orhan, On strongly $\sigma-$-regular summability methods, Bull. Cal. Math. Soc. 83 (1991) 51–56.

[40] E. Savaş, Some sequence spaces and almost convergence, Ann. Univ. Timiora 30(2-3) (1992) 6–10.

[41] E. Savaş, Matrix transformations of some generalized sequence spaces, J. Orissa Math. Soc. 4 (1985) 37–51.

[42] E. Savaş, On σ-regular dual summability methods, Indian J. Pure Appl. Math. 18(11) (1987) 1011–1015.

[43] E. Savaş, Matrix transformations and absolute almost convergence, Bull. Inst. Math. Acad. Sinica 15(3) (1987) 345–355.

[44] E. Savaş, Matrix transformations between some new sequence spaces, Tamkang J. Math. 19(4) (1988) 75–80.

[45] E. Savaş, σ-summable sequences and matrix transformations, Chinese J. Math. 18(3) (1990) 201–207.

[46] E. Savaş, Matrix transformations and almost convergence, Math. Student 59(1-4) (1991) 170–176.

[47] E. Savaş, Matrix transformations of X_p into Cs, Punjab Univ. J. Math. (Lahore) 24 (1991) 59–66.

[48] E. Savaş, Mursaleen, Matrix transformations in some sequence spaces, Istanbul Univ. Fen Fak. Mat. Derg. 52 (1993) 1–5.

[49] E. Savaş, On some generalized sequence spaces defined by a modulus, Indian J. Pure Appl. Math. 30(5) (1999) 459–464.

[50] E. Savaş, On asymptotically lacunary statistical equivalent functions via ideals, J. Math. Comp. Sci. 19(1) (2019) 35–40.

[51] E. Savaş, Double ideal lacunary statistical convergence in topological groups, Adv. Stud. Contemp. Math. (Kyungshang) 28(4) (2018) 635–642.

[52] P. Schaefer, Infinite matrices and invariant means, Proc. Amer. Math. Soc. 36 (1972) 104–110.

Chapter 11

Carathéodory Theory of Dynamic Equations on Time Scales

Sanket Tikare

Ramniranjan Jhunjhunwala College, sankettikare@rjcollege.edu.in

11.1 Introduction and Preliminaries

The time scales calculus and the theory of dynamic equations on time scales appears in the literature in the past three decades. It provides a unified framework in which one can study both continuous and discrete dynamical process simultaneously. Hence, it is realistic to model a hybrid phenomenon involving continuous and discrete data simultaneously, using dynamic equations on time scales. Thus, this topic has gained significant attention from many researchers. However, in the literature, little study has been done for dynamic equations on time scales involving Carathéodory function. In this chapter, we will investigate Carathéodory solutions of dynamic initial value problem of the type

$$\begin{cases} x^{\Delta} = f(t, x), & t \in [a, b]_{\mathbb{T}}^{\kappa}; \\ x(a) = A, \end{cases} \tag{11.1}$$

where $A \in \mathbb{R}$ and $f : [a, b]_{\mathbb{T}} \times \mathbb{R} \to \mathbb{R}$ is a Δ-Carathéodory function, and x^{Δ} is the delta derivative $x : [a, b)_{\mathbb{T}} \to \mathbb{R}$.

In the Carathéodory theory, we study the solutions of dynamic equations under weak conditions that the right-hand side of the dynamic equations is not necessarily continuous. This allows us to broaden the class of dynamic

DOI: 10.1201/9781003178200-11

equations beyond those with continuous right-hand sides. We assume that the reader is familiar with the classical theory of Carathéodory. In this section, several concepts and results of the time scale calculus are briefly exposed that provide a framework for the study of dynamic equations in the sense of Carathéodory. The reader can find information on time scales calculus, for examples, in [2, 3].

A *time scale* $\mathbb{T}$ is an arbitrary nonempty closed subset of $\mathbb{R}$, with the subspace topology inherited from the standard topology of $\mathbb{R}$. For $a, b \in \mathbb{T}$ with $a < b$, the time scale interval $[a, b]_{\mathbb{T}}$ is defined as $[a, b] \cap \mathbb{T} := \{t \in \mathbb{T} : a \leq t \leq b\}$. For $t \in \mathbb{T}$, we define two operators, $\sigma : \mathbb{T} \to \mathbb{T}$ as $\sigma(t) := \inf\{s \in \mathbb{T} : s > t\}$, called *the forward jump operator* and $\rho : \mathbb{T} \to \mathbb{T}$ as $\rho(t) := \sup\{s \in \mathbb{T} : s < t\}$ called *the backward jump operator*. With the help of these two jump operators, we classify the points in a time scale $\mathbb{T}$ in the following way: A point $t \in \mathbb{T}$ is *right-scattered* if $\sigma(t) > t$; while it is *left-scattered* if $\rho(t) < t$. A point $t \in \mathbb{T}$ is *right-dense* if $\sigma(t) = t$; while it is *left-dense* if $\rho(t) = t$. A point $t \in \mathbb{T}$ is *dense* if $\rho(t) = t = \sigma(t)$; while it is *isolated* if $\rho(t) < t < \sigma(t)$. The *graininess function* $\mu : \mathbb{T} \to [0, \infty)$ is defined by $\mu(t) := \sigma(t) - t$. The metric on $\mathbb{T}$ is defined by $d(r, s) := |\mu(r, s)|$, where $\mu(t, \tau) = \mu(t)$ for a fixed $\tau \in \mathbb{T}$. This metric d generates the order topology on time scale $\mathbb{T}$.

Below we list some standard notations that are being used in this chapter. The set of continuous functions from $[a, b]_{\mathbb{T}}$ to $\mathbb{R}$ is denoted by $\mathrm{C}([a, b]_{\mathbb{T}}, \mathbb{R})$. The set of absolutely continuous functions from $[a, b]_{\mathbb{T}}$ to $\mathbb{R}$ is denoted by $\mathrm{AC}([a, b]_{\mathbb{T}}, \mathbb{R})$. The set of Δ-measurable functions from $[a, b]_{\mathbb{T}}$ to $\mathbb{R}$ is denoted by $\mathrm{M}([a, b]_{\mathbb{T}}, \mathbb{R})$. The set of bounded Δ-measurable functions from $[a, b]_{\mathbb{T}}$ to $\mathbb{R}$ is denoted by $\mathrm{BM}([a, b]_{\mathbb{T}}, \mathbb{R})$. The set of Lebesgue Δ-integrable functions from $[a, b]_{\mathbb{T}}$ to $\mathbb{R}_+$ is denoted by $\mathrm{L}_1([a, b]_{\mathbb{T}}, \mathbb{R}_+)$. We note that

$$\mathrm{AC}([a, b]_{\mathbb{T}}, \mathbb{R}) \subset \mathrm{C}([a, b]_{\mathbb{T}}, \mathbb{R}) \subset \mathrm{BM}([a, b]_{\mathbb{T}}, \mathbb{R}) \subset \mathrm{M}([a, b]_{\mathbb{T}}, \mathbb{R}).$$

For $x : [a, b]_{\mathbb{T} \to \mathbb{R}}$, we define *sup-norm* on the set $\mathrm{BM}([a, b]_{\mathbb{T}}, \mathbb{R})$ as $\|x\|_0 := \sup_{t \in [a,b]_{\mathbb{T}}} |x(t)|$. Also, we define the *generalized Bielecki's norm*, called *TZ-norm* on the space $\mathrm{BM}([a, b]_{\mathbb{T}}, \mathbb{R})$ as

$$\|x\|_\beta := \sup_{t \in [a,b]_{\mathbb{T}}} \frac{|x(t)|}{e_\beta(t, a)}.$$

We note that the sup-norm $\|\cdot\|_0$ and the TZ-norm $\|\cdot\|_\beta$ are equivalent. Since $(\mathrm{BM}([a, b]_{\mathbb{T}}, \mathbb{R}), \|\cdot\|_0)$ is a Banach space, it follows that $(\mathrm{BM}([a, b]_{\mathbb{T}}, \mathbb{R}), \|\cdot\|_\beta)$ is also a Banach space.

Definition 11.1.1. *(See [2, Definition 1.58]) A function $x : [a, b)_{\mathbb{T}} \to \mathbb{R}$ is said to be rd-continuous if it is continuous at every right-dense points in $[a, b)_{\mathbb{T}}$ and its left sided limits exist at left dense points in $[a, b)_{\mathbb{T}}$. The notation $\mathrm{C}_{\mathrm{rd}}([a, b)_{\mathbb{T}}, \mathbb{R})$ is used to denote the set of all rd-continuous functions $x : [a, b)_{\mathbb{T}} \to \mathbb{R}$.*

Definition 11.1.2. *(See [2, Definition 2.45]) A function $\beta : [a,b]_{\mathbb{T}} \to \mathbb{R}$ is said to be positively regressive if $1 + \mu(t)\beta(t) > 0$ for all $t \in [a,b]_{\mathbb{T}}$. The notation $\mathcal{R}^+$ is used to denote the set of all positively regressive rd-continuous functions $\beta : [a,b]_{\mathbb{T}} \to \mathbb{R}$.*

Definition 11.1.3. *(See [2, Definition 2.30]) For $\beta \in \mathcal{R}^+$, the exponential function $e_\beta(\cdot, a)$ is defined as*

$$e_\beta(t,a) := \begin{cases} \exp\left(\int_a^t \beta(s)ds\right) & \text{if } \mu(t) = 0; \\ \exp\left(\int_a^t \frac{Log(1+\mu(s)\beta(s))}{\mu(s)}\Delta s\right) & \text{if } \mu(t) > 0, \end{cases}$$

for all $t \in [a,b]_{\mathbb{T}}$, where Log *is the principal logarithm function.*
Since $\beta \in \mathcal{R}^+$, we note that $e_\beta(t,a) > 0$ for all $t \in [a,b]_{\mathbb{T}}$ and $e_\beta(a,a) = 1$.

Definition 11.1.4. *[4] A function $x : [a,b]_{\mathbb{T}} \to \mathbb{R}$ is said to be absolutely continuous if, for every $\varepsilon > 0$, there exists $\delta > 0$ such that*

$$\sum_{i=1}^{n} |x(d_i) - x(c_i)| < \varepsilon,$$

whenever $\{[c_i, d_i]_{\mathbb{T}} : 1 \le i \le n\}$ is a finite family of pairwise disjoint subintervals of $[a,b]_{\mathbb{T}}$ obeying

$$\sum_{i=1}^{n} (d_i - c_i) < \delta.$$

A function $x : [a,b]_{\mathbb{T}} \to \mathbb{R}$ is called an arc if it is absolutely continuous.

Theorem 11.1.5. *(Fundamental theorem of calculus for vector valued functions, [4, Theorem 4.1]) A function $f : [a,b]_{\mathbb{T}} \to \mathbb{R}$ is an arc if and only if the following hold.*

(i) *for Δ-a.e. $t \in [a,b)_{\mathbb{T}}$, the function f is Δ-differentiable and $f^\Delta \in \mathrm{L}_1([a,b)_{\mathbb{T}}, \mathbb{R})$;*

(ii) *for each $t \in [a,b]_{\mathbb{T}}$, we have*

$$f(t) = f(a) + \int_a^t f^\Delta(s)\Delta s.$$

Theorem 11.1.6. *(See [8, Proposition 2.4]) Let $\{x_i\}$ be a sequence of arcs $x_i : \mathbb{T} \to \mathbb{R}$ and let $\phi : \mathbb{T} \to [0, +\infty)$ be in $L_1([a,b)_{\mathbb{T}}, \mathbb{R}_+)$ such that for each $i \in \mathbb{N}$, we get*

$$\|x_i^\Delta(s)\| \le \phi(s) \quad \Delta\text{-a.e. } s \in [a,b)_{\mathbb{T}}. \tag{11.2}$$

If $\{x_i(a)\}$ is a bounded sequence, then there exists a subsequence $\{x_{i_k}\} \subset \{x_i\}$ and an arc $x : \mathbb{T} \to \mathbb{R}$ such that $x_{i_k} \rightrightarrows x$. Furthermore, if $c, d \in \mathbb{T}$ with $c < d$, then

$$\lim_{k\to\infty} \int_c^d x_{i_k}^\Delta(s)\Delta s = \int_c^d x^\Delta(s)\Delta s.$$

Remark 11.1.7. *The above theorem remains valid if the condition given in* (11.2) *is replaced by*

$$\left|\int_c^d x_i^{\Delta}(s)\Delta s\right| \leq \int_c^d \phi(s)\Delta s$$

for each $c, d \in \mathbb{T}$ *with* $c < d$.

Theorem 11.1.8. *(Measure continuity [8, Proposition 1.11]) Let* $f : [a,b]_{\mathbb{T}} \to [0, +\infty)$ *be a Lebesgue* Δ*-integrable function. For every* $\varepsilon > 0$ *there exists* $\delta > 0$ *such that, if* A *is a* Δ*-measurable subset of* $[a,b]_{\mathbb{T}}$ *with* $\mu_{\Delta}(A) < \delta$*, then* $\displaystyle\int_A f(s)\Delta s < \varepsilon$.

Theorem 11.1.9. *(See [5, Proposition 4.1]) Let* $f : \mathbb{T} \to \mathbb{R}$ *be given function and* $\tilde{f} : [a,b] \to \mathbb{R}$ *be its extension defined as*

$$\tilde{f}(t) := \begin{cases} f(t) & \text{if } t \in \mathbb{T}, \\ f(t_i) & \text{if } t \in (t_i, \sigma(t_i)), \text{ for some } i \in I \subset \mathbb{N}. \end{cases} \tag{11.3}$$

Then, f *is* Δ*-measurable if and only if* $\tilde{f}$ *is Lebesgue measurable.*

Lemma 11.1.10. *(See [5, Lemma 3.2]) Let* $E \subset [a,b)_{\mathbb{T}}$ *be such that* E *does not contain any right-scattered points. Then* $m^*(E) = \mu^*(E)$.

Theorem 11.1.11. *(See [5, Theorem 5.1]) Let* $E \subset \mathbb{T}$ *be a* Δ*-measurable set such that* $b \notin E$*. Define the set* $\tilde{E}$ *as*

$$\tilde{E} := E \cup \bigcup_{i \in I_E} (t_i, \sigma(t_i)),$$

where $I_E := \{i \in I : t_i \in E \cap R\}$ *with* $I \subset \mathbb{N}$ *and* $R := \{t \in \mathbb{T} : t < \sigma(t)\}$*. Let* $f : \mathbb{T} \to \mathbb{R}$ *be a* Δ*-measurable function and* $\tilde{f} : [a,b] \to \mathbb{R}$ *be its extension defined in* (11.3)*. Then,* $f \in \mathrm{L}_1(E, \mathbb{R})$ *if and only if* $\tilde{f} \in \mathrm{L}_1(\tilde{E}, \mathbb{R})$*. Further*

$$\int_E f(s)\Delta s = \int_{\tilde{E}} \tilde{f}(s) ds.$$

Definition 11.1.12. *(See [6, Definition 2.30]) A function* $f : [a,b]_{\mathbb{T}} \to \mathbb{R}$ *is said to be* Δ*-Carathéodory function if it satisfies the following conditions.*

(C-i) *The map* $t \mapsto f(t,x)$ *is* Δ*-measurable for every* $x \in \mathbb{R}$.

(C-ii) *The map* $x \mapsto f(t,x)$ *is continuous* Δ*-a.e.* $t \in [a,b]_{\mathbb{T}}$.

(C-iii) *For every* $r > 0$*, there exists* $h_r \in \mathrm{L}_1([a,b]_{\mathbb{T}}, \mathbb{R}_+)$ *such that*

$$|f(t,x)| \leq h_r(t)$$

for Δ*-a.e.* $t \in [a,b]_{\mathbb{T}}$ *and* $|x| < r + |x_0|$.

Definition 11.1.13. *An arc* $x : [a,b)_{\mathbb{T}} \to \mathbb{R}$ *is said to be a solution of dynamic problem* (11.1)*, if it satisfy* (11.1)*. A solution of* (11.1) *is* Δ*-Carathéodory solution if the function* f *in* (11.1) *is* Δ*-Carathéodory function.*

11.2 Carathéodory Solutions

We shall deal with the Carathéodory solutions of dynamic problem (11.1). We start with the following useful lemma.

Lemma 11.2.1. *Let $x : \mathbb{T} \to \mathbb{R}$ be an arc and $f \in \mathrm{L}_1([a, b)_{\mathbb{T}}, \mathbb{R})$ be such that*

$$x(t) = \int_c^t f(s)\Delta s,$$

for each $t \in [c, d]_{\mathbb{T}}$, where $c, d \in \mathbb{T}$ with $c < d$. Then

$$x^{\Delta}(t) = f(t) \quad \Delta\text{-a.e. } t \in [c, d)_{\mathbb{T}}.$$

Proof. Consider all those points $t \in [c, d)_{\mathbb{T}}$ where $x^{\Delta}(t)$ exists. If $\sigma(t) > t$, then by definition

$$x^{\Delta}(t) = \frac{\int_c^{\sigma(t)} f(s)\Delta s - \int_c^t f(s)\Delta s}{\mu(t)} = \frac{\int_t^{\sigma(t)} f(s)\Delta s}{\mu(t)} = f(t).$$

If $\sigma(t) = t$ and t is a Lebesgue point of $\tilde{f}$, then there is a sequence $\{t_i\}_{i \in \mathbb{N}}$ in $[c, d)_{\mathbb{T}}$ such that $t_i \to t$. Now, using Theorem 11.1.11, we can write

$$x^{\Delta}(t) = \lim_{i \to \infty} \frac{x(t_i) - x(t)}{t_i - t} = \lim_{i \to \infty} \frac{\int_t^{t_i} f(s)\Delta s}{t_i - t} = \lim_{i \to \infty} \frac{\int_t^{t_i} \tilde{f}(s)ds}{t_i - t} = \tilde{f}(t) = f(t).$$

Thus, if $D := \{t \in [c, d)_{\mathbb{T}} : x^{\Delta}(t) \neq f(t)\}$, then $D \subset R \cap E$, where $R := \{t \in [c, d)_{\mathbb{T}} : \sigma(t) = t\}$ and E is the set of points $t \in [c, d)_{\mathbb{T}}$ such that t is not a Lebesgue point of $\tilde{f}$. From Lemma 11.1.10, we find that

$$m^*(D) \leq m^*(R \cap E) = \mu^*(R \cap E) \leq \mu^*(E) = 0.$$

Hence, $x^{\Delta}(t) = f(t)$ Δ-a.e. $t \in [c, d)_{\mathbb{T}}$. □

Lemma 11.2.2. *Let $f : [a, b]_{\mathbb{T}} \times \mathbb{R} \to \mathbb{R}$ be an integrable function.*

1. *If an arc $x : [a, b]_{\mathbb{T}} \to \mathbb{R}$ is a solution of dynamic problem (11.1), then it follows that*

$$x(t) = A + \int_a^t f(s, x(s))\Delta s \quad \forall\, t \in [a, b]_{\mathbb{T}}. \tag{11.4}$$

2. *Every arc $x : [a, b]_{\mathbb{T}} \to \mathbb{R}$ obeying (11.4) is a solution of dynamic problem (11.1).*

Theorem 11.2.3. *Let $f : [a,b]_{\mathbb{T}} \times \mathbb{R} \to \mathbb{R}$ be a Δ-Carathéodory function. Then there exists $b_1 \in \mathbb{T} \setminus \{a\}$ such that the dynamic problem* (11.1) *defined on $[a,b_1)_{\mathbb{T}}$ has a Carathéodory solution.*

Proof. First, we assume that a is right-scattered point, $\sigma(a) > a$. Then take $b_1 = \sigma(a)$. Define an arc $x : [a,b_1]_{\mathbb{T}} \to \mathbb{R}$ by

$$x(t) := \begin{cases} A, & \text{if } t = a, \\ \mu(a)f(a,x(a)) + x(a), & \text{if } t = b_1. \end{cases}$$

Then, by definition

$$x^{\Delta}(a) = \frac{x(\sigma(a)) - x(a)}{\sigma(a) - a} = \frac{x(\sigma(a)) - x(a)}{\mu(a)} = \frac{\mu(a)f(a,x(a))}{\mu(a)} = f(a,x(a)).$$

Thus, $x^{\Delta}(t) = f(t,x(t))$ for $t \in [a,b_1)_{\mathbb{T}}$. Hence, x is a Carathéodory solution to dynamic problem (11.1). Next, we assume that a is right-dense point. That is, $\sigma(a) = a$. Let $r > 0$ be arbitrarily fixed. Then, by Theorem 11.1.8, there exists $b_1 \in \mathbb{T} \setminus \{a\}$ such that

$$\int_a^{b_1} h_r(s)\Delta s \le r.$$

We define the set $\mathcal{A}$ to be the set of all arcs $x : \mathbb{T} \to \mathbb{R}$ satisfying $|x(t) - A| \le r$ for all $t \in [a,b_1]_{\mathbb{T}}$ and

$$\left| \int_c^d x^{\Delta}(s)\Delta s \right| \le \int_c^d h_r(s)\Delta s$$

for each $c, d \in \mathbb{T}$ such that $c < d$. Using Theorem 11.1.6, we conclude that the set $\mathcal{A}$ is a compact set. Also, $\mathcal{A}$ is a convex set. Now, for each $x \in \mathcal{A}$, we define the function $y_x : \mathbb{T} \to \mathbb{R}$ by

$$y_x(t) := \begin{cases} x(t), & \text{if } t \in [a,b_1]_{\mathbb{T}}, \\ x(b_1), & \text{if } t \in \mathbb{T} \setminus [a,b_1]_{\mathbb{T}}. \end{cases}$$

Note that the function y_x is a continuous function. Therefore, the function $f(s, y_x(s))$ is Δ-measurable. Define the mapping $F : \mathcal{A} \to \mathrm{C}(\mathbb{T},\mathbb{R})$ by

$$F[x](t) := A + \int_a^t f(s, y_x(s))\Delta s \tag{11.5}$$

for all $t \in \mathbb{T}$ and all $x \in \mathcal{A}$. Then, by Theorem 11.1.8, Fx is an arc, for arbitrarily fixed $x \in \mathcal{A}$. From Lemma 11.2.1, it follows that

$$(F[x](t))^{\Delta} = f(t, y_x(t)) \quad \Delta\text{-a.e. } t \in [a,b)_{\mathbb{T}}.$$

Therefore, for each $c, d \in \mathbb{T}$ such that $c < d$, we have

$$\left| \int_c^d (F[x](s))^{\Delta} \Delta s \right| = \left| \int_c^d f(s, y_x(s)) \Delta s \right| \leq \int_c^d h_r(s) \Delta s.$$

Now, for each $t \in [a, b_1]_{\mathbb{T}}$, we deduce that

$$\begin{aligned} |F[x](t) - A| = \left| \int_a^t f(s, y_x(s)) \Delta s \right| \leq \int_a^t |f(s, y_x(s))| \Delta s &\leq \int_a^t h_r(s) \Delta s \\ &\leq \int_a^{b_1} h_r(s) \Delta s \\ &\leq r. \end{aligned}$$

Hence, we obtain $F(\mathcal{A}) \subset \mathcal{A}$. Let $\bar{x} \in \mathcal{A}$ be arbitrary. Then, there is a sequence $\{x_i\}_{i \in \mathbb{N}}$ in $\mathcal{A}$ such that $\|x_i - \bar{x}\|_{\infty} \to 0$ as $i \to \infty$. Thus, for each $s \in \mathbb{T}$,

$$\lim_{i \to \infty} |f(s, y_{x_i}(s)) - f(s, y_{\bar{x}}(s))| = 0$$

and

$$|f(s, y_{x_i}(s)) - f(s, y_{\bar{x}}(s))| \leq 2h_r(s)$$

for all $i \in \mathbb{N}$. By Lebesgue dominated convergence theorem, we can write

$$\lim_{i \to \infty} \int_a^b |f(s, y_{x_i}(s)) - f(s, y_{\bar{x}}(s))| \, \Delta s = 0$$

and then $\lim_{i \to \infty} |Fx_i - F\bar{x}| = 0$. Hence, the mapping F is continuous at $\bar{x}$. Therefore, by Schaefer's fixed point theorem [12], there exists $x^* \in \mathcal{A}$ such that $F(x^*) = x^*$. From Lemma 11.2.1, we may conclude that $x^{*\Delta}(t) = f(t, x^*(t))$ Δ- a.e. $t \in [a, b_1)_{\mathbb{T}}$. Hence, x^* is a solution of dynamic problem (11.1). □

Theorem 11.2.4. *Let $f : [a, b]_{\mathbb{T}} \times \mathbb{R} \to \mathbb{R}$ be a Δ-Carathéodory function. If there is a sequence $\{f_k\}$ of Δ-Carathéodory functions such that*

$$\int_a^t f_k(s, z_k(s)) \Delta s \to \int_a^t f(s, x(s)) \Delta s,$$

where $\{z_k\}$ is the sequence of arcs converges uniformly to x on $[a, b)_{\mathbb{T}}$, then there exists $b_1 \in [a, b)_{\mathbb{T}} \setminus \{a\}$ such that the dynamic problem (11.1) has a Δ-Carathéodory solution x on $[a, b_1)_{\mathbb{T}}$.

Proof. If a is right scattered point in $[a, b]_{\mathbb{T}}$, then take $b_1 = \sigma(a)$ and whence $[a, b_1)_{\mathbb{T}} = \{a\}$. Define the arc $x : [a, b_1]_{\mathbb{T}} \to \mathbb{R}$ by

$$x(t) := \begin{cases} A & \text{for } t = a, \\ f(a, x(a))\mu(a) + x(a) & \text{for } t = b_1. \end{cases}$$

Then

$$x(b_1) - x(a) = f(a, x(a))\mu(a) \text{ implies } \frac{x(b_1) - x(a)}{\mu(a)} = f\big(a, x(a)\big).$$

That is, $x^{\Delta}(t) = f(t, x(t))$ for $t \in [a, b_1)_{\mathbb{T}} = \{a\}$. Thus, x is a Δ-Carathéodory solution to dynamic problem (11.1). If $\sigma(a) = a$, then by Theorem 11.1.8, there exists $b_1 \in [a, b)_{\mathbb{T}} \setminus \{a\}$ such that

$$\int_a^{b_1} h_r(s)\Delta s < r \tag{11.6}$$

for arbitrarily fixed $r > 0$. Since $\{f_i\}$ is a sequence of Δ-Carathéodory functions, by Theorem 11.2.3, there is a sequence of $\{x_i\}$ of arcs $x_i : [a, b_1)_{\mathbb{T}} \to \mathbb{R}$, $b_1 \in [a, b)_{\mathbb{T}} \setminus \{a\}$ obeying (11.1). Therefore, each x_i is a Δ-Carathéodory solution of dynamic problem(11.1) on $[a, b_1)_{\mathbb{T}}$. By Lemma 11.2.2, for $t \in [a, b_1)_{\mathbb{T}}$ and for each $i \in \mathbb{N}$, we obtain

$$x_i(t) = A + \int_a^t f_i(s, x_i(s))\Delta s.$$

Then

$$|x_i(t)| \leq |A| + \int_a^t |f_i(s, x_i(s))|\Delta s \leq |A| + \int_a^{b_1} h_r(s)\Delta s$$

which, by (11.6), yields $|x_i(t)| \leq |A| + r$ $(r > 0)$. Hence, $\{x_i\}$ is uniformly bounded on $[a, b_1)_{\mathbb{T}}$. Now, for $t_1, t_2 \in [a, b_1)_{\mathbb{T}}$, we have

$$\begin{aligned}
|x_i(t_1) - x_i(t_2)| &= \left|\int_a^{t_1} f_i(s, x_i(s))\Delta s - \int_a^{t_2} f_i(s, x_i(s))\Delta s\right| \\
&\leq \int_{t_1}^{t_2} |f_i(s, x_i(s))|\Delta s \\
&\leq \int_{t_1}^{t_2} h_r(s)\Delta s.
\end{aligned}$$

But, from Theorem 11.1.8, for every $\varepsilon > 0$, there exists $\delta > 0$ such that $t_1, t_2 \in [a, b_1)_{\mathbb{T}}$ and $|t_1 - t_2| \leq \delta$ imply

$$\int_{t_1}^{t_2} h_r(s)\Delta s < \varepsilon.$$

Consequently, $|x_i(t_1) - x_i(t_2)| < \varepsilon$. Therefore, $\{x_i\}$ is equicontinuous on $[a, b_1)_{\mathbb{T}}$. By the Arzelà–Ascoli theorem [1], there is a subsequence $\{y_j\}$ of $\{x_i\}$ which converges uniformly on $[a, b_1)_{\mathbb{T}}$ to an arc $x : [a, b_1)_{\mathbb{T}} \to \mathbb{R}$. We show that this x is solution of dynamic problem (11.1). Since $x_i(a) = A$, $x(a) = \lim_{j\to\infty} y_j(a) = A$. Let $t \in [a, b_1)_{\mathbb{T}}$ be fixed. Then, by hypothesis,

$$\lim_{k\to\infty} \int_a^t f_k(s, z_k(s))\Delta s = \int_a^t f(s, x(s))\Delta s,$$

where $\{z_k\}$ is a subsequence of $\{y_j\}$. Since $z_k(t) = A + \int_a^t f_k(s, z_k(s))\Delta s$ and $\lim_{k\to\infty} z_k(t) = x(t)$, it follows that

$$\lim_{k\to\infty}\left(A + \int_a^t f_k(s, z_k(s))\Delta s\right) = x(t),$$

and we obtain $A + \int_a^t f(s, x(s))\Delta s = x(t)$ which, by Lemma 11.2.2, proves that x is a solution of dynamic problem (11.1). □

Theorem 11.2.5. *Let $f : [a,b]_{\mathbb{T}} \times \mathbb{R} \to \mathbb{R}$ be a Δ-Carathéodory function. If there exists a positively regressive and rd-continuous function $\beta : [a,b]_{\mathbb{T}} \to \mathbb{R}$ such that $|f(t,u) - f(t,v)| \leq \beta(t)|u - v|$ for Δ-a.e. $t \in [a,b]_{\mathbb{T}}$ and for all $u, v \in \mathbb{R}$, then the dynamic problem* (11.1) *has a unique solution x such that $\|x\|_\beta \leq |A| + M\|h_r\|_\beta$ for some positive constant M.*

Proof. Define the TZ-norm $\|\cdot\|_\beta$ on the space $BM([a,b]_{\mathbb{T}}, \mathbb{R})$ as

$$\|x\|_\beta = \sup_{t\in[a,b]_{\mathbb{T}}} \frac{|x(t)|}{e_\beta(t,a)},$$

where $\beta : [a,b]_{\mathbb{T}} \to \mathbb{R}$ is regressive, rd-continuous function. Then $|x(t)| \leq \|x\|_\beta\, e_\beta(t,a)$ for all $t \in [a,b]_{\mathbb{T}}$. Let $F : \mathrm{BM}([a,b]_{\mathbb{T}}, \mathbb{R}) \to \mathrm{BM}([a,b]_{\mathbb{T}}, \mathbb{R})$ be defined by

$$F[x](t) = A + \int_a^t f(s, x(s))\Delta s, \quad t \in [a,b]_{\mathbb{T}}.$$

Let $x_1, x_2 \in \mathrm{BM}([a,b]_{\mathbb{T}}, \mathbb{R})$. Then, for all $t \in [a,b]_{\mathbb{T}}$,

$$F[x_1](t) = A + \int_a^t f(s, x_1(s))\Delta s \text{ and } F[x_2](t) = A + \int_a^t f(s, x_2(s))\Delta s.$$

Therefore,

$$\begin{aligned} |F[x_1](t) - F[x_2](t)| &= \left|\int_a^t [f(s, x_1(s)) - f(s, x_2(s))]\,\Delta s\right| \\ &\leq \int_a^t \beta(s)|x_1(s) - x_2(s)|\Delta s. \end{aligned}$$

Then

$$\begin{aligned} \frac{|F[x_1](t) - F[x_2](t)|}{e_\beta(t,a)} &\leq \frac{1}{e_\beta(t,a)} \int_a^t \beta(s)\,|x_1(s) - x_2(s)|\,\Delta s \\ &\leq \frac{1}{e_\beta(t,a)} \int_a^t \beta(s) e_\beta(s,a)\|x_1 - x_2\|_\beta \Delta s \\ &= \frac{1}{e_\beta(t,a)} \int_a^t e_\beta^\Delta(s,a)\|x_1 - x_2\|_\beta \Delta s \\ &= \left[1 - \frac{1}{e_\beta(t,a)}\right] \|x_1 - x_2\|_\beta. \end{aligned}$$

This yields

$$\sup_{t\in[a,b]_{\mathbb{T}}} \frac{|F[x_1](t) - F[x_2](t)|}{e_\beta(t,a)} \leq \sup_{t\in[a,b]_{\mathbb{T}}} \left[1 - \frac{1}{e_\beta(t,a)}\right] \|x_1 - x_2\|_\beta$$
$$= \left[1 - \frac{1}{e_\beta(b,a)}\right] \|x_1 - x_2\|_\beta = \alpha\|x_1 - x_2\|_\beta$$

where $\alpha = 1 - \dfrac{1}{e_\beta(b,a)} < 1$. Hence,

$$\|Fx_1 - Fx_2\|_\beta \leq \alpha\|x_1 - x_2\|_\beta \quad \text{and} \quad 0 < \alpha < 1 \ (b \neq a).$$

The Banach's fixed point theorem [12] assures that the function F has a unique fixed point in $\mathrm{BM}([a,b]_{\mathbb{T}}, \mathbb{R})$. This yields that the dynamic problem (11.1) has a unique solution. Now, for $t < b$, dividing throughout by $e_\beta(t,a)$ to (11.4), we get

$$\frac{|x(t)|}{e_\beta(t,a)} \leq \frac{|A|}{e_\beta(t,a)} + \frac{1}{e_\beta(t,a)} \int_a^t |f(s,x(s))|\,\Delta s$$

and then

$$\begin{aligned}
\frac{|x(t)|}{e_\beta(t,a)} &\leq \frac{|A|}{e_\beta(t,a)} + \frac{1}{e_\beta(t,a)} \int_a^t h_r(s)\Delta s \\
&\leq \frac{|A|}{e_\beta(t,a)} + \frac{1}{e_\beta(t,a)} \|h_r\|_\beta \int_a^t e_\beta(s,a)\Delta s \\
&\leq \frac{|A|}{e_\beta(t,a)} + \frac{1}{e_\beta(a,a)} \|h_r\|_\beta \int_a^t e_\beta(s,a)\Delta s \\
&\leq \frac{|A|}{e_\beta(t,a)} + \|h_r\|_\beta \int_a^t e_\beta(s,a)\Delta s \\
&\leq \frac{|A|}{e_\beta(t,a)} + \|h_r\|_\beta\,(b-a)e_\beta(b,a).
\end{aligned}$$

That is,

$$\frac{|x(t)|}{e_\beta(t,a)} \leq \frac{|A|}{e_\beta(t,a)} + \|h_r\|_\beta\,(b-a)e_\beta(b,a) \leq |A| + M\|h_r\|_\beta,$$

where $M := (b-a)e_\beta(b,a)$. Therefore $\|x\|_\beta \leq |A| + M\|h_r\|_\beta$ which shows that the solution x is bounded with respect to $\|\cdot\|_\beta$. □

Theorem 11.2.6. *If $f : [a,b]_{\mathbb{T}} \times \mathbb{R} \to \mathbb{R}$ is a continuous function such that there exist a constant $L > 0$ and a function $c \in L_1([a,b]_{\mathbb{T}}, \mathbb{R}_+)$ satisfying*

$$|f(t,x)| \leq L|x| + c(t)$$

for Δ-a.e. $t \in [a,b)_{\mathbb{T}}$ and for all $x \in \mathbb{R}$, then the dynamic problem (11.1) *has at least one continuous solution.*

Proof. Define the operator $F : C([a,b]_{\mathbb{T}}, \mathbb{R}) \to C([a,b]_{\mathbb{T}}, \mathbb{R})$ by

$$F[x](t) := A + \int_a^t f(s, x(s))\Delta s, \quad t \in [a,b]_{\mathbb{T}}. \tag{11.7}$$

We use Schaefer's fixed point theorem [12]. In view of Lemma 11.2.2, the fixed points of F will be solutions to the dynamic problem (11.1). We first show that F is continuous. For this, take a sequence $\{u_k\}$ in $C([a,b]_{\mathbb{T}}, \mathbb{R})$ such that $\|u_k - u\|_\beta \to 0$. Thence $\|u_k - u\|_0 \to 0$. Now, For $t \in [a,b]_{\mathbb{T}}$, we have

$$\begin{aligned}|F[u_k](t) - F[u](t)| &= \left| \int_a^t f(s, u_k(s))\Delta s - \int_a^t f(s, u(s))\Delta s \right| \\ &\leq \int_a^t |f(s, u_k(s)) - f(s, u(s))|\, \Delta s \\ &\leq \int_a^b |f(s, u_k(s)) - f(s, u(s))|\, \Delta s\end{aligned}$$

and then

$$\|Fu_k - Fu\|_0 \leq \int_a^b |f(s, u_k(s)) - f(s, u(s))|\, \Delta s.$$

Since

$$|f(t, u_k(t)) - f(t, u(t))| \to 0 \quad \Delta\text{-a.e. } t \in [a,b]_{\mathbb{T}},$$

we may conclude that $\|Fu_k - Fu\|_0 \to 0$, and then $\|Fu_k - Fu\|_\beta \to 0$. Thus, F is a continuous map from $C([a,b]_{\mathbb{T}}, \mathbb{R})$ to itself. Let Ω be a bounded subset of $C([a,b]_{\mathbb{T}}, \mathbb{R})$. So, there exists a constant $k > 0$ obeying $\|u\|_0 \leq k$, for all $u \in \Omega$. For $u \in \Omega$, we have

$$\begin{aligned}|F[u](t)| &= \left| A + \int_a^t f(s, u(s))\Delta s \right| \\ &\leq |A| + \int_a^b |f(s, u(s))|\Delta s \\ &\leq |A| + \int_a^b (L|u(s)| + c(s))\, \Delta s \\ &\leq |A| + \int_a^b L\|u\|_0 \Delta s + \int_a^b c(s)\Delta s \\ &\leq |A| + Lk(b-a) + \int_a^b c(s)\Delta s\end{aligned}$$

and then

$$\|Fu\|_0 \leq |A| + Lk(b-a) + \int_a^b c(s)\Delta s.$$

Therefore, $F(\Omega)$ is uniformly bounded. Next, for $t_1, t_2 \in [a,b]_{\mathbb{T}}$ with $t_1 < t_2$, we have

$$\begin{aligned}
|F[u](t_1) - F[u](t_2)| &\leq \int_{t_1}^{t_2} |f(s, u(s))|\,\Delta s \\
&\leq \int_{t_1}^{t_2} (L|u(s)| + c(s))\,\Delta s \\
&\leq \int_{t_1}^{t_2} L\|u\|_0\,\Delta s + \int_{t_1}^{t_2} c(s)\Delta s \\
&\leq Lk(t_2 - t_1) + \int_{t_1}^{t_2} c(s)\Delta s.
\end{aligned}$$

Let $\varepsilon > 0$ be given. Then, by Theorem 11.1.8, there exists a $\delta > 0$ such that $t_1, t_2 \in [a,b]_{\mathbb{T}}$ and $|t_1 - t_2| < \delta$ imply $|F[u](t_1) - F[u](t_2)| < \varepsilon$ for all $u \in \Omega$. Hence, $F(\Omega)$ is equicontinuous set in $\mathrm{C}([a,b]_{\mathbb{T}}, \mathbb{R})$ and the Arzelà–Ascoli theorem [1] assures that the set $F(\Omega)$ is relatively compact. Therefore, the map $F : \mathrm{C}([a,b]_{\mathbb{T}}, \mathbb{R}) \to \mathrm{C}([a,b]_{\mathbb{T}}, \mathbb{R})$ is completely continuous. Now, consider the set $\Gamma \subset \mathrm{C}([a,b]_{\mathbb{T}}, \mathbb{R})$ defined by

$$\Gamma := \big\{x \in \mathrm{C}([a,b]_{\mathbb{T}}, \mathbb{R}) : x = \lambda F x \text{ for some } \lambda \in [0,1]\big\}.$$

We note that

$$\int_a^t L e_L(s,a)\,\Delta s = e_L(t,a) - 1 \text{ and } \|A\|_L = \sup_{t\in[a,b]_{\mathbb{T}}} \frac{|A|}{e_L(t,a)} = |A|.$$

If $x \in \Gamma$, it follows that

$$x(t) = \lambda A + \lambda \int_a^t f\big(s, x(s)\big)\Delta s.$$

Then $\|x\|_L = \lambda\|Fx\|_L \leq \|Fx\|_L$ and so

$$\begin{aligned}
\|x\|_L &\leq \sup_{t\in[a,b]_{\mathbb{T}}} \frac{\left|A + \int_a^t f\big(s,x(s)\big)\Delta s\right|}{e_L(t,a)} \\
&\leq |A| + \sup_{t\in[a,b]_{\mathbb{T}}} \frac{\int_a^t \big|f\big(s,x(s)\big)\big|\,\Delta s}{e_L(t,a)} \\
&\leq |A| + \sup_{t\in[a,b]_{\mathbb{T}}} \frac{\int_a^t \{L|x(s)| + c(s)\}\,\Delta s}{e_L(t,a)} \\
&= |A| + \sup_{t\in[a,b]_{\mathbb{T}}} \frac{\int_a^t L\, e_L(s,a)\frac{|x(s)|}{e_L(s,a)}\Delta s + \int_a^t c(s)\Delta s}{e_L(t,a)}
\end{aligned}$$

$$\le |A| + \sup_{t\in[a,b]_{\mathbb{T}}} \frac{\displaystyle\int_a^t L\, e_L(s,a)\|x\|_L\, \Delta s + \int_a^t c(s)\Delta s}{e_L(t,a)}$$

$$= |A| + \sup_{t\in[a,b]_{\mathbb{T}}} \frac{\|x\|_L(e_L(t,a)-1) + \displaystyle\int_a^t c(s)\Delta s}{e_L(t,a)}$$

$$= |A| + \sup_{t\in[a,b]_{\mathbb{T}}} \left\{ \|x\|_L\left(1 - \frac{1}{e_L(t,a)}\right) + \frac{\displaystyle\int_a^t c(s)\Delta s}{e_L(t,a)} \right\}$$

$$\le |A| + \|x\|_L\left(1 - \frac{1}{e_L(b,a)}\right) + \frac{\displaystyle\int_a^b c(s)\Delta s}{e_L(a,a)}$$

$$= |A| + \|x\|_L - \frac{\|x\|_L}{e_L(b,a)} + \int_a^b c(s)\Delta s.$$

Hence,

$$\|x\|_L \le |A|\; e_L(b,a) + e_L(b,a)\int_a^b c(s)\Delta s$$

and therefore Γ is bounded. As a consequence of Schaefer's fixed point theorem [12], we deduce that, F has a fixed point u in $C([a,b]_{\mathbb{T}};\mathbb{R})$, equivalently, the dynamic problem (11.1) has a solution in $C\big([a,b]_{\mathbb{T}},\mathbb{R}\big)$. □

11.3 Generalized Dynamic Equations

11.3.1 Henstock–Kurzweil Δ-integral

In this section, the solutions of Carathéodory dynamic problem (11.1) is discussed, based on the concept of Henstock–Kurzweil delta integral.

Definition 11.3.1. *[11] A function $x : [a,b]_{\mathbb{T}} \to \mathbb{R}$ is said to be absolutely continuous in the restricted sense on $E \subset [a,b]_{\mathbb{T}}$ if, for every $\varepsilon > 0$, there exists $\delta > 0$ such that*

$$\sum_{i=1}^{n} \operatorname{osc}(x, [c_i, d_i]_{\mathbb{T}}) < \varepsilon,$$

whenever $\{[c_i,d_i]_{\mathbb{T}} : 1 \le i \le n\}$ is a finite family of pairwise disjoint subintervals of $[a,b]_{\mathbb{T}}$ with $c_i, d_i \in E$ satisfying

$$\sum_{i=1}^{n} \mu_\Delta([c_i,d_i]_{\mathbb{T}}) < \delta.$$

We write $x \in \Delta\text{-AC}_*$. A function x is said to be generalized absolutely continuous in the restricted sense if x is continuous on $E \subset \mathbb{T}$ and E can be written as countable union of sets on each of which x is $\Delta\text{-AC}_*$. In this case, we write $x \in \Delta\text{-ACG}_*$. If x is uniformly continuous on E and E can be written as countable union of sets on each of which x is $\Delta\text{-AC}_*$, then we say that x is uniformly $\Delta\text{-ACG}_*$ and write $x \in \Delta\text{-UACG}_*$.

Definition 11.3.2. *A pair of functions (δ_L, δ_R) is said to be a Δ-gauge for $[a,b]_{\mathbb{T}}$ if $\delta_L, \delta_R : [a,b]_{\mathbb{T}} \to \mathbb{R}$ be two functions such that $\delta_L(t) > 0$ for $t \in (a,b]_{\mathbb{T}}$, $\delta_L(a) \geq 0$, $\delta_R(t) > 0$ for $t \in [a,b)_{\mathbb{T}}$, $\delta_R(b) \geq 0$ and $\delta_R(t) \geq \mu(t)$ for all $t \in [a,b)_{\mathbb{T}}$. We write $\delta = (\delta_L, \delta_R)$.*

Definition 11.3.3. *A division of $[a,b]_{\mathbb{T}}$ defined by*

$$\mathcal{P} := \{a = t_0 \leq \xi_1 \leq t_1 \leq \xi_2 \leq t_2 \dots t_{n-1} \leq \xi_n \leq t_n = b\}$$

with $t_i > t_{i-1}$ for $1 \leq i \leq n$ and $t_i,\ \xi_i \in \mathbb{T}$ is called as a partition of $[a,b]_{\mathbb{T}}$. The point ξ_i is called tag point associated with the subinterval $[t_{i-1}, t_i]_{\mathbb{T}}$ of $[a,b]_{\mathbb{T}}$. We denote such partition by $\mathcal{P} = \{\xi_i, [t_{i-1}, t_i]_{\mathbb{T}}\}_{i=1}^n$.

Definition 11.3.4. *A partition $\mathcal{P}$ of $[a,b]_{\mathbb{T}}$ is δ-fine if*

$$\xi_i - \delta_L(\xi_i) \leq t_{i-1} < t_i \leq \xi_i + \delta_R(\xi_i)$$

for $1 \leq i \leq n$, where $\delta := (\delta_L, \delta_R)$ is a Δ-gauge for $[a,b]_{\mathbb{T}}$.

Definition 11.3.5. *A function $f : [a,b]_{\mathbb{T}} \to \mathbb{R}$ is said to be Henstock–Kurzweil Δ-integrable on $[a,b]_{\mathbb{T}}$ provided there is a real number I such that for every $\varepsilon > 0$, there is a Δ-gauge δ, for $[a,b]_{\mathbb{T}}$ such that for all δ-fine partitions $\mathcal{P} = \{\xi_i, [t_{i-1}, t_i]_{\mathbb{T}}\}_{i=1}^n$ of $[a,b]_{\mathbb{T}}$,*

$$\left| I - \sum_{i=1}^{n} f(\xi_i)(t_i - t_{i-1}) \right| < \varepsilon.$$

The number I is called as Henstock–Kurzweil Δ-integral of f on $[a,b]_{\mathbb{T}}$ and we write

$$I = (HK) \int_a^b f(t)\Delta t.$$

Using Henstock–Kurzweil Δ-integral, we generalize the Δ-Carathéodory dynamic equation (11.1) as follows. The dynamic initial value problem (11.1) is said to be *generalized Δ-Carathéodory dynamic problem* if the function $f : [a,b]_{\mathbb{T}} \times \mathbb{R} \to \mathbb{R}$ satisfies the following conditions.

(C1) The map $t \mapsto f(t,x)$ is Δ-measurable for every $x \in \mathbb{R}$.

(C2) The map $x \mapsto f(t,x)$ is continuous Δ-a.e. $t \in [a,b]_{\mathbb{T}}$.

(C3) For every real number $r > 0$, there exist two Henstock–Kurzweil Δ-integrable functions g_r and h_r on $[a,b]_{\mathbb{T}}$ such that $g_r(t) \leq f(t,x) \leq h_r(t)$ for Δ-a.e. $t \in [a,b]_{\mathbb{T}}$ and for $x \in \overline{B}_r(A) := \{x \in \mathbb{R} : |x - A| \leq r\}$.

In this case, the function f is said to be the *generalized Δ-Carathéodory function.*

Definition 11.3.6. *A function $x : [a,b]_{\mathbb{T}} \to \mathbb{R}$ is said to be a solution of generalized Δ-Carathéodory dynamic problem* (11.1) *if x satisfies the following conditions.*

(i) *x is Δ-ACG$_*$ function on each compact time scale subinterval $\mathbb{J}$ of $[a,b]_{\mathbb{T}}$.*

(ii) *$x^{\Delta} = f(t,x)$ Δ-a.e. $t \in [a,b]_{\mathbb{T}}$ and $x(a) = A$.*

The above condition (C3) can be written as

$$0 \leq f(t,x) - g_r(t) \leq h_r(t) - g_r(t) \ \ (\Delta\text{-a.e. } t \in [a,b]_{\mathbb{T}} \text{ and } x \in \overline{B}_r(A)).$$

Then, $h_r - g_r$ is nonnegative Henstock–Kurzweil Δ-integrable function and by Theorem 2.19 of [7], it is a Δ-Lebesgue integrable function. Also, $f(t,x) - g_r(t)$ satisfies condition (C-iii). Thus, our generalized Δ-Carathéodory dynamic equation (11.1) is the essentially first-order dynamic equation of the form $x^{\Delta} = F(t,x) + G(t)$, where $F(t,x)$ is Δ-Lebesgue integrable function and $G(t)$ is Henstock–Kurzweil Δ-integrable function on $\mathbb{T}$, that is, *a first-order dynamic equation perturbed by Henstock–Kurzweil Δ-integrable function.*

11.3.2 Existence and uniqueness of solutions

Theorem 11.3.7. *If $f : [a,b]_{\mathbb{T}} \times \mathbb{R} \to \mathbb{R}$ be a generalized Δ-Carathéodory function, then the dynamic problem* (11.1) *has generalized Δ-Carathéodory solution in $[a,b]_{\mathbb{T}}$.*

Proof. Since f is a generalized Δ-Carathéodory function, by condition (C3), there exist two Henstock–Kurzweil Δ-integrable functions g_r and h_r on $\mathbb{T}$ such that for all $x \in \overline{B}_r(A)$,

$$g_r(t) \leq f(t,x) \leq h_r(t) \quad \Delta\text{-a.e. } t \in \mathbb{T}.$$

The function $h_r - g_r$ is nonnegative and Henstock–Kurzweil Δ-integrable on $[a,b]_{\mathbb{T}}$, it follows that, $h_r - g_r$ is Δ-Lebesgue integrable on $[a,b]_{\mathbb{T}}$. Define $F : [a,b]_{\mathbb{T}} \times \overline{B}_r(A) \to \mathbb{R}$ by

$$F(t,x) := f\left(t, x + (\text{HK}) \int_a^t g_r(s)\Delta s\right) - g_r(t).$$

Then F is a Δ-Carathéodory function satisfying

$$0 \leq F(t,x) \leq h_r(t) - g_r(t) \text{ for all } (t,x) \in \Omega,$$

where $\Omega := \left\{(t,x) \in [a,b]_{\mathbb{T}} \times \overline{B}_r(A) : \left| x + (\text{HK}) \int_a^t g_r(\tau)\Delta\tau - A \right| \leq r\right\}$. By

Theorem 11.2.3, there is a function $y : [a,b]_{\mathbb{T}} \to \mathbb{R}$ such that $y^{\Delta}(t) = F(t,y)$ Δ-a.e. $t \in [a,b]_{\mathbb{T}}$ and $y(a) = A$. Define

$$\phi(t) := y(t) + (\text{HK}) \int_a^t g_r(\tau)\Delta\tau \text{ for } t \in [a,b]_{\mathbb{T}}.$$

Then, for Δ-a.e. $t \in [a,b]_{\mathbb{T}}$,

$$\begin{aligned}\phi^{\Delta}(t) &= y^{\Delta}(t) + g_r(t)\\ &= F(t,y) + g_r(t)\\ &= f\left(t, y + (\text{HK}) \int_a^t g_r(\tau)\Delta\tau\right) - g_r(t) + g_r(t)\\ &= f\left(t, y + (\text{HK}) \int_a^t g_r(\tau)\Delta\tau\right)\end{aligned}$$

and $\phi(a) = y(a) = A$. Here, we note that the function ϕ is Δ-ACG$_*$ on $[a,b]_{\mathbb{T}}$ because y is Δ-ACG$_*$ on $[a,b]_{\mathbb{T}}$ and g_r is Henstock–Kurzweil Δ-integrable there. □

Example 11.3.8. *Let* $\mathbb{T} := \left\{t = \frac{1}{n} : n \in \mathbb{N}\right\} \cup \{0\}$. *Let* $f(t,x) := h(t,x) + g(t)$, *where* $|h(t,x)| \leq H(t)$, $\forall\, t \in [0,1]_{\mathbb{T}}$ *and* $x \in \overline{B}_r(A)$ *and* $H(t)$ *is Lebesgue* Δ*-integrable on* $\mathbb{T}$. *Define* $g : \mathbb{T} \to \mathbb{R}$ *by*

$$g(t) := \begin{cases} (-1)^n\, n & \text{if } t = \dfrac{1}{n}, \\ L & \text{if } t = 0, \end{cases}$$

where L *is any constant. The function* g *is neither* Δ*-integrable nor Lebesgue* Δ*-integrable on* $\mathbb{T}$. *But, it is Henstock–Kurzweil* Δ*-integrable on* $\mathbb{T}$. *Note that* $\forall\, t \in [0,1]_{\mathbb{T}}$ *and* $x \in \overline{B}_r(A)$, $g(t) - H(t) \leq f(t,x) \leq g(t) + H(t)$. *Thus,* $f(t,x)$ *is generalized* Δ*-Carathéodory function and by Theorem 11.3.7 the generalized dynamic problem* (11.1) *has a solution in* $\mathbb{T}$. *When* $h(t,x) = 0$, *the function* $x : \mathbb{T} \to \mathbb{R}$ *defined by*

$$x(t) := \begin{cases} 0 & \text{if } t = 1, \\ \displaystyle\sum_{k=2}^{n} \frac{(-1)^{k+1}}{k-1} & \text{if } t = \dfrac{1}{n}, \\ -\ln(2) & \text{if } t = 0 \end{cases}$$

is a solution of the generalized dynamic problem (11.1) *with* f *as defined above and* $x(0) = A = -\ln(2)$.

Theorem 11.3.9. *Let* $f : [a,b]_{\mathbb{T}} \times \mathbb{R} \to \mathbb{R}$ *be a* Δ*-Carathéodory function. Define*

$$F_n(t) = (HK) \int_a^t \phi_n(\tau)\Delta\tau,$$

where $\{\phi_n\}$ *is a sequence of* Δ*-measurable simple functions. If* $\{F_n\}$ *is a sequence of* Δ-UACG$_*$ *functions on* $[a,b]_{\mathbb{T}}$ *such that* $F_n \to x$ *uniformly* Δ*-a.e. on* $[a,b]_{\mathbb{T}}$*, then* x *is generalized* Δ*-Carathéodory solution to the dynamic problem* (11.1).

Proof. Since $f(t,x)$ is Δ-measurable in t, there exists a nondecreasing sequence of Δ-measurable simple functions $\{\phi_n\}$ on $[a,b]_{\mathbb{T}}$ such that $\phi_n(t) \to f(t,x(t))$ for all $t \in [a,b]_{\mathbb{T}}$. Since each ϕ_n is Henstock–Kurzweil Δ-integrable and $\phi_n(t) \leq \phi_{n+1}(t)$ Δ-a.e. on $[a,b]_{\mathbb{T}}$ and for all $n \in \mathbb{N}$, by monotone convergence theorem, it follows that $f(t,x(t))$ is Henstock–Kurzweil Δ-integrable on $[a,b]_{\mathbb{T}}$ and

$$\text{(HK)} \int_a^t f(\tau, x(\tau))\Delta\tau = \lim_{n\to\infty} \text{(HK)} \int_a^t \phi_n(\tau)\Delta\tau.$$

Define

$$F_n(t) := \text{(HK)} \int_a^t \phi_n(\tau)\Delta\tau + A, \quad \text{for } t \in [a,b]_{\mathbb{T}}.$$

Then, $\{F_n\}$ is a sequence of Δ-UACG$_*$ functions and $F_n(t) \to x(t)$, $t \in [a,b]_{\mathbb{T}}$. Therefore

$$\lim_{n\to\infty} F_n(t) = \text{(HK)} \int_a^t f(\tau, x(\tau))\Delta\tau + A.$$

This gives

$$x(t) = \text{(HK)} \int_a^t f(\tau, x(\tau))\Delta\tau + A, \quad t \in [a,b]_{\mathbb{T}}$$

which gives the dynamic problem (11.1). Now, since $\{F_n\}$ is a sequence of Δ-UACG$_*$ functions converges to x and x is a primitive of Henstock–Kurzweil Δ-integrable function f on $[a,b]_{\mathbb{T}}$, it follows that x is Δ-ACG$_*$ on $[a,b]_{\mathbb{T}}$. □

Below we give an inequality which is useful for further analysis.

Theorem 11.3.10. *Suppose* $g : [0,\infty) \to [0,\infty)$ *is continuous nondecreasing function and* $y : [a,b]_{\mathbb{T}} \to [0,\infty)$ *is bounded function such that* $g \circ y : [a,b]_{\mathbb{T}} \to [0,\infty)$ *is rd-continuous. If* $G : [a,b]_{\mathbb{T}} \to \mathbb{R}$ *be a strictly increasing function such that*

$$G^{\Delta}(u) = \frac{1}{g(u)} \neq 0 \text{ for } u \in \mathbb{T},$$

then for all $t \in [a,b]_{\mathbb{T}}$ *and for some constant* $l \geq 0$

$$y(t) \leq l + (HK) \int_a^t g(y(\tau))\Delta\tau \tag{11.8}$$

implies

$$y(t) \leq G^{-1}(G^{\sigma}(l) + t - a),$$

where $G^{-1} : \widetilde{\mathbb{T}} = G([a,b]_{\mathbb{T}}) \to [a,b]_{\mathbb{T}}$ *is the inverse of* $G : [a,b]_{\mathbb{T}} \to \mathbb{R}$.

Proof. Assume for all $t \in [a,b]_{\mathbb{T}}$,

$$y(t) \leq l + (\text{HK}) \int_a^t g\big(y(\tau)\big)\Delta\tau$$

for some constant $l \geq 0$. Denote $G(a) = \alpha$, $G(b) = \beta$, where $\alpha = \min \widetilde{\mathbb{T}}$.

Let $T \in [a,b]_{\mathbb{T}}$ be such that $\alpha - (t-a) < G(T) < \beta - (t-a)$ for all $t \in [a,b]_{\mathbb{T}}$. For $s \in [\alpha - G(T), \beta - G(T)]_{\widetilde{\mathbb{T}}}$, set $G^{-1}(G(T)+s) = u$. Then, $u \in [a,b]_{\mathbb{T}}$ and by hypothesis

$$G^{\Delta}(G^{-1}(G(T)+s)) = \frac{1}{g(G^{-1}(G(T)+s))}.$$

But, by derivative of the inverse, we have $(G^{-1}(p))^{\Delta} = (\frac{1}{G^{\Delta}} \circ G^{-1})(p)$ for $p \in [\alpha, \beta]_{\widetilde{\mathbb{T}}}$. So, $(G^{-1}(p))^{\Delta} = (g \circ G^{-1})(p)$. Now, for $t \in [a,b]_{\mathbb{T}}$

$$\begin{aligned}
&(\text{HK}) \int_a^t g(G^{-1}(G(T)+\tau-a))\Delta\tau \\
&\quad = (\text{HK}) \int_a^t (g \circ G^{-1})(G(T)+\tau-a)\Delta\tau \\
&\quad = (\text{HK}) \int_a^t (G^{-1})^{\Delta}(G(T)+\tau-a)\Delta\tau \\
&\quad = G^{-1}(G(T)+t-a) - G^{-1}(G(T)+a-a) \\
&\quad = G^{-1}(G(T)+t-a) - T.
\end{aligned}$$

Hence,

$$T + (\text{HK}) \int_a^t g(G^{-1}(G(T)+\tau-a))\Delta\tau = G^{-1}(G(T)+t-a).$$

Now, let $T = \sigma(l)$ for some $l \in [a,b]_{\mathbb{T}}^{\kappa}$ such that $G^{\sigma}(l) < \alpha - b + a$. Then,

$$\sigma(l) + (\text{HK}) \int_a^t g(G^{-1}(G^{\sigma}(l)+\tau-a))\Delta\tau = G^{-1}(G^{\sigma}(l)+t-a). \quad (11.9)$$

Since G^{-1} is bounded, combing (11.8) and (11.9), we obtain

$$\begin{aligned}
&y(t) - G^{-1}(G^{\sigma}(l)+t-a) \\
&\quad \leq l + (\text{HK}) \int_a^t g(y(\tau))\Delta\tau - \sigma(l) - (\text{HK}) \int_a^t g(G^{-1}(G^{\sigma}(l)+\tau-a))\Delta\tau \\
&\quad = l - \sigma(l) + (\text{HK}) \int_a^t [g(y(\tau)) - g(G^{-1}(G^{\sigma}(l)+\tau-a))]\Delta\tau.
\end{aligned}$$

Since g is nondecreasing, and y and G^{-1} are bounded, we arrive at

$$\big|g(y(\tau)) - g(G^{-1}(G^{\sigma}(l)+\tau-a))\big| < M$$

for some $M > 0$ and for all $\tau \in [a,b]_{\mathbb{T}}$. Therefore, for $t \in [a,b]_{\mathbb{T}}$

$$y(t) - G^{-1}\big(G^{\sigma}(l) + t - a\big) \le l - \sigma(l) + M(t-a).$$

If $\Lambda := \big\{t \in [a,b]_{\mathbb{T}} : y(t) < G^{-1}(G^{\sigma}(l) + \tau - a) \text{ for } a \le \tau \le b\big\}$, then one can see that $\sup \Lambda = \max \mathbb{T} = b$. Hence, $y(t) \le G^{-1}\big(G^{\sigma}(l) + t - a\big)$ for $t \in [a,b]_{\mathbb{T}}$. □

The local uniqueness of solution of dynamic problem (11.1) involving an Osgood-type condition is given below.

Theorem 11.3.11. *Assume that $f : [a,b]_{\mathbb{T}} \times \overline{B}_r(A) \to \mathbb{R}$ satisfy the condition*

$$|f(t,x) - f(t,y)| \le g(|x-y|),$$

where $g : [a,b]_{\mathbb{T}} \to [0,\infty)$ is such that its extension $\overline{g} : [0,\infty) \to [0,\infty)$ is an increasing function with $\overline{g}(0) = 0$ and $\overline{g}(s) > 0$ for $s \in (0,\infty)$ and for every $k > 0$,

$$(HK) \int_0^k \frac{1}{\overline{g}(s)} ds = \infty.$$

Then, the generalized Δ-Carathéodory dynamic problem (11.1) *has unique solution in $[a,b]_{\mathbb{T}}$.*

Proof. Assume that x and y are two solutions of the generalized Δ-Carathéodory dynamic problem (11.1). Then, $x, y : [a,b]_{\mathbb{T}} \to \mathbb{R}$ are such that

$$x(t) = A + (\mathrm{HK}) \int_a^t f\big(\tau, x(\tau)\big)\Delta\tau$$

and

$$y(t) = A + (\mathrm{HK}) \int_a^t f\big(\tau, y(\tau)\big)\Delta\tau.$$

If $z(t) := x(t) - y(t)$ for all $t \in [a,b]_{\mathbb{T}}$, then

$$\begin{aligned} |z(t)| &= \left|(\mathrm{HK}) \int_a^t [f(\tau, x(\tau)) - f(\tau, y(\tau))]\Delta\tau\right| \\ &\le (\mathrm{HK}) \int_a^t |f(\tau, x(\tau)) - f(\tau, y(\tau))|\,\Delta\tau \\ &\le (\mathrm{HK}) \int_a^t g(|z(\tau)|)\Delta\tau \\ &\le (\mathrm{HK}) \int_a^b g(|z(\tau)|)\Delta\tau. \end{aligned}$$

If $G^{\Delta}(u) = \dfrac{1}{g(u)}$, then by Theorem 11.3.10, we can write

$$\begin{aligned}|z(t)| &\leq G^{-1}(G^{\sigma}(0) + b - a)\\ G(|z(t)|) &\leq G^{\sigma}(0) + b - a\\ G(|z(t)|) - G^{\sigma}(0) &\leq b - a\\ (\text{HK})\int_{\sigma(0)}^{|z(t)|} G^{\Delta}(r)\Delta r &\leq b - a\\ (\text{HK})\int_{\sigma(0)}^{|z(t)|} \frac{1}{g(r)}\Delta r &\leq b - a.\end{aligned}$$

But, from Theorems 19 and 20 of [13],

$$(\text{HK})\int_{\sigma(0)}^{|z(t)|} \frac{1}{g(r)}\Delta r = (\text{HK})\int_{0}^{|z(t)|} \frac{1}{\overline{g}(s)}ds.$$

Therefore,

$$(\text{HK})\int_{0}^{|z(t)|} \frac{1}{\overline{g}(s)}ds \leq b - a$$

which gives $\infty \leq b - a$, a contradiction. Hence, we must have $|z(t)| = 0$ and therefore the dynamic problem (11.1) has unique solution in $[a, b]_{\mathbb{T}}$. □

Remark 11.3.12. *If the conditions given in Theorem 11.3.11 do not hold, then the dynamic problem* (11.1) *may have more than one solutions. This is shown in the following example:*

Let $\mathbb{T} := \left\{t = \dfrac{1}{n} : n \in \mathbb{N}\right\} \cup \{0\}$ *and define* $f : [0, 1]_{\mathbb{T}} \times \overline{B}_r(0) \to \mathbb{R}$*by*

$$f(t, x) := \begin{cases} \dfrac{1}{2(\sqrt{t} + \sqrt{\sigma(t)})} & \text{for } x < t,\\ \dfrac{1}{(\sqrt{t} + \sqrt{\sigma(t)})} & \text{for } x \geq t,\end{cases}$$

where $\overline{B}_r(0) = \{x \in \mathbb{R} : |x| \leq r\}$. *Then, we find that*

$$(\text{HK})\int_{0}^{1} f\big(t, x(t)\big)\Delta t = 1.$$

So, f *does not satisfy conditions given in Theorem 11.3.11. In fact,* $x_1(t) = \sqrt{t}$ *and* $x_2(t) = \dfrac{\sqrt{t}}{2}$ *are two solutions of dynamic problem* (11.1).

Remark 11.3.13. *The conditions given in the Theorem 11.3.11 do not guarantee the existence of a solution of the dynamic problem* (11.1), *as shown in the next example:*

Let $[0,1)_{\mathbb{T}}$ be a time scale interval that contains a countable infinite subset $\cup_{i=1}^{\infty}\{r_i\}$ with $\sigma(r_i)=r_i$. Let $f:[0,1]_{\mathbb{T}}\times\overline{B}_r(0)\to\mathbb{R}$ defined by

$$f(t,x) := \begin{cases} 0 & \text{for } x\neq 0,\ t\neq r_i \\ 1 & \text{for } x\neq 0,\ t=r_i \quad \text{or } x=0,\ t\neq r_i, \\ 2 & \text{for } x=0,\ t=r_i. \end{cases}$$

This f is generalized Δ-Carathéodory function and satisfy condition given in Theorem 11.3.11 with $\overline{g}\equiv 0$. For this function f, dynamic problem (11.1) has no solution.

11.4 Dependency and Convergence of Solutions

Theorem 11.4.1. *Assume that $f:[a,b]_{\mathbb{T}}\times\overline{B}_r(A)\to\mathbb{R}$ satisfy the condition*

$$|f(t,x)-f(t,y)|\leq g(|x-y|),$$

where $g:[a,b]_{\mathbb{T}}\to[0,\infty)$ is an increasing continuous function such that $\dfrac{1}{g(u)}\neq 0$ for $u\in[a,b]_{\mathbb{T}}$. Let x and y be such that $x^{\Delta}(t)=f(t,x)$, $x(a)=A$ and $y^{\Delta}(t)=f(t,y)$, $y(a)=B$. Then

$$|x(t)-y(t)|\leq G^{-1}(G(|A-B|)+t-a) \quad \text{for all } t\in[a,b]_{\mathbb{T}},$$

where $G:[a,b]_{\mathbb{T}}\to\mathbb{R}$ is a strictly increasing function such that $G^{\Delta}(u)=\dfrac{1}{g(u)}$ for all $u\in[a,b]_{\mathbb{T}}$.

Proof. By hypothesis, we have

$$x(t)=A+(\text{HK})\int_a^t f(\tau,x(\tau))\Delta\tau \quad \text{and} \quad y(t)=B+(\text{HK})\int_a^t f(\tau,y(\tau))\Delta\tau.$$

Then for all $t\in[a,b]_{\mathbb{T}}$,

$$\begin{aligned} |x(t)-y(t)| &\leq |A-B|+(\text{HK})\int_a^t |f(\tau,x(\tau))-f(\tau,y(\tau))|\,\Delta\tau \\ &\leq |A-B|+(\text{HK})\int_a^t g\left(|x(\tau)-y(\tau)|\right)\Delta\tau. \end{aligned}$$

By Theorem 11.3.10, we have

$$|x(t) - y(t)| \le G^{-1}\left(G(|A - B|) + t - a\right) \quad \text{for all } t \in [a, b]_{\mathbb{T}},$$

which shows that the solutions of dynamic problem (11.1) depend continuously on the initial condition. To emphasize the dependence on the initial condition, we denote the solutions of (11.1) by $\phi(t, a, A)$. Thus,

$$|\phi(t, a, A) - \phi(t, a, B)| \le G^{-1}\left(G(|A - B|) + t - a\right)$$

for all $t \in [a, b]_{\mathbb{T}}$. □

Corollary 11.4.2. *Under the assumption of the above theorem we have the following.*

$$\begin{aligned}|\phi(t, a, A) - \phi(s, \bar{a}, B)| \le\ & G^{-1}\left(G(|A - B|) + t - a\right) \\ & + G^{-1}\left(G(M\,|\bar{a} - a|) + t - \bar{a}\right) + M\,|t - s|,\end{aligned}$$

where $M := \sup_{t \in \mathbb{T}} |f(t, \bar{a}, B)|$ *and* $\bar{a} \in (a, b]_{\mathbb{T}}$.

Proof. We write

$$\begin{aligned}|\phi(t, a, A) - \phi(s, \bar{a}, B)| \le\ & |\phi(t, a, A) - \phi(t, a, B)| + |\phi(t, a, B) - \phi(t, \bar{a}, B)| \\ & + |\phi(t, \bar{a}, B) - \phi(s, \bar{a}, B)|.\end{aligned} \tag{11.10}$$

Consider

$$\begin{aligned}&|\phi(t, a, A) - \phi(t, a, B)| \\ &\quad \le |A - B| + (\mathrm{HK}) \int_a^t |f(\tau, \phi(\tau, a, A)) - f(\tau, \phi(\tau, a, B))|\, \Delta\tau \\ &\quad \le |A - B| + (\mathrm{HK}) \int_a^t g\left(|\phi(\tau, a, A) - \phi(\tau, a, B)|\right) \Delta\tau,\end{aligned}$$

which yields

$$|\phi(t, a, A) - \phi(t, a, B)| \le G^{-1}\left(G(|A - B|) + t - a\right). \tag{11.11}$$

Assuming $0 \leq a < \bar{a} < t$,

$$\begin{aligned}
&|\phi(t,a,B) - \phi(t,\bar{a},B)| \\
&\leq (\mathrm{HK}) \int_a^t |f(\tau,\phi(\tau,a,B)) - f(\tau,\phi(\tau,\bar{a},B))|\,\Delta\tau \\
&= (\mathrm{HK}) \int_a^{\bar{a}} |f(\tau,\phi(\tau,a,B)) - f(\tau,\phi(\tau,\bar{a},B))|\,\Delta\tau \\
&\quad + (\mathrm{HK}) \int_{\bar{a}}^t |f(\tau,\phi(\tau,a,B)) - f(\tau,\phi(\tau,\bar{a},B))|\,\Delta\tau \\
&\leq (\mathrm{HK}) \int_a^{\bar{a}} |f(\tau,\phi(\tau,a,B))|\,\Delta\tau \\
&\quad + (\mathrm{HK}) \int_{\bar{a}}^t |f(\tau,\phi(\tau,a,B)) - f(\tau,\phi(\tau,\bar{a},B))|\,\Delta\tau \\
&\leq M\,|\bar{a} - a| + (\mathrm{HK}) \int_{\bar{a}}^t |f(\tau,\phi(\tau,a,B)) - f(\tau,\phi(\tau,\bar{a},B))|\,\Delta\tau \\
&\leq M\,|\bar{a} - a| + (\mathrm{HK}) \int_{\bar{a}}^t g\left(|\phi(\tau,a,B) - \phi(\tau,\bar{a},B)|\right)\Delta\tau,
\end{aligned}$$

which yields

$$|\phi(t,a,B) - \phi(t,\bar{a},B)| \leq G^{-1}\left(G(M\,|\bar{a} - a|) + t - \bar{a}\right). \tag{11.12}$$

Now,

$$\begin{aligned}
|\phi(t,\bar{a},B) - \phi(s,\bar{a},B)| &= \left|(\mathrm{HK}) \int_s^t f(\tau,\bar{a},B)\Delta\tau\right| \\
&\leq (\mathrm{HK}) \int_s^t |f(\tau,\bar{a},B)|\Delta\tau.
\end{aligned}$$

That is,

$$|\phi(t,\bar{a},B) - \phi(s,\bar{a},B)| \leq M\,|t - s|, \tag{11.13}$$

Hence, using (11.11), (11.12) and (11.13), (11.10) becomes

$$\begin{aligned}
|\phi(t,a,A) - \phi(s,\bar{a},B)| \leq\ & G^{-1}\left(G(|A - B|) + t - a\right) \\
& + G^{-1}\left(G(M\,|\bar{a} - a|) + t - a\right) + M\,|t - s|\,.
\end{aligned} \tag{11.14}$$

□

Now, to investigate the convergence of solutions of dynamic problem (11.1), we provide an analogous concept of equi-integrability and a related basic result.

Definition 11.4.3. *A family of functions $\{f_n\}$ is said to be Henstock–Kurzweil Δ-equi-integrable on $[a,b]_{\mathbb{T}}$, if for every $\varepsilon > 0$, there exists a Δ-gauge,*

δ for $[a,b]_{\mathbb{T}}$ such that for all $n \in \mathbb{N}$

$$\left|(HK)\int_a^b f_n - \sum_{i=1}^k f_n(\xi_i)(t_i - t_{i-1})\right| < \varepsilon$$

for all δ-fine partitions $\mathcal{P} = \{\xi_i, [t_{i-1}, t_i]_{\mathbb{T}}\}_{i=1}^k$ of $[a,b]_{\mathbb{T}}$.

Theorem 11.4.4. *If $\{f_n\}$ is a sequence of Henstock–Kurzweil Δ-equi-integrable functions on $[a,b]_{\mathbb{T}}$ such that $f_n \to f$ on $[a,b]_{\mathbb{T}}$, then f is Henstock–Kurzweil Δ-integrable on $[a,b]_{\mathbb{T}}$ and*

$$\lim_{n\to\infty}(HK)\int_a^b f_n(t)\Delta t = (HK)\int_a^b f(t)\Delta t.$$

Proof. Since $\{f_n\}$ is sequence of Henstock–Kurzweil Δ-equi-integrable functions on $[a,b]_{\mathbb{T}}$, for every $\varepsilon > 0$, there is a Δ-gauge, δ for $[a,b]_{\mathbb{T}}$ such that for all $n \in \mathbb{N}$

$$\left|(\mathrm{HK})\int_a^b f_n - \sum_{i=1}^k f_n(\xi_i)(t_i - t_{i-1})\right| < \varepsilon$$

for all δ-fine partitions $\mathcal{P} = \{\xi_i, [t_{i-1}, t_i]_{\mathbb{T}}\}_{i=1}^k$ of $[a,b]_{\mathbb{T}}$.

We claim that the sequence $\left\{(HK)\int_a^b f_n(t)\Delta t\right\}$ is convergent. Consider

$$\begin{aligned}
&\left|(HK)\int_a^b f_n(t)\Delta t - (HK)\int_a^b f_m(t)\Delta t\right| \\
&= \left|(HK)\int_a^b f_n(t)\Delta t - \sum_{i=1}^k f_n(\xi_i)(t_i - t_{i-1}) + \sum_{i=1}^k f_n(\xi_i)(t_i - t_{i-1})\right. \\
&\qquad \left. - \sum_{i=1}^k f_m(\xi_i)(t_i - t_{i-1}) + \sum_{i=1}^k f_m(\xi_i)(t_i - t_{i-1}) - (HK)\int_a^b f_m(t)\Delta t\right| \\
&\le \left|(HK)\int_a^b f_n(t)\Delta t - \sum_{i=1}^k f_n(\xi_i)(t_i - t_{i-1})\right| \\
&\qquad + \left|\sum_{i=1}^k f_n(\xi_i)(t_i - t_{i-1}) - \sum_{i=1}^k f_m(\xi_i)(t_i - t_{i-1})\right| \\
&\qquad + \left|\sum_{i=1}^k f_m(\xi_i)(t_i - t_{i-1}) - (HK)\int_a^b f_m(t)\Delta t\right| \\
&< \varepsilon + \varepsilon + \left|\sum_{i=1}^k f_n(\xi_i)(t_i - t_{i-1}) - \sum_{i=1}^k f_m(\xi_i)(t_i - t_{i-1})\right| \\
&= 2\varepsilon + \sum_{i=1}^k \left|f_n(\xi_i) - f_m(\xi_i)\right| |t_i - t_{i-1}|.
\end{aligned}$$

Since $f_n \to f$ on $[a,b]_{\mathbb{T}}$, for every $\varepsilon > 0$, there exists $n_0 \in \mathbb{N}$ such that for all $m,\ n \geq n_0$ and for all $\xi_i \in [a,b]_{\mathbb{T}}$, we have

$$\left|f_n(\xi_i) - f_m(\xi_i)\right| < \frac{\varepsilon}{(b-a)}.$$

Then

$$\begin{aligned}\left|(HK)\int_a^b f_n(t)\Delta t - (HK)\int_a^b f_m(t)\Delta t\right| &< 2\varepsilon + \sum_{i=1}^{k} \frac{\varepsilon}{(b-a)}\left|t_i - t_{i-1}\right| \\ &< 2\varepsilon + \varepsilon = 3\varepsilon.\end{aligned}$$

This implies that $\left\{(HK)\int_a^b f_n(t)\Delta t\right\}$ is a Cauchy sequence in $\mathbb{R}$ and hence it is convergent. Let

$$\lim_{n\to\infty}(HK)\int_a^b f_n(t)\Delta t = L.$$

Then

$$\begin{aligned}&\left|\sum_{i=1}^{k} f(\xi_i)(t_i - t_{i-1}) - L\right| \\ &= \left|\sum_{i=1}^{k} f(\xi_i)(t_i - t_{i-1}) - \sum_{i=1}^{k} f_n(\xi_i)(t_i - t_{i-1}) + \sum_{i=1}^{k} f_n(\xi_i)(t_i - t_{i-1}) - L\right| \\ &\leq \left|\sum_{i=1}^{k} f(\xi_i)(t_i - t_{i-1}) - \sum_{i=1}^{k} f_n(\xi_i)(t_i - t_{i-1})\right| + \left|\sum_{i=1}^{k} f_n(\xi_i)(t_i - t_{i-1}) - L\right| \\ &< \varepsilon + \varepsilon = 2\varepsilon.\end{aligned}$$

This yields that

$$(HK)\int_a^b f(t)\Delta t = L.$$

Hence, f is Henstock–Kurzweil Δ-integrable on $\mathbb{T}$ and

$$\lim_{n\to\infty}(HK)\int_a^b f_n(t)\Delta t = (HK)\int_a^b f(t)\Delta t.$$

□

Using the above result we have the following.

Theorem 11.4.5. *Assume that $\{f_n\}$ is a sequence of Henstock–Kurzweil Δ-equi-integrable functions on $[a,b]_{\mathbb{T}}$ such that $f_n \to f$ on $[a,b]_{\mathbb{T}}$ and for each $n \in \mathbb{N}$, $f_n : [a,b]_{\mathbb{T}} \times \overline{B}_r(A) \to \mathbb{R}$ satisfy*

$$\left|f_n(t,x) - f_n(t,y)\right| \leq g(|x-y|),$$

where $g : [a,b]_{\mathbb{T}} \to [0,\infty)$ *is a continuous increasing function with* $g(0) = 0$ *and* $g(t) > 0$ *for* $t \in [a,b]_{\mathbb{T}}$. *If, for each* $n \in \mathbb{N}$, x_n *is a solution of*

$$\begin{cases} x_n^{\Delta}(t) = f_n(t, x_n), & t \in [a,b]_{\mathbb{T}}^{\kappa}; \\ x_n(a) = A, \end{cases} \tag{11.15}$$

then there is a subsequence $\{y_n\}$ *of* $\{x_n\}$ *and a function* $x : [a,b]_{\mathbb{T}} \to \mathbb{R}$ *such that* $y_n \to x$ *and* x *is solution of dynamic problem* (11.1) *in* $[a,b]_{\mathbb{T}}$.

Proof. From the hypothesis, we have

$$x_n(t) = x_n(a) + (HK)\int_a^t f_n(\tau, x_n)\Delta\tau,$$

which yields

$$|x_n(t) - x_n(a)| \le (HK)\int_a^t |f_n(\tau, x_n)|\,\Delta\tau. \tag{11.16}$$

Now,

$$\begin{aligned} |f_n(\tau, x_n(\tau))| &= |f_n(\tau, x_n(a)) + f_n(\tau, x_n(\tau)) - f_n(\tau, x_n(a))| \\ &\le |f_n(\tau, x_n(a))| + |f_n(\tau, x_n(\tau)) - f_n(\tau, x_n(a))| \\ &\le |f_n(\tau, x_n(a))| + g\left(|x_n(\tau) - x_n(a)|\right). \end{aligned}$$

Then

$$(HK)\int_a^t |f_n(\tau, x_n(\tau))|\,\Delta\tau \le M + (HK)\int_a^t g\left(|x_n(\tau) - x_n(a)|\right)\Delta\tau,$$

where $M := (HK)\int_a^b |f_n(\tau, x_n(a))|\,\Delta\tau$. By Theorem 11.3.10, we have

$$(HK)\int_a^t |f_n(\tau, x_n(\tau))|\,\Delta\tau \le G^{-1}(G(M) + t - a),$$

where $G^{\Delta}(u) = \dfrac{1}{g(u)}$. So, equation (11.16) becomes

$$|x_n(t) - x_n(a)| \le G^{-1}(G(M) + t - a).$$

Therefore

$$\begin{aligned} |x_n(t)| &\le |x_n(t) - x_n(a)| + |x_n(a)| \\ &\le G^{-1}\left(G(M) + t - a\right) + x_0 \\ &= \alpha, \end{aligned}$$

where $\alpha := G^{-1}(G(M) + b - a) + x_0$. Therefore, $\{x_n\}$ is uniformly bounded on $[a,b]_{\mathbb{T}}$.

Now, for given any $\varepsilon > 0$, we can find $\delta > 0$ such that if $|t - t_1| < \delta$, then

$$\begin{aligned}|x_n(t) - x_n(t_1)| &= \left| x_n(a) + (HK)\int_a^t f_n(\tau, x_n(\tau))\Delta\tau - x_n(a) \right. \\ &\qquad \left. - (HK)\int_a^{t_1} f_n(\tau, x_n(\tau))\Delta\tau \right| \\ &= \left| (HK)\int_{t_1}^t f_n(\tau, x_n(\tau))\Delta\tau \right| \\ &\leq (HK)\int_{t_1}^t |f_n(\tau, x_n(\tau))|\,\Delta\tau \\ &\leq L\,|t - t_1| < L\delta.\end{aligned}$$

That is,

$$|x_n(t) - x_n(t_1)| < \varepsilon,$$

where $\varepsilon := L\delta$. Therefore $\{x_n\}$ is equicontinuous on $[a, b]_{\mathbb{T}}$. By Arzelá–Ascoli's theorem [1], there exists a subsequence $\{y_n\}$ of $\{x_n\}$ such that $y_n \to x$ uniformly on $\mathbb{T}$. Also, by hypothesis, we have

$$y_n(t) = A + (HK)\int_a^t f_n(\tau, y_n(\tau))\Delta\tau.$$

Since $f_n \to f$ on $[a, b]_{\mathbb{T}}$ and $\{f_n\}$ is Henstock–Kurzweil Δ-equi-integrable on $\mathbb{T}$, by Theorem 11.3.10, it follows that

$$\begin{aligned}\lim_{n\to\infty} y_n(t) &= A + \lim_{n\to\infty} (HK)\int_a^t f_n\big(\tau, y_n(\tau)\big)\Delta\tau \\ x(t) &= A + (HK)\int_a^t \lim_{n\to\infty} f_n\big(\tau, y_n(\tau)\big)\Delta\tau \\ x(t) &= A + (HK)\int_a^t f\big(\tau, x(\tau)\big)\Delta\tau.\end{aligned}$$

That is,

$$\begin{cases} x^{\Delta}(t) = f\big(t, x(t)\big), \ t \in \mathbb{T}, \\ x(a) = A. \end{cases}$$

□

Bibliography

[1] R. Agarwal, M. Bohner, P. Řehák, Half-linear dynamic equations, Nonlinear analysis and applications: to V. Lakshmikantham on his 80th birthday. Vol. 1, 2, pages 1–57, Kluwer Acad. Publ., Dordrecht (2003).

[2] M. Bohner, A. C. Peterson, Dynamic equations on time scales: An introduction with applications, Birkhäuser, Boston (2001).

[3] M. Bohner, A. C. Peterson, Advances in dynamic equations on time scales, Birkhäuser, Boston (2004).

[4] A. Cabada, D. R. Vivero, Criterions for absolute continuity on time scales, J. Differ. Equ. Appl. 11(11) (2005) 1013–1028.

[5] A. Cabada, D. R. Vivero, Expression of the Lebesgue Δ-integral on time scales as a usual Lebesgue integral: application to the calculus of Δ antiderivatives, Math. Comput. Model. 43 (2006) 194–207.

[6] H. Gilbert, Existence theorems for first-order equations on time scales with Δ-Carathéodory functions, Adv. Difference Equ. 2010 (2010) Article ID 650827, 20 pages.

[7] A. C. Peterson, B. Thompson, Henstock–Kurzweil delta and nabla integrals, J. Math. Anal. Appl. 323 (2006) 162–178.

[8] I. L. D. Santos, G. N. Silva, Absolute continuity and existence of solutions to dynamic inclusions in time scales, Math. Ann. 356(1) (2013) 373–399.

[9] I. L. D. Santos, Discontinuous dynamic equations on time scales, Rend. Circ. Mat. Palermo 64(3) (2015) 383–402.

[10] I. L. D. Santos, On qualitative and quantitative results for solutions to first-order dynamic equations on time scales, Bol. Soc. Mat. Mex. 21 (2015) 205–218.

[11] B. Satco, A Cauchy problem on time scales with applications, An. Stiint. Univ. Al. I. Cuza Iasi. Mat. (N.S.) Matematica, Tomul LVII (2011) 221–234.

[12] D. Smart, Fixed point theorems, Cambridge Tracts in Mathematics, Cambridge University Press (1974).

[13] B. S. Thomson, Henstock–Kurzweil integrals on time scales, Panamer. Math. J. 18(1) (2008) 1–19.

[14] S. Tikare, Generalized first order dynamic equations on time scales with Δ-Carathéodory functions, Differ. Equ. Appl. 11(1) (2019) 167–182.

[15] S. Tikare, I. L. D. Santos, Existence results for solutions to discontinuous dynamic equations on time scales, Differ. Equ. Appl. 12(1) (2020) 89–100.

Chapter 12

Vector Valued Ideal Convergent Generalized Difference Sequence Spaces Associated with Multiplier Sequences

Binod Chandra Tripathy

Tripura University, tripathybc@gmaiil.com

12.1 Introduction

A sequence space is a vector space whose elements are infinite sequences of real or complex numbers. Equivalently, it is a function space whose elements are functions from the natural numbers to the field K of real or complex numbers. The set of all such functions is naturally identified with the set of all possible infinite sequences with elements in K and can be turned into a vector space under the operations of point wise addition of functions and

DOI: 10.1201/9781003178200-12

point wise scalar multiplication. All sequence spaces are linear subspaces of this space.

The class of all sequence w contains sequences those are unbounded even the subsequential limits do not exist. In order to have some investigation about their nature, subset of w of such type of sequences, the researcher around the globe developed different methods such as matrix maps on them, considered their differences, multiplied by a sequence of scalars etc. This is the origin of the concept of summability theory. The arithmetic mean i.e., $(C, 1)$ mean is one of the beautiful example, by which some bounded non-convergent sequences can have their sequence of means converging to a limit.

The sequence spaces are being investigated from different aspects. The summability theory brought a significant change and opened a lot of scope for investigations on sequence spaces. Frobenius introduced the method of summability by arithmetic mean, which was generalized by Cesàro (1880) as the (C, k)-method of summability. The most important sequences spaces in analysis are the ℓ^p spaces, consisting of the pth-power summable sequences, with the p-norm. Other important classes of sequences like convergent sequences or null sequences form sequence spaces, respectively denoted c and c_0.

The sequence spaces were investigated by G. Köthe, O. Toeplitz, T. Kojima, I. Schur, H. Steinhaus, G.H. hardy, K. Knopp, B. Kuttner, J.A. Fridy, A. Wilansky, K. Zeller, I.J. Maddox, G. G. Lorentz, A. Peyerimhoff, G. M. Petersen, T. Salat, B. E. Rhoades, T. Pati, J. Boos, S. Mazur, W. Orlicz are a few to be named.

The other as, Section of this article contains the materials given as section 2 Definitions and Preliminaries contains the basic definitions. It also contains subsections on difference sequences, matrix characterizations, basics of vector valued sequences spaces (VVSS). Section 3 is on ideal convergence of sequences, it also contains its particular cases such as statistical convergence of sequences. Section 4 is on Sequence spaces associated with multiplier sequences. We introduce the definitions of vector valued generalized difference ideal convergent sequences spaces associated with multiplier sequences. In section 5, contains the main results of this chapter.

12.2 Definitions and Preliminaries

Throughout ℓ_∞, c, c_0 denote bounded, convergent and null sequence spaces of real or complex numbers. These are normed linear spaces, normed by the sup-norm,

$$||(x_k)|| = \sup_{k \in N} |x_k|.$$

Definition 12.2.1. *A subset E of the whole sequences w is said to be solid or normal if $(x_k) \in E$ implies $(\alpha_k x_k) \in E$, for all (α_k) with $| \alpha_k | \leq 1$, for all $k \in N$.*

Remark 12.2.2. *Geometrically, $E \subset w$ is said to be solid or normal if for each $x = (x_k) \in E$, the closed ball B with origin as center and x on the circumference is contained in E.*

Definition 12.2.3. *A subset E of the whole sequences w is said to be monotone, if it contains the canonical pre-images of all its step spaces.*

Remark 12.2.4. *In the definition of solid, if one takes $\alpha_k = 1$ or $\alpha_k = 0$, for all $k \in N$, then the definition of normal can be obtained. Hence subset E of all sequences w is solid implies it is monotone, but not necessarily conversely (one may refer to Kamthan and Gupta [23]).*

Definition 12.2.5. *A subset E of the whole sequences w is said to be symmetric, if $(x_{\pi(k)}) \in E$, whenever $(x_k) \in E$, where π is a permutation of N.*

Definition 12.2.6. *A subset E of the whole sequences w is said to be sequence algebra, if $(x_k).(y_k) = (x_k * y_k) \in E$, whenever if $(x_k), (y_k) \in E$.*

Definition 12.2.7. *A subset E of the whole sequences w is said to be convergence free, if $(y_k) \in E$, whenever $(x_k) \in E$ and $x_k = 0$ implies $y_k = 0$.*

Remark 12.2.8. *In most of the cases the authors have considered the E to be a sequence space, instead of a subset of all sequences. For all these definitions, one may refer to research monographs listed in the references, in particular Kamthan and Gupta [23].*

12.2.1 Difference sequence spaces

The notion of generalized difference sequences is introduced by Tripathy et al. [47] as follows:

Definition 12.2.9. *Let $m \geq 0, n \geq 0$ be non-negative natural numbers, then the generalized difference operator Δ_m^n on sequences $= (x_k)$ is defined as follows:*

$$\Delta_m^n x_k = x_k - x_{k+nm}, \text{ for all } k \in N.$$

Remark 12.2.10. *On taking $m = 1, n = 1$, the difference operator reduces to the difference operator introduced and studied by Kizmaz [24]. When $m = 1$, the difference operator reduces to the generalized difference operator introduced and studied by Et and Colak [13]. When $m = 1$, the difference operator reduces to the difference operator introduced and studied by Tripathy and Esi [46].*

Remark 12.2.11. *When $m = 0$ or $m = n = 0$ or $n = 0$, the difference operator Δ_0^0 is represented as, $\Delta_0^0 x_k = \Delta_0^n x_k = \Delta_m^0 x_k = x_k$, for all $k \in N$.*

The difference sequence spaces have been studied from different aspects by many others, one may refer to Altinok et al. [1], Tripathy and Dutta [44], Tripathy and Mahanta [59].

12.2.2 Matrix transformation between sequence spaces

Let $A = (a_{n,k})$ be infinite matrix of real or complex numbers. Then we say that A transforms from the sequence space E into the sequence space F, if $Ax \in F$, for each sequence $(x_k) \in E$ that is, $Ax = (A_n x) \in F$, where $A_n x = \sum_{k=1}^{\infty} a_{nk} x_k$, provided that the infinite series converges for each $n \in N$.

The necessary and sufficient condition(s) under which an infinite matrix operator transforms sequences in the space E in to the sequences space F known as the characterization result.

Lemma 12.2.12. *A matrix $A = (a_{n,k})$ is said to be conservative (which transforms a convergent sequence into a convergent sequence) if and only if it satisfies the following well-known Kojima-Schur conditions*

(i) $||A|| = \sup_{n \in N} \sum_{k=1}^{\infty} |a_{nk}| < \infty$,

(ii) $\lim_{n \to \infty} a_{nk} = \alpha_k$, *exists for each* $k \in N$,

(iii) $\lim_{n \to \infty} \sum_{k=1}^{\infty} a_{nk} = \alpha$.

Lemma 12.2.13. *A matrix $A = (a_{n,k})$ is said to be regular (or limit preserving transformation) if and only if it satisfies the following well-known Silverman-Toeplitz conditions i.e. $A = (a_{n,k})$ satisfies the above Kojima-Schur conditions with $\alpha_k = 0$, for all $k \in N$ in (ii) and $\alpha = 1$ in (iii).*

For the characterization of different matrix classes, one may refer to Basar [3], Cooke [9], Hardy [20], Kamthan and Gupta [23], Maddox [26], Peterson [28], Peyerimhoff [29], Ruckle [32], Wilansky [61], Boos [4] and others. Using the concept of matrix maps, Chandra and Tripathy [5] generalized the concept of duals of the sequence spaces and investigated their different properties.

12.2.3 Vector valued sequence spaces

The vector valued sequence spaces (VVSS) was studied by Kamthan [22], Gupta [19], Ratha and Srivastava [30], Tripathy and Sen [53,54], Tripathy and Mahanta [51], Tripathy et al. [41], Tripathy et al. [48], Tripathy et al. [43], Tripathy and Dutta [45] and many others. For the basics on VVSS, one may refer to Kamthan and Gupta [23].

Let X be a seminormed space, seminormed by q. By $w(X)$, we denote the class of all vector valued sequences spaces defined over X. The vector valued sequence spaces $c(X)$, $c_0(X)$ and $\ell_\infty(X)$ are defined as follows:

$$c(X) = \{(x_k) \in w(X) : \lim_{k\to\infty} q(x_k - L) = 0, \text{ for some } L \in X\}.$$
$$\ell_\infty(X) = \{(x_k) \in w(X) : \sup_{k\in n} q(x_k) < \infty\}.$$
$$c_0(X) = \{(x_k) \in w(X) : \lim_{k\to\infty} q(x_k) = 0\}.$$

The above classes of sequences are seminormed spaces, seminormed by

$$f(x) = f((x_k)) = \sup_{k\in N} q(x_k).$$

Ratha and Srivastava [30] considered different Banach spaces for each term of the sequence i.e. the Banach spaces E_k, normed by f_k, for all $k \in N$. Tripathy and Mahanta [51] considered the terms of the sequence from different seminormed spaces E_k, seminormed by f_k, for all $k \in N$. Later on many researcher have investigated VVSS form different aspects.

12.3 Ideal Convergence of Sequences

The concept of ideal was introduced by K. Kuratowski, defined as follows.

Definition 12.3.1. *Let X be a nonempty set, then a class I of subsets of X is said to be an ideal if*

(i) $A \in I$ and $B \subset A$ implies $B \in I$ (Hereditary property)

(ii) $A, B \in I$ implies $A \cup B \in I$ (Additive property)

Example 12.3.2. *The class of all finite subsets of N, the set of natural numbers is an example of ideal.*

Example 12.3.3. *Let $I = \{A \subset N : \sum_{a\in A} \frac{1}{a} < \infty\}$ be a collection of subsets of the set N of natural numbers. Then it can be verified that I is an ideal of N.*

Definition 12.3.4. *Let X be a nonempty set, then a class of subsets $\mathcal{F}$ of X, not containing the empty set $\emptyset$ is said to be a filter, if the following conditions are satisfied.*

(i) $A \in \mathcal{F}$ and $B \supset A$ implies $B \in \mathcal{F}$.

(ii) $A, B \in \mathcal{F}$ implies $A \cap B \in \mathcal{F}$.

Example 12.3.5. *The class of all subsets of N of the type $\{N - A :$ A is a fine subset of $N\}$ is an example of a filter of N.*

The works on I-convergence was studied at the initial stage by Kostyrko et al. [25]. Later on it was further investigated by Salat et al. [34,35], Demirci [10], Demirci and Orhan [11], Tripathy and Hazarika [50], Tripathy et al. [55], Tripathy and Sen [56,57] and others. The idea depends on the notion of ideal of subsets of N.

Definition 12.3.6. *A given complex sequence $x = (x_k)$ is said to be I-convergent to L, if for any $\varepsilon > 0$, we have $\{k \in N :| x_k - L |\geq \varepsilon\} \in I$. We write, $x_k \rightarrow_I L$ or $I - \lim x_k = L$.*

We write $m^I = c^I \cap \ell_\infty$ and $m_0^I = c_0^I \cap \ell_\infty$.
The spaces $c^I(\Lambda), c_0^I(\Lambda)$ are defined as follows:

$$c^I(\Lambda) = \{(x_k) \in \omega : I - \lim(\lambda_k x_k - L) = 0, \text{ for some } L \in C\},$$

$$c_0^I(\Lambda) = \{(x_k) \in \omega : I - \lim(\lambda_k x_k) = 0\}.$$

Also, we write $m^I(\Lambda) = c^I(\Lambda) \cap \ell_\infty(\Lambda)$ and $m_0^I(\Lambda) = c_0^I(\Lambda) \cap \ell_\infty(\Lambda)$.

12.3.1 Statistically convergent sequence space

In order to extend the notion of convergence of sequences, statistical convergence of sequences was introduced by Fast [15], Buck [2] and Schoenberg [36], independently. It is also found in Zygmund [62]. Later on it was studied from sequence space point of view and linked with summability theory by Connor [6], Connor and Kline [7], Fridy [16, 17], Salat [33], Rath and Tripathy [31], Tripathy [38], Tripathy and Sen [53,54], Tripathy and Nath [52], Tripathy and Sarma [58] and many others.

The notion of statistical convergence depends on the notion of asymptotic density of subsets of the set N of natural numbers (one may refer to Niven et al. [27]).

Definition 12.3.7. *A given complex sequencex $= (x_k)$ is said to be statistically convergent to L, if for any $\varepsilon > 0$, we have $\delta(\{k \in N :| x_k - L |\geq \varepsilon\}) = 0$. We write, $x_k \rightarrow_{stat} L$ or $stat - lim x_k = L$.*

By $\overline{c}$ and $\overline{c_0}$ we denote the space of all statistically convergent and statistically null sequence respectively. Clearly $\overline{c}$, $\overline{c_0}$ are linear spaces. We write $m = \overline{c} \cap l_\infty$ and $m_0 = \overline{c_0} \cap \ell_\infty$. Salat [33, Theorem 2.1] showed m and m_0 are close subspace of ℓ_∞ under the sup-norm.

Fridy [16], Salat [33] and Connor [6] established some relations between statistical convergence and convergence of sequences. Those results are known as decomposition theorems. Later on these decomposition results have been generalized for ideal and statistical version of notions such as monotonic property etc, researcher have proved the respective decomposition results.

Definition 12.3.8. *A sequence $x = (x_k)$ is said to be a Cauchy sequence if for a given $\varepsilon > 0$, there exists n_0 an natural number such that the set $\{k \in N : |x_k - x_{n_0}| \geq \varepsilon\}$ is a finite set.*

Based on the above idea of Cauchy sequence, Fridy [16] introduced the notion of statistically Cauchy as follows.

Definition 12.3.9. *A sequence $x = (x_k)$ is said to be statistically Cauchy, if for every $\varepsilon > 0$, there exists $n_0 = n_0(\varepsilon) \in N$ $\delta(\{k \in N : | x_k - x_{k_0} | \geq \varepsilon\}) = 0$.*

A necessary and sufficient condition for a sequence to be statistically convergent was given by Schoenberg [36], which was improved by Salat [33].

Lemma 12.3.10. *(Tripathy and Mahanta [51, Proposition 4]). If a space E is bounded and solid then $(\lambda_k) \in M(E, E)$ if and only if $(\lambda_k) \in \ell_\infty$.*

Lemma 12.3.11. *(Salat et al. [35, Lemma 2.5]) Let $K \in \Im(I)$ and $M \subseteq N$. If $M \notin I$, then $M \cap K \notin I$.*

12.4 Sequence Spaces Associated with the Multiplier Sequences

Definition 12.4.1. *Let $\Lambda = (\lambda_k)$ be a general multiplier sequence of non-zero scalars. Then for a sequence space E, the multiplier sequence space $E(\Lambda)$, associated with the multiplier sequence Λ is defined by $E(\Lambda) = \{(x_k) \in w : (\lambda_k x_k) \in E\}$.*

Example 12.4.2. *Let $\Lambda = (\lambda_k) = (k^{-1})$, then for the sequence space c, the sequence $(x_k) = (k) \in c(\Lambda)$. It means that some unbounded sequences are also included in the class of convergent sequences associated with the multiplier sequence (k^{-1}).*

Remark 12.4.3. *The studies on sequence spaces can be considered as a matrix A having the main diagonal as the sequence (λ_k), which transforms a class of sequences into the known sequence space E. That is the domain of the matrix A is to be determined. However in the case of characterization theorems, the matrix is characterized.*

Remark 12.4.4. *In the case of characterization of matrix classes, we find necessary and sufficient conditions for an infinite matrix to transform a sequence space E into another sequence space F. In the case of the sequence spaces associated with the multiplier sequences, we consider an infinite matrix with the main diagonal as the sequence $\Lambda = (\lambda_k)$ and then find the subset of w for which it transforms into the sequence space E.*

12.4.1 Relation with real-life problems

The role of the multiplier sequence is quite similar with the role of the catalysts in chemical reaction. It accelerates or retards the process of chemical reaction without effecting the chemical reaction.

12.4.2 Advantages

The class of sequence is enlarged or condensed based on the type of multiplier sequence considered.

The multiplier problem was solved for the first time by Tripathy and Mahanta [51]. It deals with the characterization of the domain of the matrix having the multiplier sequence as the diagonal elements, so that it will map from a class of sequence E into E i.e. $(\lambda_k x_k) \in E$, for all $(x_k) \in E$. They obtained the characterization of the domain based on the characterization of the matrix class (E, E). Solution to multiplier problem is possible, provided the matrix characterization of (E, E) is known.

In the case of the sequence space associated with multiplier sequence, we find a subset F of the all sequence w such that $(\lambda_k x_k) \in E$, for all $(x_k) \in F$.

The multiplier sequences has been investigated from different aspects by Tripathy and Sen [60], Et et al. [14], Hazarika and Tripathy [41], Tripathy [39, 40], Tripathy and Chandra [42] and many others.

12.4.3 Vector valued generalized difference ideal convergent sequence spaces associated with the multiplier sequences

In this chapter, we introduce the following classes of sequences. Throughout we consider (X, q), a seminormed space, seminormed by q, $\Lambda = (\lambda_k)$ a generalized multiplier sequence, $m \geq 0$ and $n \geq 0$, be integers, then for I an ideal of N, we introduce the following classes of VVSS in this chapter. The classes of sequences $c^I(\Lambda, \Delta_m^n, q), c_0^I(\Lambda, \Delta_m^n, q)$ are defined as follows:

$$
\begin{aligned}
c^I(\Lambda, \Delta_m^n, q) &= \{(x_k) \in \omega : I - \lim q(\lambda_k \Delta_m^n x_k - L) = 0, \text{ for some } L \in X\}, \\
c_0^I(\Lambda, \Delta_m^n, q) &= \{(x_k) \in \omega : I - \lim q(\lambda_k \Delta_m^n x_k) = 0\}, \\
\ell_\infty^I(\Lambda, \Delta_m^n, q)) &= \{(x_k) \in \omega : \{q(\lambda_k \Delta_m^n x_k) \geq M\} \in I, \text{ for some} M \in R\}, \\
m^I(\Lambda, \Delta_m^n, q) &= c^I(\Lambda, \Delta_m^n, q) \cap \ell_\infty(\Lambda, \Delta_m^n, q) \\
m_0^I(\Lambda, \Delta_m^n, q) &= c_0(\Lambda, \Delta_m^n, q)^I \cap \ell_\infty(\Lambda, \Delta_m^n, q).
\end{aligned}
$$

Remark 12.4.5. *On taking $X = C$, the filed of complex numbers, $m = n = 0$ in the above classes of sequences, one will get the classes of multiplier sequences $c^I(\Lambda)$, $c_0^I(\Lambda)$, $m^I(\Lambda)$, $m_0^I(\Lambda)$, introduced and investigated by Tripathy and Hazarika [49].*

12.5 Main Results

In this section, we establish the results of this article. We state the following results without proof, which can be established using standard technique.

Theorem 12.5.1. *Let $\Lambda = (\lambda_k)$ be a given multiplier sequence, (X, q) be a a seminormed space, $m \geq 0, n \geq 0$ integers, I an ideal of N. Then the class of sequences $Z^I(\Lambda, \Delta^n_m, q)$, for $Z = c$, c_0, ℓ_∞, m and m_0 are linear spaces.*

Theorem 12.5.2. *Let $\Lambda = (\lambda_k)$ be a given multiplier sequence, (X, q) be a complete seminormed space, $m \geq 0, n \geq 0$ be integers, I an ideal of N. Then the sequence spaces $Z^I(\Lambda, \Delta^n_m, q)$, for $Z = m, m_0$ are complete seminormed space, seminormed by*

$$f(x) = f((x_k)) = \sum_{i=1}^{nm} \lambda_i q(x_i) + \sup_k \sum_{k=1}^{\infty} q(\lambda_k \Delta^n_m x_k). \tag{12.1}$$

Proof. Clearly $m^I(\Lambda, \Delta^n_m, q)$ and $m^I_0(\Lambda, \Delta^n_m, q)$ are linear spaces. Now we show that $m^I(\Lambda, \Delta^n_m, q)$ is complete with respect to the seminorm f, given by (12.1). Let $(x^{(r)})$ be a Cauchy sequence in $m^I(\Lambda, \Delta^n_m, q)$. Then for each $\varepsilon > 0$, there exists a number n_0 such that $f(x_k^{(r)} - x_k^{(s)}) < \varepsilon$ for all $r, s \geq n_0$.

$$\Rightarrow \sum_{i=1}^{nm} \lambda_i q(x_k^{(r)} - x_k^{(s)}) + \sup_k \sum_{k=1}^{\infty} q(\lambda_k \Delta^n_m (x_k^{(r)} - x_k^{(s)})) < \varepsilon, \text{ for all } r, s \geq n_0. \tag{12.2}$$

From (12.2), we have

$\Rightarrow q(\lambda_k(x_k^{(r)} - x_k^{(s)})) < \varepsilon$, for all $r, s \geq n_0$, for each $k = 1, 2, 3, ..., mn$.

$\Rightarrow (\lambda_k(x_k^{(i)})$ is a Cauchy sequence in X.

Let $(y_k^{(r)}) = (\lambda_k(x_k^{(r)})$, for all $k \in N$ and for each $r \in N$. Since X is complete seminormed space, so there exists $y_k \in X$ such that $y_k^{(i)} \to y_k$, as $i \to \infty$, for each $k \in N$. Since $m^I(\Lambda)$ is a linear space, so we can find a sequence (x_k) such that $(x_k^{(r)})$ convergence to (x_k), as $r \to \infty$. We can write y_k as $y_k = \lambda_k x_k$, for each $k \in N$. Thus, $(x_k) \in m^I(\Lambda, \Delta^n_m, q)$. Therefore,

$$\lim_{i \to \infty} (\lambda_k x_k^{(i)}) = y_k, \text{ for each } k \in N. \tag{12.3}$$

Now we have from (12.2),

$$\lim_{j \to \infty} q(\lambda_k(\Delta^n_m (x_k^{(r)} - x_k^{(s)})) < \varepsilon, \text{ for all } r, s \geq n_0, \text{ for each } k \in N.$$

Therefore, we get

$$(\lambda_k(\Delta^n_m (x_k^{(r)} - x_k^{(s)}) \tag{12.4}$$

is a Cauchy sequence in X. Using Equation (12.3), from Equation (12.4), since X is a complete seminormed space, applying standard technique, one can show that $(x_k^{(r)})$ converges in X. This shows that $m^I(\Lambda, \Delta_m^n, q)$ is complete seminormed space. Similarly, it can be established that $m_0^I(\Lambda, \Delta_m^n, q)$ is a complete seminormed space. □

Remark 12.5.3. *When $(X, q) = (X, ||.||)$ is a Banach space, the above class of sequences are Banach spaces, normed by f, q being replaced by the $||.||$.*

Theorem 12.5.4. *The sequence spaces $c_0^I(\Lambda, \Delta_m^n, q)$ and $m_0^I(\Lambda, \Delta_m^n, q)$ are solid.*

Proof. Let $(x_k) \in c_0^I(\Lambda, \Delta_m^n, q)$ and (α_k) be a sequences of scalars with $|\alpha_k| \leq 1$, for all $k \in N$. The result follows from the following relation:

$$\{k \in N : q(\lambda_k \Delta_m^n x_k) \geq \varepsilon\} \supseteq q(\alpha_k \lambda_k \Delta_m^n x_k) \geq \varepsilon\}.$$

□

Proposition 12.5.5. *The spaces $c^I(\Lambda, \Delta_m^n, q)$ and $m^I(\Lambda, \Delta_m^n, q)$ are not solid in general.*

Proof. The proof of the result follows from the following example.

Example 12.5.6. *Let $I = I_\delta$, the subsets of N having natural density zero. Let $\Lambda = (\lambda_k)$ such that $\lambda_k = k$, for all $k \in N$, $X = C$, the set of complex numbers, $m = 1, n = 1$. Consider the sequence (x_k) defined by*

$$x_k = \begin{cases} \frac{1}{k^{-2}}, & \text{if } k = i^2, i \in N \\ 1 & \text{otherwise.} \end{cases}$$

Let $\alpha_k = (-1)^k$, for all $k \in N$. Then, $(\Lambda_k x_k) \in c^I(\Lambda, \Delta_m^n, q)$, but $(\alpha_k \lambda_k x_k) \notin c^I(\Lambda, \Delta_m^n, q)$.

□

Theorem 12.5.7. *Let $I \neq I_f$, then the spaces $c^I(\Lambda, \Delta_m^n, q)$ and $c_0^I(\Lambda, \Delta_m^n, q)$ are not symmetric in general.*

Proof. The proof of the result follows from the following example.

Example 12.5.8. *Let $I = I_\delta, m = 1, n = 1$. Consider the sequence (x_k) defined by*

$$x_k = \begin{cases} k, & \text{if } k = i^2, i \in N \\ \frac{1}{k^2} & \text{otherwise.} \end{cases}$$

Let $\lambda_k = k^{-1}$, for all $k \in N$. Consider the rearrangement (y_k) of (x_k) defined by

$$(y_k) = (x_1,\ x_2,\ x_4,\ x_3,\ x_9,\ x_5,\ x_{16},\ x_6,\ x_{25},\ x_7,\ x_{36},\ x_8,\ \ldots).$$

Then $(\lambda_k y_k)$ neither belongs to $c^I(\Lambda, \Delta_m^n, q)$ nor to $c_0^I(\Lambda, \Delta_m^n, q)$; where as $(\lambda_k x_k)$ belongs to both $c^I(\Lambda, \Delta_m^n, q)$ as well as to $c_0^I(\Lambda, \Delta_m^n, q)$.

We may consider the following example too.

Example 12.5.9. *For $I = I_c$. Let $A = \{k : k = i^2$ or j^3, for $i, j \in N\}$, then $\sum_{a \in A} < \infty$. Consider the sequence (x_k) defined by*

$$x_k = \begin{cases} k, & \text{if } k = j^3, j \in N \\ 1 & \text{otherwise.} \end{cases}$$

Let $\lambda_k = 1$, for each $k \in N$. Then the rearrangement (z_k) of (x_k) defined by

$$(z_k) = (x_1,\ x_2,\ x_3,\ x_8,\ x_4,\ x_5,\ x_{27},\ x_6,\ x_7,\ x_{64},\ x_8,\ x_{10},\ \ldots).$$

Then $(\lambda_k z_k)$ neither belongs to $c^I(\Lambda)$ nor to $c_0^I(\Lambda)$; but $(\lambda_k x_k)$ belongs to both $c_0^I(\Lambda)$ and $c^I(\Lambda)$.

□

Theorem 12.5.10. *The spaces $c^I(\Lambda, \Delta_m^n, q)$ and $c_0^I(\Lambda, \Delta_m^n, q)$ are not sequence algebra in general.*

Proof. The result follows from the following example.

Example 12.5.11. *Let $X = R$, the set of real numbers. Consider the sequences $x_k = k^2, y_k = k^3$ and $\lambda_k = k^{-4}$, for all $k \in N$, then $(x_k), (y_k) \in Z(\Lambda, \Delta_m^n, q)$, but $(x_k * y_k) \notin Z(\Lambda, \Delta_m^n, q)$, where $Z = c^I, c_0^I$.*

□

Theorem 12.5.12. *The spaces $c^I(\Lambda, \Delta_m^n, q)$ and $c_0^I(\Lambda, \Delta_m^n, q)$ are not convergence free.*

Proof. The proof of this result follows from the following example.

Example 12.5.13. *Consider $P \in I$ and define the sequence (x_k) by*

$$x_k = \begin{cases} \frac{1}{k}, & \text{if } k \in P \\ k & \text{if } k \notin P. \end{cases}$$

Let $Q \subseteq N$ and define the sequence (y_k) by

$$y_k = \begin{cases} k, & \text{if } k = i^2, i \in Q \\ \frac{1}{k} & \text{if } k = i^2, i \notin Q. \end{cases}$$

Then, the result follows from the Lemma 12.3.11. □

Theorem 12.5.14. *If $I \neq I_f$, then the spaces $c^I(\Lambda, \Delta_m^n, q)$ and $c_0^I(\Lambda, \Delta_m^n, q)$ are separable.*

Proof. The proof of this result is easy, so omitted. □

Theorem 12.5.15. $(\lambda_k) \in (c_0^I(\Lambda, \Delta_m^n, q), c_0^I(\Lambda, \Delta_m^n, q))(= (m_0^I(\Lambda, \Delta_m^n, q), m_0^I(\Lambda, \Delta_m^n, q)))$ *if and only if* $(\lambda_k) \notin (c^I(\Lambda, \Delta_m^n, q), c^I(\Lambda, \Delta_m^n, q))$.

Proof. Suppose $(\lambda_k) \in \ell_\infty^I$. Then there exists a $M > 0$ such that

$$R = \{k \in N : |\lambda_k| \geq M\} \in I.$$

$$S = \left\{k \in N : q(\Delta_m^n x_k) \geq \frac{\varepsilon}{M}\right\} \in I.$$

Then $R \cup S = \{k \in N : |\lambda_k| \Delta_m^n x_k| \geq \varepsilon\} \in I$. Thus, $(\lambda_k x_k) \in c_0^I(\Lambda, \Delta_m^n, q)(= m_0^I(\Lambda, \Delta_m^n, q))$. Hence $(\lambda_k) \in (c_0^I(\Lambda, \Delta_m^n, q), c_0^I(\Lambda, \Delta_m^n, q))(= (m_0^I(\Lambda, \Delta_m^n, q), m_0^I(\Lambda, \Delta_m^n, q)))$. The converse part can be easily established using the standard technique, so omitted. □

Theorem 12.5.16. *The sequence spaces* $c^I(\Lambda, \Delta_m^n, q)$ *is not solid, then* $(\lambda_k) \notin (c^I(\Lambda, \Delta_m^n, q), c^I(\Lambda, \Delta_m^n, q))$.

Proof. Proof of the result follows from Lemma 12.3.10. □

Theorem 12.5.17. *If* $\Omega\Lambda^{-1} = (\omega_k \lambda_k^{-1}) \in \ell_\infty^I$, *then* $Z(\Omega, \Delta_m^n, q) \subset Z(\Lambda, \Delta_m^n, q)$ *and the inclusion is proper, where* $Z = c^I$, c_0^I, m_0^I, m^I.

Proof. The proof is a routine verification and the inclusion is proper follows from the following example.

Example 12.5.18. *Let* $X = R$, *the set of real numbers,* $n = m = 1$. *Consider the sequence* $(x_k) = (k^4)$ *and* $\omega_k = k^{-4}, \lambda_k = k^{-3}$, *for all* $k \in N$. *Then* $(x_k) \in Z(\Omega)$ *and* $(\omega_k \lambda_k^{-1})$ *is bounded, but* $(x_k) \notin Z(\Lambda)$.

□

12.6 Conclusion

In this chapter we have introduced the notion of vector valued generalized difference ideal convergent sequence spaces associated with the multiplier sequences, considering the seminormed space. We have investigated its different algebraic and topological properties. Some researcher have introduced the sequences spaces associated with multiplier sequences from different spaces, such as modular functions, Orlicz function, from the point of view of paranormed sequence spaces following Simons [37] etc. The list of some articles have been mentioned in the references. Further combining the existing ideas and direction of the sequence spaces associated with the multiplier sequences, many new classes of VVSS can be introduced and their different properties can be investigated.

Bibliography

[1] H. Altinok, Y. Altin, M. Et, Lacunary almost statistical convergence of fuzzy numbers, Thai Jour. Math. 2(2) (2004) 265–274.

[2] R. C. Buck, Generalized asymptotic density, Amer. Jour. Math. 75 (1953) 335–346.

[3] F. Başar, Summability Theory and its Applications, Bentham Science Publishers, Monograph (2011).

[4] J. Boos, Classical and Modern Methods in Summability, Oxford Mathematical Monographs, Oxford University Press (2000).

[5] P. Chandra, B. C. Tripathy, On generalized Köthe-Toeplitz duals of some sequence spaces, Indian J. Pure Appl. Math. 33(8) (2002) 1301–1306.

[6] J. S. Connor, The statistical and strong p-Cesàro Convergence of sequences, Analysis 8 (1988) 47–63.

[7] J. S. Connor, J. Kline, On Statistical limit points and the Consistency of Statistical convergence, J. Math. Anal. Appl. 197 (1996) 392–399.

[8] E. T. Copson, Metric Spaces, Cambridge Tracts in Mathematics, No. 57 Cambridge University Press (1992).

[9] R. G. Cook, Infinite Matrices and Sequence Spaces, Dover Publications (1955).

[10] K. Demirci, On lacunary statistical limit points, Demonstr. Math. 35(1) (2002) 93–101.

[11] K. Demirci, S. Orhan, Statistically relatively uniform convergence of positive linear operators, Results Math. 69 (2016) 359–367.

[12] A. Esi, B. C. Tripathy, Strongly almost convergent generalized difference sequences associated with multiplier sequences; Math. Slovaca 57(4) (2007) 339–348.

[13] M. Et, R. Çolak, On some generalized difference sequence Spaces; Soochow Journal of Mathematics; 21(4) (1995), 377–386.

[14] M. Et, P. Y. Lee, B. C. Tripathy, Strongly almost $(V, \lambda)(\Delta^r)$-summable sequences defined by Orlicz function, Hokkaido Math. J. 35 (2006) 197–213.

[15] H. Fast, Surla convergence statistique, Colloq. Math. 2 (1951) 241–244.

[16] J. A. Fridy, On statistical convergence, Analysis 5 (1985) 301–313.

[17] J. A. Fridy, Statistical limit points, Proc. Amer. Math. Soc. 118(4) (1993) 1187–1192.

[18] G. Goes, S. Goes, Sequences of bounded variation and sequences of Fourier coefficients, Math. Z. 118 (1970) 93–102.

[19] M. Gupta, The generalized sequence spaces $\ell^1(X)$ and $m_0(X)$, J. Math. Anal. Appl. 78(2) (1980) 357–366.

[20] G. H. Hardy, Divergent Series, Oxford University (1949).

[21] B. Hazarika, B. C. Tripathy, Some new type of generalized difference paranormed sequence spaces defined by sequences of Orlicz functions associated with multiplier sequences, Pacific J. Sci. Tech. 10(1) (2009) 161–169.

[22] P. K. Kamthan, Bases in a certain class of Frechet spaces, Tamkang J. Math. 7 (1976) 41–49.

[23] P. K. Kamthan, M. Gupta, Sequence Spaces and Series, Marcel Dekker Inc., New York (1981).

[24] H. Kizmaz, On certain sequence spaces, Canad. Math. Bull. 24 (1981) 169–176.

[25] P. Kostyrko, T. Šalát, W. Wilczyński, I-convergence, Real Anal. Exchange 26(2000-2001) 669–686.

[26] I. J. Maddox, Elements of Functional Analysis, Cambridge University (1970).

[27] I. Niven,, H. S. Zuckerman, H. L. Montgomery, An Introduction to the Theory of Numbers, John Wiley and Sons. Inc. (1991).

[28] G. M. Peterson, Regular Matrix and Transformations, McGraw-Hill Publishing Company Limited (1966).

[29] A. Peyerimhoff, Lectures on summability, Lecture Notes in Mathematics 107, Springer (1969).

[30] A. Ratha, P. D. Srivastava, On some vector valued sequence space $\mathcal{L}_\infty^{(p)}(E_k, \Lambda)$, Ganita 47(1) (1996) 1–12.

[31] D. Rath, B. C. Tripathy, Matrix maps on sequence spaces associated with sets of integers, Indian J. Pure Appl. Math. 27(2) (1996) 197–206.

[32] W. H. Ruckle, Sequence Spaces, Pitman Advanced Publishing Program (1981).

[33] T. Šalát, On statistically convergent sequences of real numbers, Math. Slovaca 30(2) (1980) 139–150.

[34] T. Šalát, B. C. Tripathy, M. Ziman, On some properties of I-convergence, Tatra Mountain Publ. Math. 28 (2004) 279–286.

[35] T. Šalát, B. C. Tripathy, M. Ziman, On I-convergence field, Italian J. Pure Appl. Math. 17 (2005) 45–54.

[36] I. J. Schoenberg, The integrability of certain functions and related summability methods, Amer. Math. Monthly 66 (1959) 361–375.

[37] S. Simons, The sequence space $\ell(p_\nu)$ and $m(p_\nu)$, Proc. London Math. Soc. 15(3) (1965) 422–436.

[38] B. C. Tripathy, Matrix transformations between some classes of sequences, J. Math. Anal. Appl. 206 (1997) 448–450.

[39] B. C. Tripathy, On a class of difference sequences related to the p-normed space ℓ^p, Demonstr. Math. 36(4) (2003) 867–872.

[40] B. C. Tripathy, On a new class of sequences, Demonstr. Math. 37(2) (2004) 377-381.

[41] B. C. Tripathy, Y. Altin, M. Et, Generalized difference sequences spaces on seminormed spaces defined by Orlicz functions, Math. Slovaca 58(3) (2008) 315–324.

[42] B. C. Tripathy, P. Chandra, On some generalized difference paranormed sequence spaces associated with multiplier sequences defined by modulus function, Anal. Theory Appl. 27(1) (2011) 21–27.

[43] B. C. Tripathy, R. Dey, N. R. Das, Ordered vector valued statistically convergent sequence space, Afr. Mat. 26 (2015) 433–441.

[44] B. C. Tripathy, H. Dutta, On some new paranormed difference sequence spaces defined by Orlicz functions, Kyungpook Math. J. 50(1) (2010) 59–69.

[45] B. C. Tripathy, H. Dutta, On some lacunary difference sequence spaces defined by a sequence of Orlicz functions and q-lacunary Δ_m^n-statistical convergence, An. Ştiinţ. Univ. "Ovidius" Constanţa Ser. Mat. 20(1) (2012) 417—430.

[46] B. C. Tripathy, A. Esi, A new type of difference sequence spaces, Internat. J. Sci. Tech. 1 (2006) 11–14.

[47] B. C. Tripathy, A. Esi, B. K. Tripathy, On some new type of generalized difference Cesàro sequence spaces, Soochow J. Math. 31(3) (2005) 333–340.

[48] B. C. Tripathy, M. Et, Y. Altin, S. Mahanta, Generalized difference paranormed sequence spaces defined by Orlicz function in a locally convex space, Indian J. Math. 48(2) (2006) 187–199.

[49] B. C. Tripathy, B. Hazarika, I-convergent sequence spaces associated with multiplier sequence spaces, Math. Ineq. Appl. 11(3) (2008) 543–548.

[50] B. C. Tripathy, B. Hazarika, I-convergent sequences spaces defined by Orlicz function, Acta Math. Appl. Sin. 27(1) (2011) 149–154.

[51] B.C. Triapthy, S. Mahanta, On a class of vector valued sequences associated with multiplier sequences, Acta Math. Appl. Sin. Eng. Ser. 20(3) (2004) 487–494.

[52] B. C. Tripathy, P. K. Nath, Statistical convergence of complex uncertain sequences, New Math. Natural Comput. 13 (2017) 359–374.

[53] B. C. Tripathy, M. Sen, Characterization of some matrix classes involving paranormed sequence spaces, Tamkang J. Math. 37(2) (2006) 155–162.

[54] B. C. Tripathy, M. Sen, Paranormed I-convergent double sequence spaces associated with multiplier sequences, Kyungpook Math. J. 54(2) (2014) 321–332.

[55] B. C. Tripathy, M. Sen, S. Nath, I-convergence in probabilistic n-normed space, Soft Comput. 16 (2012) 1021–1027.

[56] B. C. Tripathy, M. Sen, Paranormed I-convergent double sequence spaces associated with multiplier sequences, Kyungpook Math. J. 54(2) (2014) 321–332.

[57] B. C. Tripathy, M. Sen, Vector valued paranormed bounded and null sequences associated with multiplier sequences; Soochow J. Math. 29(3) (2003) 313–325.

[58] B. C. Tripathy, B. Sarma, Statistically null vector valued paranormed sequence space associated with multiplier sequences, Internat. J. Math. Sci. 3(2) (2004) 497–506.

[59] B. C. Tripathy, S. Mahanta, On a class of difference sequences related to the ℓ^p space defined by Orlicz functions, Math. Slovaca 57(2) (2007) 171–178.

[60] B. C. Tripathy, M. Sen, Paranormed I-convergent double sequence spaces associated with multiplier sequences, Kyungpook Math. J. 54(2) (2014) 321–332.

[61] A. Wilansky, Summability Through Functional Analysis, Vol. 85 1st Edition, North Holland (1984).

[62] A. Zygmund, Trigonometric Series, vol. II, Cambridge University Press, Cambridge (1993).

Chapter 13

Domain of Generalized Riesz Difference Operator of Fractional Order in Maddox's Space $\ell(p)$ and Certain Geometric Properties

Taja Yaying

Dera Natung Government College, tajayaying20@gmail.com

Bipan Hazarika

Gauhati University, bh_rgu@yahoo.co.in

S. A. Mohiuddine

King Abdulaziz University, mohiuddine@gmail.com

13.1 Introduction

Throughout this article $\Gamma(x)$ will denote the gamma function of a real number $x \notin \{0, -1, -2, \ldots\}$. It is defined as

$$\Gamma(x) = \int_0^\infty t^{x-1} e^{-t} dt. \tag{13.1}$$

Clearly $\Gamma(x+1) = x!$ for $x \in \mathbb{N}_0 \in \{0, 1, 2, \ldots\}$. Also $\Gamma(x+1) = x\Gamma(x)$ for any real number $x \notin \{0, -1, -2, \ldots\}$. Let w denotes the space of all real valued sequences. Any linear subspace of w is called sequence space. Let ℓ_∞, c and c_0 denote the spaces of all bounded, convergent and null sequences, respectively. Moreover, ℓ_1, bs and cs will denote the spaces of all absolutely summable, bounded and convergent series, respectively.

DOI: 10.1201/9781003178200-13

A Banach space V is called a BK space if each of the mappings $\pi_i : V \to \mathbb{C}$ defined by $\pi_i(\mathbf{v}) = v_i$ for all $\mathbf{v} = (v_i) \in V$ and $i \in \mathbb{N}_0$, is continuous, where $\mathbb{C}$ is the field of complex numbers. The sequence spaces ℓ_∞, c_0 and c are $BK-$spaces equipped with the norm $\|\mathbf{v}\|_{\ell_\infty} = \sup_i |v_i|$. Here and henceforth we shall use the notation that the summation symbol without limit runs from 0 to ∞. Furthermore, we shall use the notation $e = (1, 1, 1, \ldots)$ and $e^{(i)}$ is the sequence with 1 in the ith position and 0 elsewhere. Moreover, S_V and B_V shall denote the open and closed unit ball in V.

Definition 13.1.1. *A linear topological space V over the real field $\mathbb{R}$ is said to be a paranormed space if there exists a subadditive function $g : V \to \mathbb{R}$ such that $g(\theta) = 0$, $g(\boldsymbol{v}) = g(-\boldsymbol{v})$ and the scalar multiplication is continuous, i.e. $|\xi_i - \xi| \to 0$ and $g(v_i - \boldsymbol{v}) \to 0$ for all $\xi \in \mathbb{R}$ and $\boldsymbol{v} \in V$, where $\theta \in V$ is the zero vector.*

Let $p = (p_i)$ be a bounded sequence of strictly positive real numbers with $M = \max\{1, H\}$ where $H = \sup_i p_i$. Then, Maddox [40] defined the sequence space $\ell(p)$ as follows:

$$\ell(p) = \left\{ \mathbf{v} = (v_i) \in w : \sum_i |v_i|^{p_i} < \infty \right\}$$

which is a complete space paranormed by $h(\mathbf{v}) = \left(\sum_i |v_i|^{p_i}\right)^{1/M}$.

Let $\lambda = (\lambda_{ij})$ be an infinite matrix of real entries. By λ_i, we denote the sequence in the ith row of the matrix λ. The matrix λ is called a matrix mapping from a sequence space V to sequence space W if the λ-transform of the sequence $\mathbf{v} = (v_i)$ i.e. $\lambda\mathbf{v} = \{\lambda_i(\mathbf{v})\} = \{\sum_j \lambda_{ij} v_j\}$ exists and belongs to W for all $\mathbf{v} \in V$. We denote the class of all such matrices that map from V to W by $(V : W)$.

Define the set V_λ by

$$V_\lambda = \{\mathbf{v} \in w \colon \lambda\mathbf{v} \in V\}. \tag{13.2}$$

Then V_λ is a sequence space and is called the matrix domain of λ in the space V. There are several publictations in the literature dealing with the construction of sequence spaces using the domain of some special triangular matrices. For relevant literature, we refer to [4–6, 24, 41, 46]. For some recent high quality papers in this field one may refer to [3, 18, 26–30, 52, 53, 60, 61].

The following definitions are fundamental in our investigation:

Definition 13.1.2. *The α-, β- and γ-duals of the subset $V \subset w$ are defined by*

$$\begin{aligned} V^\alpha &= \{\boldsymbol{t} = (t_i) \in w : \boldsymbol{tv} = (t_i v_i) \in \ell_1 \textit{ for all } \boldsymbol{v} \in V\}, \\ V^\beta &= \{\boldsymbol{t} = (t_i) \in w : \boldsymbol{tv} = (t_i v_i) \in cs \textit{ for all } \boldsymbol{v} \in V\}, \\ V^\gamma &= \{\boldsymbol{t} = (t_i) \in w : \boldsymbol{tv} = (t_i v_i) \in bs \textit{ for all } \boldsymbol{v} \in V\}, \end{aligned}$$

respectively.

Definition 13.1.3. *A sequence $\boldsymbol{v} = (v_j)$ of a normed space $(V, \|\cdot\|)$ is called a Schauder basis of the space V if for every $\nu \in V$ there exists a unique sequence of scalars (a_j) such that*

$$\lim_{i\to\infty}\left\|\nu - \sum_{j=0}^{i} a_j v_j\right\| = 0.$$

The matrix domains $c_0(\Delta^F)$, $c(\Delta^F)$ and $\ell_\infty(\Delta^F)$ of forward difference matrix Δ^F are first studied by Kızmaz [38]. Since then several other authors [1,2,17,20–23,32,42,43,47,63] generalized the notion of difference operator Δ^F and studied difference sequence spaces of integer order. The difference sequence spaces of integer order were further generalized by Baliarsingh [6] to fractional order. Since then various authors, for instance Baliarsingh and Dutta [7,8], Dutta and Baliarsingh [19], Kadak and Baliarsingh [37], Baliarsingh and Kadak [9], Baliarsingh and Nayak [12], Meng and Mei [46], Yaying and Hazarika [60], Yaying et al. [56,57,59,62], Nayak et al. [48], Kadak [34–36], Furkan [24], Mohiuddine et al. [31,44], Özger [50,51] etc. studied forward and backward difference sequence spaces of fractional order.

Definition 13.1.4. *[7] Let $q > 0$ be a proper fraction. Then the difference operator Δ^{Bq} of fractional order q is defined as*

$$(\Delta^{Bq}\boldsymbol{v})_i = \sum_{l}(-1)^l \frac{\Gamma(q+1)}{l!(q-l+1)} v_{i-l}.$$

The difference operator Δ^{Bq} can also be expressed in the form of triangle as follows:

$$(\Delta^{Bq})_{ij} = \begin{cases} (-1)^{i-j}\frac{\Gamma(q+1)}{(i-j)!\Gamma(q-i+j+1)} & (0 \le j \le i), \\ 0 & (j > i). \end{cases}$$

Now, we give the definition of Riesz mean matrix and a short survey of literature concerning sequence spaces obtained by the domain of Riesz mean matrix in different spaces.

Definition 13.1.5. *Let (t_k) be a sequence of positive numbers and define $T_i = \sum_{l=0}^{i} t_l$, $i \in \mathbb{N}_0$. Then the Riesz mean matrix $R^t = (r^t_{ij})$ is defined as*

$$r^t_{ij} = \begin{cases} \frac{t_j}{T_i} & (0 \le j \le i), \\ 0 & (j > i). \end{cases}$$

Malkowsky [41] introduced the sequence spaces $\mathbf{r}^t_\infty$, $\mathbf{r}^t_c$ and $\mathbf{r}^t_0$ as the set of all sequences whose R^t-transforms are in the spaces ℓ_∞, c and c_0, respectively. Altay and Başar [4] studied the sequence space $\mathbf{r}^t(p)$ defined by

$$\mathbf{r}^t(p) = \left\{ v = (v_j) \in w : \sum_{i\in\mathbb{N}_0}\left|\frac{1}{T_i}\sum_{j=0}^{i} t_j v_j\right|^{p_i} < \infty \right\},$$

where $p = (p_j)$ is a bounded sequence of positive real numbers and studied their topological structures, obtained α-, β- and γ-duals, and characterized certain class of matrices related to $\mathbf{r}^t(p)$. The same authors in [5] introduced the sequence spaces $\mathbf{r}^t_\infty(p)$, $\mathbf{r}^t_c(p)$ and $\mathbf{r}^t_0(p)$ as the set of all sequences whose R^t- transforms are in the spaces $\ell_\infty(p)$, $c(p)$ and $c_0(p)$, respectively. Başarır and Öztürk [15] introduced the Riesz difference sequence space

$$\mathbf{r}^t(p, \Delta) = \{\mathbf{v} = (v_i) : (\Delta \mathbf{v})_i \in \mathbf{r}^t(p)\}$$

and studied their topological structures, obtained Schauder basis and β- and γ-duals. Başarır and Kayıkçı [14] introduced more generalized Riesz difference sequence space $\mathbf{r}^t(p, B^m) = \{\mathbf{v} = (v_j) \in w : B^m\mathbf{v} \in \mathbf{r}^t(p)\}$ and studied their topological properties, α- and β-duals and determined the Schauder basis, Besides certain matrix transformations and geometric properties related to the space $\mathbf{r}^t(p, B^m)$ are examined. Başarır [13] introduced the generalized Riesz difference seqence spaces $\mathbf{r}^t_\infty(B,p)$, $\mathbf{r}^t_c(B,p)$ and $\mathbf{r}^t_0(B,p)$ as the set of all sequences whose R^tB-transforms are in the spaces $\ell_\infty(p)$, $c(p)$ and $c_0(p)$, respectively. Başarır and Öztürk [16] generalized the results of Başarır [13], thereby introducing Riesz difference seqence spaces $\mathbf{r}^t_\infty(p, B^m)$, $\mathbf{r}^t_c(p, B^m)$ and $\mathbf{r}^t_0(p, B^m)$ as the set of all sequences whose R^tB^m-transforms are in the spaces $\ell_\infty(p)$, $c(p)$ and $c_0(p)$, respectively. Besides Başarır and Kara in [11] studied some classes of compact operators on Riesz difference sequence spaces $\mathbf{r}^t(B^m)$, and $\mathbf{r}^t_\infty(B^m)$, $\mathbf{r}^t_c(B^m)$ and $\mathbf{r}^t_0(B^m)$, respectively, using Hausdorff measure of noncompactness. Sengönül and Kayaduman [54] introduced the sequence spaces $\hat{f}$ and $\hat{f}_0$ as the set of all sequences whose R^t-transforms are in the spaces f and f_0 of almost convergent and almost null sequences, respectively. Recently Yaying [58] introduced the paranormed Riesz difference sequence spaces $\mathbf{r}^t_\infty(p, \Delta^{Bq})$, $\mathbf{r}^t_c(p, \Delta^{Bq})$ and $\mathbf{r}^t_0(p, \Delta^{Bq})$ as the set of all sequences whose $R^t\Delta^{Bq}$-transforms are in the spaces $\ell_\infty(p)$, $c(p)$ and $c_0(p)$, respectively.

13.2 Paranormed Riesz Difference Sequence Space $\mathbf{r}^t(p, \Delta^{Bq})$ of Fractional Order

Before constructing the sequence space $\mathbf{r}^t(p, \Delta^{Bq})$, we recall the following notions and definitions which can be found in [58]:

Definition 13.2.1. *[58] Let $q > 0$ be a proper fraction. Then the product matrix $R^t\Delta^{Bq}$ of Riesz mean R^t and backward difference operator Δ^{Bq} is defined by:*

$$\left(R^t\Delta^{Bq}\right)_{ij} = \begin{cases} \sum\limits_{l=j}^{i}(-1)^{l-j}\frac{\Gamma(q+1)}{(l-j)!\Gamma(q-l+j+1)}\frac{t_l}{T_i} & (0 \leq j \leq i), \\ 0 & (j > i). \end{cases}$$

In the rest of the paper, for brevity, we shall denote $R^t\Delta^{Bq}$ by $\mathcal{R}$.

Definition 13.2.2. *[58, Lemma 2.1] The inverse of the product matrix $\mathcal{R} = R^t\Delta^{Bq}$ is given by:*

$$\left(R^t\Delta^{Bq}\right)^{-1}_{ij} = \begin{cases} (-1)^{i-j}\sum\limits_{l=j}^{j+1}\frac{\Gamma(-q+1)}{(i-l)!\Gamma(-\alpha-i+l+1)}\frac{T_j}{t_l} & (0 \leq j < i), \\ \frac{T_i}{t_i} & (j = i), \\ 0 & (j > i). \end{cases}$$

Let us define the $R^t\Delta^{Bq} = \mathcal{R}$-transform of a sequence $\mathbf{v} = (v_j)$ as follows:

$$u_i = (\mathcal{R}\mathbf{v})_i = \sum_{j=0}^{i-1}\left[\sum_{l=j}^{i}(-1)^{l-j}\frac{\Gamma(q+1)}{(l-j)!\Gamma(q-l+j+1)}\frac{t_l}{T_i}\right]v_j + \frac{t_i}{T_i}v_i, \quad (i \in \mathbb{N}_0). \tag{13.3}$$

Now, we introduce paranormed Riesz difference sequence space $\mathbf{r}^t(p, \Delta^{Bq})$ of fractional order q as follows:

$$\begin{aligned} \mathbf{r}^t(p, \Delta^{Bq}) &= \{\mathbf{v} = (v_i) \in w : \mathcal{R}\mathbf{v} \in \ell(p)\} \\ &= \left\{\mathbf{v} = (v_i) \in w : \sum_i |(\mathcal{R}\mathbf{v})_i|^{p_i} < \infty\right\}. \end{aligned}$$

The above sequence space can be expressed in the form of the notation (13.2) as follows:

$$\mathbf{r}^t(p, \Delta^{Bq}) = (\ell(p))_{R^t\Delta^{Bq}} = (\ell(p))_{\mathcal{R}}.$$

The sequence space $\mathbf{r}^t(p, \Delta^{Bq})$ may be reduced to the following classes of sequence spaces in the special cases of q and $p = (p_i)$:

1. If $q = 0$ then the sequence space $\mathbf{r}^t(p, \Delta^{Bq})$ reduces to $\mathbf{r}^t(p)$ as defined by Altay and Başar [4].

2. If $q = 1$ then the sequence space $\mathbf{r}^t(p, \Delta^{Bq})$ reduces to $\mathbf{r}^t(p, \Delta^{B})$ where $(\Delta^B\mathbf{v})_j = v_j - v_{j-1}$.

3. If $q = m \in \mathbb{N}$ then the sequence space $\mathbf{r}^t(p, \Delta^{Bq})$ reduces to $\mathbf{r}^t(p, \Delta^{Bm})$ where $(\Delta^{Bm}\mathbf{v})_j = \sum\limits_{l=0}^{m}(-1)^l\binom{m}{l}v_{j-l}$ (cf. [45]).

4. If $p = e = (1, 1, 1, \ldots)$ then the sequence space $\mathbf{r}^t(p, \Delta^{Bq})$ reduces to $\mathbf{r}^t_p(\Delta^{Bq})$ (cf. [62]).

We begin with the following theorem:

Theorem 13.2.3. *The sequence space $\boldsymbol{r}^t(p, \Delta^{Bq})$ is a complete linear metric space paranormed by $\mathcal{P}_q$ defined by*

$$\mathcal{P}_q\boldsymbol{v} = \|\mathcal{R}\boldsymbol{v}\|_{\ell(p)} = \left(\sum_i \left|\sum_{j=0}^{i-1}\left[\sum_{l=j}^{i}(-1)^{l-j}\frac{\Gamma(q+1)}{(l-j)!\Gamma(q-l+j+1)}\frac{t_l}{T_i}\right]v_j + \frac{t_i}{T_i}v_i\right|^{p_i}\right)^{1/M} \tag{13.4}$$

with $0 < p_i \le H < \infty$.

Proof. The proof is a routine verification and hence omitted. □

Theorem 13.2.4. *The Riesz difference sequence space $\boldsymbol{r}^t(p, \Delta^{Bq})$ is linearly isomorphic to $\ell(p)$ where $0 < p_i \le H < \infty$.*

Proof. Define a mapping $\psi : \mathbf{r}^t(p, \Delta^{Bq}) \to \ell(p)$ by $\mathbf{v} \mapsto \mathbf{u} = \psi\mathbf{v} = \mathcal{R}\mathbf{v} = R^t\Delta^{Bq}\mathbf{v}$. It is easy to see that ψ is linear and $\mathbf{v} = 0$ whenever $\psi\mathbf{v} = 0$. Thus ψ is injective.

Let $\mathbf{u} = (u_j) \in \ell(p)$ and define the sequence $\mathbf{v} = (v_j)$ by

$$v_j = \sum_{l=0}^{j-1}\left[\sum_{m=l}^{l+1}(-1)^{j-l}\frac{\Gamma(-q+1)}{(j-m)!\Gamma(-q-j+m+1)}\frac{T_l}{t_m}u_l\right] + \frac{T_j}{t_j}u_j, \quad (j \in \mathbb{N}_0). \tag{13.5}$$

Then

$$\begin{aligned}
\mathcal{P}_q\mathbf{v} &= \left(\sum_j |(\mathcal{R}\mathbf{v})_j|^{p_j}\right)^{1/M} \\
&= \left(\sum_j \left|\sum_{l=0}^{j-1}\left(\sum_{m=l}^{j}(-1)^{m-l}\frac{\Gamma(q+1)}{(m-l)!\Gamma(q-m+l+1)}\frac{t_m}{T_j}\right)v_l + \frac{t_j}{T_j}v_j\right|^{p_j}\right)^{1/M} \\
&= \left(\sum_j \left|\sum_{l=0}^{j}\delta_{jl}u_l\right|^{p_j}\right)^{1/M} = \left(\sum_j |u_j|^{p_j}\right)^{1/M} < \infty,
\end{aligned}$$

where

$$\delta_{ij} = \begin{cases} 1 & \text{if } i = j, \\ 0 & \text{if } i \neq j. \end{cases}$$

Thus $\mathbf{v} \in \mathbf{r}^t(p, \Delta^{Bq})$. Consequently ψ is surjective and paranorm preserving. Thus $\mathbf{r}^t(p, \Delta^{Bq}) \cong \ell(p)$. □

We now construct sequence of points in the space $\mathbf{r}^t(p, \Delta^q)$ which will form Schauder basis for that space.

Theorem 13.2.5. *Let $\mathcal{R}_j(t) = (\mathcal{R}\boldsymbol{v})_j$ for all $j \in \mathbb{N}_0$, where $\mathcal{R} = R^t\Delta^{Bq}$ and $0 < p_j \leq H < \infty$. Define the sequence $b^{(j)}(t) = (b_i^{(j)}(t))$ of the elements of the space $\boldsymbol{r}^t(p, \Delta^{Bq})$ for every fixed $j \in \mathbb{N}_0$ by*

$$b_i^{(j)}(t) = \begin{cases} \sum_{l=j}^{j+1} (-1)^{i-j} \frac{\Gamma(-q+1)}{(i-l)!\Gamma(-q-i+l+1)} \frac{T_j}{t_l} & (j < i), \\ \frac{T_j}{t_j} & (j = i), \\ 0 & (j > i). \end{cases} \tag{13.6}$$

Then, the sequence $\{b^{(j)}(t)\}$ is basis for the space $\boldsymbol{r}^t(p, \Delta^{Bq})$ and every $\boldsymbol{v} \in \boldsymbol{r}^t(p, \Delta^{Bq})$ has a unique representation of the form

$$\boldsymbol{v} = \sum_j \mathcal{R}_j(t) b^{(j)}(t). \tag{13.7}$$

Proof. Using the definition of $\mathcal{R}$ and $b^{(j)}(t)$, we observe that

$$\mathcal{R}b^{(j)}(t) = e^{(j)} \in \ell(p) \tag{13.8}$$

for $0 < p_j \leq H < \infty$. Let $\mathbf{v} \in \mathbf{r}^t(p, \Delta^{Bq})$ and we denote

$$v^{[m]} = \sum_{j=0}^{m} \mathcal{R}_j(t) b^{(j)}(t) \tag{13.9}$$

for every non-negative integer m. Now, using equations (13.8) and (13.9), we obtain

$$\begin{aligned} \mathcal{R}v^{[m]} &= \sum_{j=0}^{m} \mathcal{R}_j(t) \mathcal{R}b^{(j)}(t) \\ &= (\mathcal{R}v)_j e^{(j)}, \end{aligned}$$

and

$$\left(\mathcal{R}(v - v^{[m]})\right)_i = \begin{cases} 0 & (0 \leq i \leq m), \\ (\mathcal{R}v)_i & (i > m). \end{cases}$$

Now, for a given $\varepsilon > 0$, there exists an integer m_0 such that

$$\left(\sum_{i \geq m} |(\mathcal{R}v)_i|^{p_i} \right)^{1/M} < \frac{\varepsilon}{2}$$

for all $m \geq m_0$. Hence

$$\mathcal{P}_q\left(v - v^{[m]}\right) = \left(\sum_{i\geq m} |(\mathcal{R}v)_i|^{p_i}\right)^{1/M}$$
$$\leq \left(\sum_{i\geq m_0} |(\mathcal{R}v)_i|^{p_i}\right)^{1/M} < \frac{\varepsilon}{2} < \varepsilon$$

for all $m \geq m_0$. This establishes the representation (13.7). To prove the uniqueness of the representation (13.7), we consider another representation of $\mathbf{v}$ as

$$\mathbf{v} = \sum_j \mathcal{R}'_j(t) b^{(j)}(t).$$

Then, we have

$$(\mathcal{R}v)_i = \sum_j \mathcal{R}'_j(t)\left(\mathcal{R}b^{(j)}(t)\right)_i = \sum_j \mathcal{R}'_j(t) e_n^{(j)} = \mathcal{R}'_i(t), \quad (i \in \mathbb{N}_0).$$

Thus, the representation (13.7) is unique. □

13.3 The α-, β- and γ-Duals

In this section, we obtain α-, β- and γ-duals of $\mathbf{r}^t(p, \Delta^{Bq})$. First, we quote certain lemmas which will be used to obtain the duals. Throughout the paper $\mathcal{F}$ will denote the collection of all finite subsets of $\mathbb{N}_0$ and $s_j = \frac{p_j}{p_j - 1}$ for all $j \in \mathbb{N}_0$.

Lemma 13.3.1. *[25, Theorem 5.1.0]*

(a) Let $1 < p_j \leq H < \infty$ for every $j \in \mathbb{N}_0$. Then $\lambda \in (\ell(p), \ell_1)$ if and only if there exists an integer $B > 1$ such that

$$\sup_{I\in\mathcal{F}} \sum_j \left|\sum_{i\in I} \lambda_{ij} B^{-1}\right|^{s_j} < \infty. \tag{13.10}$$

(b) Let $0 < p_j \leq 1$ for every $j \in \mathbb{N}_0$. Then $\lambda \in (\ell(p), \ell_1)$ if and only if

$$\sup_{I\in\mathcal{F}} \sup_{j\in\mathbb{N}_0} \left|\sum_{i\in I} \lambda_{ij}\right|^{p_j} < \infty. \tag{13.11}$$

Lemma 13.3.2. *[39, Theorem 1]*

(a) Let $1 < p_j \leq H < \infty$ for every $j \in \mathbb{N}_0$. Then $\lambda \in (\ell(p), \ell_\infty)$ if and only if there exists an integer $B > 1$ such that

$$\sup_{i \in \mathbb{N}_0} \sum_j \left|\lambda_{ij} B^{-1}\right|^{s_j} < \infty. \tag{13.12}$$

(b) Let $0 < p_j \leq 1$ for every $j \in \mathbb{N}_0$. Then $\lambda \in (\ell(p), \ell_\infty)$ if and only if

$$\sup_{i,j \in \mathbb{N}_0} |\lambda_{ij}|^{p_j} < \infty. \tag{13.13}$$

Lemma 13.3.3. *[39, Theorem 1] Let $0 < p_j \leq H < \infty$ for every $j \in \mathbb{N}_0$. Then $\lambda \in (\ell(p), c)$ if and only if (13.12), (13.13) hold, and*

$$\lim_{i \to \infty} \lambda_{ij} = \alpha_j \text{ for } j \in \mathbb{N}_0 \tag{13.14}$$

also holds.

Theorem 13.3.4. *Let $1 < p_j \leq H < \infty$ for every $j \in \mathbb{N}_0$. Define the set*

$$\mathcal{D}_1(p) = \bigcup_{B>1} \left\{ \mathbf{a} = (a_j) \in w : \sup_{I \in \mathcal{F}} \sum_j \left| \sum_{i \in I} \left[\sum_{l=j}^{j+1} (-1)^{i-j} \right.\right.\right.$$
$$\left.\left.\left. \times \frac{\Gamma(-q+1)}{(i-l)!\Gamma(-q-i+l+1)} \frac{T_j}{t_l} a_j + \frac{T_i}{t_i} a_i \right] B^{-1} \right|^{s_j} < \infty \right\}.$$

Then $\left[\boldsymbol{r}^t(p, \Delta^{Bq})\right]^\alpha = \mathcal{D}_1(p)$.

Proof. Consider the sequence $\mathbf{a} = (a_j) \in w$ and $\mathbf{v} = (v_j)$ is as defined in (13.5), then we have

$$\begin{aligned} a_i v_i &= \sum_{j=0}^{i-1} \left[\sum_{l=j}^{j+1} (-1)^{i-j} \frac{\Gamma(-q+1)}{(i-l)!\Gamma(-q-i+l+1)} \frac{T_j}{t_l} a_i u_j \right] + \frac{T_i}{t_i} a_i u_i \\ &= (Gu)_i, \text{ for each } i \in \mathbb{N}_0, \end{aligned} \tag{13.15}$$

where $G = (g_{ij})$ is a matrix defined by

$$g_{ij} = \begin{cases} \sum\limits_{l=j}^{j+1} (-1)^{i-j} \frac{\Gamma(-q+1)}{(i-l)!\Gamma(-q-i+l+1)} \frac{T_j}{t_l} a_i & (0 \leq j < i), \\ \frac{T_i}{t_i} a_i & (j = i), \\ 0 & (j > i). \end{cases}$$

Thus, we observe from (13.15) that $\mathbf{av} = (a_i v_i) \in \ell_1$ whenever $\mathbf{v} = (v_j) \in \mathbf{r}^t(p, \Delta^q)$ if and only if $G\mathbf{u} \in \ell_1$ whenever $\mathbf{u} = (u_j) \in \ell(p)$. This implies that $\mathbf{a} = (a_i) \in \left[\mathbf{r}^t(p, \Delta^{Bq})\right]^\alpha$ if and only if $G \in (\ell(p), \ell_1)$. Thus by using Part (a) of Lemma 13.3.1, we conclude that $\left[\mathbf{r}^t(p, \Delta^{Bq})\right]^\alpha = \mathcal{D}_1(p)$. ∎

Theorem 13.3.5. *Let $1 < p_j \leq H < \infty$ for every $j \in \mathbb{N}_0$. Define the set $\mathcal{D}_2(p)$ as follows:*

$$\mathcal{D}_2(p) = \bigcup_{B>1} \left\{ \boldsymbol{a} = (a_k) \in w : \sum_j \left| \Delta^q \left(\frac{a_j}{t_j} \right) T_j B^{-1} \right|^{s_j} < \infty \right.$$
$$\left. \text{and } \left\{ \left(\frac{a_j}{t_j} T_j B^{-1} \right)^{s_j} \right\} \in \ell_\infty \right\},$$

where

$$\Delta^{Bq}\left(\frac{a_j}{t_j}\right) = \frac{a_j}{t_j} + \sum_{l=j+1}^{i} (-1)^{l-j} a_l \sum_{m=j}^{j+1} \frac{\Gamma(-q+1)}{(l-m)!\Gamma(-\alpha-l+m+1)t_m}. \tag{13.16}$$

Then $\left[\boldsymbol{r}^t(p, \Delta^{Bq})\right]^\beta = \mathcal{D}_2(p)$ and $\left[\boldsymbol{r}^t(p, \Delta^{Bq})\right]^\gamma = \mathcal{D}_2(p)$.

Proof. Let $\mathbf{a} = (a_j) \in w$ and $\mathbf{v} = (v_j)$ is as defined in (13.5). Consider the following equation

$$\begin{aligned}
\sum_{j=0}^{i} a_j v_j &= \sum_{j=0}^{i} a_j \left[\sum_{l=0}^{j-1} \left(\sum_{m=l}^{l+1} (-1)^{j-l} \frac{\Gamma(-q+1)}{(j-m)!\Gamma(-q-j+m+1)} \frac{T_l}{t_m} u_l \right) \right. \\
&\quad \left. + \frac{T_j}{t_j} u_j \right] \\
&= \sum_{j=0}^{i-1} u_j T_j \left(\frac{a_j}{t_j} + \sum_{l=j+1}^{i} (-1)^{l-j} a_l \sum_{m=j}^{j+1} \frac{\Gamma(-q+1)}{(l-m)!\Gamma(-\alpha-l+m+1)t_m} \right) \\
&\quad + \frac{T_i}{t_i} a_i u_i \\
&= \sum_{j=0}^{i-1} u_j T_j \Delta^{Bq}\left(\frac{a_j}{t_j}\right) + \frac{T_i}{t_i} a_i u_i \qquad (13.17) \\
&= (Hu)_i, \text{ for each } i \in \mathbb{N}_0, \qquad (13.18)
\end{aligned}$$

where $H = (h_{nk})$ is a matrix defined by

$$h_{ij} = \begin{cases} \Delta^{Bq}\left(\frac{a_j}{t_j}\right) T_j & \text{if } 0 \leq j < i, \\ \frac{T_i}{t_i} a_i & \text{if } j = i, \\ 0 & \text{if } j > i. \end{cases}$$

and $\Delta^{Bq}\left(\frac{a_j}{t_j}\right)$ is as defined in (13.16). Thus, we deduce from (13.17) that $\mathbf{av} = (a_j v_j) \in cs$ whenever $\mathbf{v} = (v_j) \in \mathbf{r}^t(p, \Delta^{Bq})$ if and only if $H\mathbf{u} \in c$ whenever $\mathbf{u} = (u_j) \in \ell(p)$. This gives the fact $\mathbf{a} = (a_j) \in [\mathbf{r}^t(p, \Delta^{Bq})]^\beta$ if and only if $H \in (\ell(p), c)$. Therefore, by using Lemma 13.3.3, we get that

$$\sum_j \left| \Delta^q \left(\frac{a_j}{t_j} \right) T_j B^{-1} \right|^{s_j} < \infty \text{ and } \sup_{j\in\mathbb{N}_0} \left| \frac{a_j}{t_j} T_j B^{-1} \right|^{s_j} < \infty.$$

Thus $\left[\mathbf{r}^t(p, \Delta^{Bq})\right]^\beta = \mathcal{D}_2(p)$.

By the similar proceedings as above, we write $\mathbf{a} = (a_j) \in [\mathbf{r}^t(p, \Delta^{Bq})]^\gamma$ if and only if $H \in (\ell(p), \ell_\infty)$. Therefore, using Part (a) of Lemma 13.3.2, we immediately get $\left[\mathbf{r}^t(p, \Delta^{Bq})\right]^\gamma = \mathcal{D}_2(p)$. □

Theorem 13.3.6. *Let $0 < p_j \leq 1$ for every $j \in \mathbb{N}_0$. Define the sets $\mathcal{D}_3(p)$ and $\mathcal{D}_4(p)$ as follows:*

$$\mathcal{D}_3(p) = \Bigg\{ \boldsymbol{a} = (a_j) \in w : \sup_{I\in\mathcal{F}} \sup_{j\in\mathbb{N}_0} \Bigg| \sum_{i\in I} \Bigg[\sum_{l=j}^{j+1} (-1)^{i-j} \times \frac{\Gamma(-q+1)}{(i-l)!\Gamma(-q-i+l+1)} \frac{T_j}{t_l} a_j + \frac{T_i}{t_i} a_i \Bigg] B^{-1} \Bigg|^{p_j} < \infty \Bigg\}$$

and

$$\mathcal{D}_4(p) = \left\{ \boldsymbol{a} = (a_j) \in w : \sup_{j\in\mathbb{N}_0} \left| \Delta^q \left(\frac{a_j}{t_j} \right) T_j \right|^{p_j} < \infty \text{ and } \sup_{j\in\mathbb{N}_0} \left| \frac{a_j}{t_j} T_j \right|^{p_j} < \infty \right\}.$$

Then $\left[\boldsymbol{r}^t(p, \Delta^{Bq})\right]^\alpha = \mathcal{D}_3(p)$, $\left[\boldsymbol{r}^t(p, \Delta^{Bq})\right]^\beta = \left[\boldsymbol{r}^t(p, \Delta^{Bq})\right]^\gamma = \mathcal{D}_4(p)$.

Proof. The proof can be established analogously to Theorems 13.3.4 and 13.3.5 except that the second parts of Lemmas 13.3.1 and 13.3.2, and Lemma 13.3.3 are employed instead of the first. So, we omit details. □

13.4 Matrix Transformations

In this section, we characterize certain matrix transformation from the space $\mathbf{r}^t(p, \Delta^{Bq})$ to the sequence space $W \in \{\ell_\infty, c, c_0\}$.

Theorem 13.4.1. *Let W be any given BK-space. Then $\lambda \in (\boldsymbol{r}^t(p, \Delta^{Bq}), W)$ if and only if $\lambda_i \in \left[\boldsymbol{r}^t(p, \Delta^{Bq})\right]^\beta$ for all $i \in \mathbb{N}_0$ and $\tilde{\lambda} \in (\ell(p), W)$, where*

$$\tilde{\lambda}_{ij} = \left(\frac{\lambda_{ij}}{t_j} + \sum_{l=j+1}^{\infty} (-1)^{l-j} \lambda_{il} \sum_{m=j}^{j+1} \frac{\Gamma(-q+1)}{(l-m)!\Gamma(-q-l+m+1)q_m} \right) T_j$$

for all $i, j \in \mathbb{N}_0$.

Proof. Let $\lambda \in (\mathbf{r}^t(p, \Delta^{Bq}), W)$ and $\mathbf{v} \in \mathbf{r}^t(p, \Delta^{Bq})$. Then $\lambda_i = (\lambda_{ij})_{j\in\mathbb{N}_0} \in \left[\mathbf{r}^t(p, \Delta^{Bq})\right]^\beta$ for all $i \in \mathbb{N}_0$. Taking $\mathbf{u} = \mathcal{R}\mathbf{v}$, we get

$$
\begin{aligned}
(\tilde{\lambda}u)_j &= \sum_j \tilde{\lambda}_{ij} u_j \\
&= \sum_j T_j \left(\frac{\lambda_{ij}}{t_j} + \sum_{l=j+1}^{\infty} (-1)^{l-j} \lambda_{il} \sum_{m=j}^{j+1} \frac{\Gamma(-q+1)}{(l-m)!\Gamma(-q-l+m+1)t_m} \right) \\
&\quad \left(\sum_{l=0}^{j-1} \left[\sum_{m=l}^{j} (-1)^{m-l} \frac{\Gamma(q+1)}{(m-l)!\Gamma(q-m+l+1)} \frac{t_m}{T_j} \right] v_l + \frac{t_j}{T_j} v_j \right) \\
&= \sum_j \lambda_{ij} v_j = (\lambda v)_j
\end{aligned} \tag{13.19}
$$

for all $j \in \mathbb{N}_0$, where the matrix $\tilde{\lambda} = (\tilde{\lambda}_{ij})$ is defined in Theorem 13.4.1. This implies that $\tilde{\lambda}_i \in \ell_1$ for each $i \in \mathbb{N}_0$ and $\tilde{\lambda}\mathbf{u} \in W$, which gives us the desired consequence $\tilde{\lambda} \in (\ell(p), W)$.

Conversely, let $\lambda_i \in \left[\mathbf{r}^t(p, \Delta^{Bq})\right]^\beta$ for each $i \in \mathbb{N}_0$ and $\tilde{\lambda} \in (\ell(p), W)$, and take $\mathbf{v} \in \mathbf{r}^t(p, \Delta^{Bq})$. Then it is clear that (13.19) holds, which in turn implies that $\lambda\mathbf{v} \in W$ for all $\mathbf{v} \in \mathbf{r}^t(p, \Delta^{Bq})$ i.e. $\lambda \in (\mathbf{r}^t(p, \Delta^{Bq}), W)$. This completes the proof. □

Now using the above result, we characterize certain matrix classes from the space $\mathbf{r}^t(p, \Delta^{Bq})$ to space W, where W is any of the spaces ℓ_∞, c or c_0.

Corollary 13.4.2. *Let $\lambda = (\lambda_{ij})$ be an infinite matrix. Then we have the following results:*

(a) $\lambda \in (\mathbf{r}^t(p, \Delta^q) : \ell_\infty)$ if and only if $\lambda_i \in \left[\mathbf{r}^t(p, \Delta^{Bq})\right]^\beta$ for all $i \in \mathbb{N}_0$ and (13.12) and (13.13) hold with $\tilde{\lambda}_{ij}$ instead of λ_{ij}.

(b) $\lambda \in (\mathbf{r}^t(p, \Delta^{Bq}) : c)$ if and only if $\lambda_i \in \left[\mathbf{r}^t(p, \Delta^q)\right]^\beta$ for all $i \in \mathbb{N}_0$, and (13.12), (13.13) and (13.14) hold with $\tilde{\lambda}_{ij}$ instead of λ_{ij}.

(c) $\lambda \in (\mathbf{r}^t(p, \Delta^{Bq}) : c_0)$ if and only if $\lambda_i \in \left[\mathbf{r}^t(p, \Delta^{Bq})\right]^\beta$ for all $i \in \mathbb{N}_0$, and (13.12), (13.13) and (13.14) hold with $\tilde{\lambda}_{ij}$ instead of λ_{ij} and $\beta_j = 0$ for all $j \in \mathbb{N}_0$.

13.5 Certain Geometric Properties

In this section, we investigate certain geometric properties of the space $\mathbf{r}^t(p, \Delta^{Bq})$. We first recall certain notions and definitions which are necessary to establish our results:

Definition 13.5.1. *A point $\boldsymbol{w} \in S_V$ is an extreme point if for every $\boldsymbol{u}, \boldsymbol{v} \in S_V$ the equality $2\boldsymbol{w} = \boldsymbol{u} + \boldsymbol{v}$ implies $\boldsymbol{u} = \boldsymbol{v}$. A Banach space V is said to be rotund if every point of S_V is an extreme point.*

Definition 13.5.2. *A Banach space V is said to have Kadec-Klee property (or property (H)) (cf. [49]) if every weakly convergent sequence on the unit sphere is convergent in norm.*

Definition 13.5.3. *Let V be a real vector space. A functional $\sigma : V \to [0, \infty)$ is called a modular if*

(a) $\sigma(\boldsymbol{v}) = 0$ if and only if $\boldsymbol{v} = \theta$;

(b) $\sigma(\alpha\boldsymbol{v}) = \sigma(\boldsymbol{v})$ for scalars $|\alpha| = 1$;

(c) $\sigma(\alpha\boldsymbol{u} + \beta\boldsymbol{v}) \leq \sigma(\boldsymbol{u}) + \sigma(\boldsymbol{v})$ for all $\boldsymbol{u}, \boldsymbol{v} \in V$ and $\alpha, \beta > 0$ with $\alpha + \beta = 1$.

The modular σ is called convex if $\sigma(\alpha\boldsymbol{u} + \beta\boldsymbol{v}) \leq \alpha\sigma(\boldsymbol{u}) + \beta\sigma(\boldsymbol{v})$ for $\boldsymbol{u}, \boldsymbol{v} \in V$ and $\alpha, \beta > 0$ with $\alpha + \beta = 1$.

We define the operator σ_p on $\mathbf{r}^t(p, \Delta^{Bq})$ by

$$\begin{aligned}\sigma_p(\mathbf{v}) &= \sum_i |(\mathcal{R}\mathbf{v})_i|^{p_i} \\ &= \sum_i \left|(R^t\Delta^q\mathbf{v})_i\right|^{p_i} \\ &= \sum_i \left(\sum_{j=0}^{i-1} \left[\sum_{l=j}^{i} (-1)^{l-j} \frac{\Gamma(q+1)}{(l-j)!\Gamma(q-l+j+1)} \frac{t_l}{T_i} \right] v_j + \frac{t_i}{Q_i} v_i \right)^{p_i}. \end{aligned} \tag{13.20}$$

It is clear that $\sigma_p(\mathbf{v})$ is a convex modular on $\mathbf{r}^t(p, \Delta^{Bq})$. Now we equip the sequence space $\mathbf{r}^t(p, \Delta^q)$ with the Luxemburg norm (cf. [49]) defined by

$$\|\mathbf{v}\| = \inf\left\{k > 0 : \sigma_p\left(\frac{\mathbf{v}}{k}\right) \leq 1\right\}.$$

Now, we give certain basic properties of the modular σ_p.

Proposition 13.5.4. *The modular σ_p on $\boldsymbol{r}^t(p, \Delta^{Bq})$ satisfies the following statements with $p_j \geq 1$ for all $j \in \mathbb{N}_0$:*

(a) If $0 < k \leq 1$, then $k^M\sigma_p\left(\frac{\boldsymbol{v}}{k}\right) \leq \sigma_p(\boldsymbol{v})$ and $\sigma_p(k\boldsymbol{v}) \leq k\sigma_p(\boldsymbol{v})$.

(b) If $k \geq 1$, then $\sigma_p(\boldsymbol{v}) \leq k^p\sigma_p\left(\frac{\boldsymbol{v}}{k}\right)$.

(c) If $k \geq 1$, then $\sigma_p(\boldsymbol{v}) \leq k\sigma_p(\boldsymbol{v}) \leq \sigma_p(k\boldsymbol{v})$.

Proposition 13.5.5. *The following statements hold for $\boldsymbol{v} \in \boldsymbol{r}^t(p, \Delta^{Bq})$:*

(a) If $\|\boldsymbol{v}\| < 1$, then $\sigma_p(\boldsymbol{v}) \leq \|\boldsymbol{v}\|$.

(b) If $\|\boldsymbol{v}\| > 1$*, then* $\sigma_p(\boldsymbol{v}) \geq \|v\|$.

(c) $\|\boldsymbol{v}\| = 1$ *if and only if* $\sigma_p(\boldsymbol{v}) = 1$.

(d) $\|\boldsymbol{v}\| < 1$ *if and only if* $\sigma_p(\boldsymbol{v}) < 1$.

(e) $\|\boldsymbol{v}\| > 1$ *if and only if* $\sigma_p(\boldsymbol{v}) > 1$.

Proof. The results can be established analogously to Proposition 17 [64, p.7] (also see [33, Proposition 3], [49, Proposition 6]). Hence, we omit details. □

Proposition 13.5.6. *Let* (v_i) *be a sequence in* $\mathbf{r}^t(p, \Delta^{Bq})$.

(a) If $\lim_{i\to\infty} \|v_i\| = 1$*, then* $\lim_{i\to\infty} \sigma_p(v_i) = 1$.

(b) If $\lim_{i\to\infty} \sigma_p(v_i) = 0$*, then* $\lim_{i\to\infty} \|v_i\| = 0$.

Proof. The proof is analogous to the proof of Theorem 10 [49, p.4]. So we omit details. □

Theorem 13.5.7. *The sequence space* $\mathbf{r}^t(p, \Delta^{Bq})$ *is a Banach space with respect to the Luxemburg norm.*

Proof. It is enough to show that every Cauchy sequence in $\mathbf{r}^t(p, \Delta^{Bq})$ is convergent in Luxemburg norm. Let $v^{(i)} = (v_j^{(i)})$ be a Cauchy sequence in $\mathbf{r}^t(p, \Delta^{Bq})$ and $\varepsilon \in (0, 1)$. Then there exists a positive integer i_0 such that $\left\|v^{(i)} - v^{(j)}\right\| < \varepsilon$ for all $i, j \geq i_0$. Using Part (a) of Proposition 13.5.5, we obtain

$$\sigma_p(v^{(i)} - v^{(j)}) < \left\|v^{(i)} - v^{(j)}\right\| < \varepsilon \tag{13.21}$$

for all $i, j \geq i_0$. This gives

$$\sum_l \left|\left(\mathcal{R}(v^{(i)} - v^{(j)})\right)_l\right|^{p_l} < \varepsilon. \tag{13.22}$$

Thus, for each fixed l and for all $i, j \geq i_0$

$$\left|\left(\mathcal{R}(v^{(i)} - v^{(j)})\right)_l\right| = \left|\left(\mathcal{R}v^{(i)}\right)_l - \left(\mathcal{R}v^{(j)}\right)_l\right| < \varepsilon.$$

Hence, the sequence $\left\{\left(\mathcal{R}v^{(i)}\right)_l\right\}$ is Cauchy sequence in $\mathbb{R}$. Since $\mathbb{R}$ is complete, there exists $(\mathcal{R}\mathbf{v})_l \in \mathbb{R}$ such that $\left\{\left(\mathcal{R}v^{(i)}\right)_l\right\} \to (\mathcal{R}\mathbf{v})_l$ as $i \to \infty$. Therefore as $i \to \infty$, using (13.22), we have

$$\sum_l \left|\left(\mathcal{R}(v^{(i)} - \mathbf{v})\right)_l\right|^{p_l} < \varepsilon$$

for all $i \geq i_0$.

It remains to show that (v_i) is an element of $\mathbf{r}^t(p, \Delta^{Bq})$. Since $\{(\mathcal{R}v^{(j)})_l\} \to (\mathcal{R}\mathbf{v})_l$ as $j \to \infty$, we have

$$\lim_{j\to\infty} \sigma_p(v^{(i)} - v^{(j)}) = \sigma_p(v^{(i)} - \mathbf{v}).$$

Thus, using equation (13.21), we get that $\sigma_p(v^{(i)} - \mathbf{v}) < \left\|v^{(i)} - \mathbf{v}\right\| < \varepsilon$ for all $i \geq i_0$. This implies that $v^{(i)} \to \mathbf{v}$ as $i \to \infty$. Thus, we have $\mathbf{v} = v^{(i)} - (v^{(i)} - \mathbf{v}) \in \mathbf{r}^t(p, \Delta^{Bq})$.

Hence, the space $\mathbf{r}^t(p, \Delta^{Bq})$ is complete under the Luxemburg norm. □

Theorem 13.5.8. *The sequence space $\mathbf{r}^t(p, \Delta^{Bq})$ equipped with the Luxemburg norm is rotund if and only if $p_j > 1$ for all $j \in \mathbb{N}_0$.*

Proof. Let the space $\mathbf{r}^t(p, \Delta^{Bq})$ be rotund and choose $j \in \mathbb{N}_0$ such that $p_j = 1$ for all $j < 3$. Now consider the following sequences for a proper fraction q

$$\mathbf{u} = (1, q - \frac{t_0}{1_1}, \frac{q(q+1)}{2!} - q\frac{t_0}{t_1}, \frac{q(q+1)(q+2)}{3!} - \frac{q(q+1)}{2!}\frac{t_0}{t_1}, \ldots)$$

and

$$\mathbf{v} = (0, \frac{T_1}{t_1}, q\frac{T_1}{t_1} - \frac{T_1}{t_2}, \frac{q(q+1)}{2!}\frac{T_1}{t_1} - qT_1\frac{t_1}{t_2}, \ldots).$$

Then $\mathbf{u} \neq \mathbf{v}$ and it can be clearly seen that

$$\sigma_p(\mathbf{u}) = \sigma_p(\mathbf{v}) = \sigma_p\left(\frac{\mathbf{u}+\mathbf{v}}{2}\right) = 1.$$

Then by Part (c) of Proposition 13.5.6, $\mathbf{u}, \mathbf{v}, \frac{\mathbf{u}+\mathbf{v}}{2} \in S_{\mathbf{r}^t(p,\Delta^{Bq})}$ which contradicts the fact that $\mathbf{r}^t(p, \Delta^{Bq})$ is not rotund. Hence, $p_j > 1$ for all $j \in \mathbb{N}_0$.

Conversely, let $\mathbf{w} \in S_{\mathbf{r}^t(p,\Delta^{Bq})}$ and $\mathbf{u}, \mathbf{v} \in S_{\mathbf{r}^t(p,\Delta^{Bq})}$ be such that $\mathbf{w} = \frac{\mathbf{u}+\mathbf{v}}{2}$. By the convexity of σ_p and using Part (c) of Proposition 13.5.5, we have

$$1 = \sigma_p(\mathbf{w}) \leq \frac{1}{2}\left[\sigma_p(\mathbf{u}) + \sigma_p(\mathbf{v})\right] \leq \frac{1}{2} + \frac{1}{2} = 1.$$

This implies that $\sigma_p(\mathbf{u}) = \sigma_p(\mathbf{v}) = 1$ and $\sigma_p(\mathbf{w}) = \frac{\sigma_p(u)+\sigma_p(\mathbf{v})}{2}$.

Thus using the definition of σ_p and from the above discussion, we get that

$$\sum_i |(\mathcal{R}\mathbf{w})_i|^{p_i} = \frac{1}{2}\sum_i |(\mathcal{R}\mathbf{u})_i|^{p_i} + \frac{1}{2}\sum_i |(\mathcal{R}\mathbf{v})_i|^{p_i}.$$

Again $\mathbf{w} = \frac{\mathbf{u}+\mathbf{v}}{2}$, we have

$$\sum_i \left|\left(\mathcal{R}\left(\frac{\mathbf{u}+\mathbf{v}}{2}\right)\right)_i\right|^{p_i} = \frac{1}{2}\sum_i |(\mathcal{R}\mathbf{u})_i|^{p_i} + \frac{1}{2}\sum_i |(\mathcal{R}\mathbf{v})_i|^{p_i}.$$

This implies that

$$\left|\left(\mathcal{R}\left(\frac{\mathbf{u}+\mathbf{v}}{2}\right)\right)_i\right|^{p_i} = \frac{1}{2}\left|(\mathcal{R}\mathbf{u})_i\right|^{p_i} + \frac{1}{2}\left|(\mathcal{R}\mathbf{v})_i\right|^{p_i}. \tag{13.23}$$

From (13.23), it follows immediately that $\mathbf{u} = \mathbf{v}$. Thus the space $\mathbf{r}^t(p, \Delta^{Bq})$ is rotund. □

Theorem 13.5.9. *The sequence space $\mathbf{r}^t(p, \Delta^{Bq})$ has the Kadec-Klee property.*

Proof. Let $\mathbf{v} \in S_{\mathbf{r}^t(p,\Delta^{Bq})}$ and $(v^{(i)}) \subset \mathbf{r}^t(p, \Delta^{Bq})$ such that $\left\|v^{(i)}\right\| \to 1$ and $v^{(i)} \to \mathbf{v}$ weakly. Using Part (a) of Proposition 13.5.6, we get

$$\sigma_p(v^{(i)}) \to 1 \text{ as } i \to \infty. \tag{13.24}$$

Also $\mathbf{v} \in S_{\mathbf{r}^t(p,\Delta^{Bq})}$ and using Part (c) of Proposition 13.5.5, we observe that

$$\sigma_p(\mathbf{v}) = 1. \tag{13.25}$$

Thus observing (13.24) and (13.25), we write

$$\sigma_p(v^{(i)}) \to \sigma_p(\mathbf{v}) \text{ as } i \to \infty.$$

Since $v^{(i)} \to \mathbf{v}$ weakly and the jth coordinate mapping $\pi_j : \mathbf{r}^t(p, \Delta^{Bq}) \to \mathbb{R}$ defined by $\pi_j(\mathbf{v}) = v_j$ is continuous, these facts imply that $v_l^{(i)} \to v_l$ as $i \to \infty$. Therefore, $v^{(i)} \to \mathbf{v}$ as $i \to \infty$. This completes the proof. □

Bibliography

[1] Z. U. Ahmad, M. Mursaleen, Köthe-Toeplitz duals of some new sequence spaces and their matrix maps, Publ. Inst. Math. (Beograd) 42 (1987) 57–61.

[2] Z. U. Ahmad, M. Mursaleen, A. H. A. Bataineh, On some generalized spaces of difference sequences, Southeast Asian Bull. Math. 29(4) (2005) 635–649.

[3] P. Z. Alp, A new paranormed sequence space defined by Catalan conservative matrix, Math. Meth. Appl. Sci. 44 (2021) 7651–7658.

[4] B. Altay, F. Başar, On the paranormed Riesz sequence spaces of non-absolute type, Southeast Asian Bull. Math. 26 (2002) 701–715.

[5] B. Altay, F. Başar, Some paranormed Riesz sequence spaces of non-absolute type, Southeast Asian Bull. Math. 30 (2006) 591–608.

[6] P. Baliarsingh, Some new difference sequence spaces of fractional order and their dual spaces, Appl. Math. Comput. 219 (2013) 9737–9742.

[7] P. Baliarsingh, S. Dutta, A unifying approach to the difference operators and their applications, Bol. Soc. Paran. Mat. 33 (2015) 49–57.

[8] P. Baliarsingh, S. Dutta, On the classes of fractional order difference sequence spaces and their matrix transformations, Appl. Math. Comput. 250 (2015) 665–674.

[9] P. Baliarsingh, U. Kadak, On matrix transformations and Hausdorff measure of noncompactness of Euler difference sequence spaces of fractional order, Quaest. Math. 43(11) (2020) 1645–1661.

[10] P. Baliarsingh, U. Kadak, M. Mursaleen, On statistical convergence of difference sequences of fractional order and related Korovkin type approximation theorems, Quaest. Math. 41(8) (2018) 1117–1133.

[11] M. Başarır, E. E. Kara, On compact operators on the Riesz B^m−difference sequence space, Iran. J. Sci. Technol. Trans. A Sci. A4 (2011) 279–285.

[12] P. Baliarsingh, L. Nayak, A note on fractional difference operators, Alexandria Eng. J. 57(2) (2018) 1051–1054.

[13] M. Başarır, On the generalized Riesz B-difference sequence spaces, Filomat 24(4) (2010) 35–52.

[14] M. Başarır, M. Kayıkçı, On the generalized B^m-Riesz difference sequence space and β-property, J. Inequal. Appl. 2009 (2009), Article ID 385029, 19 pp.

[15] M. Başarır, M. Öztürk, On the Riesz difference sequence space, Rend. Circ. Mat. Palermo, II Ser. 57 (2008) 377–389.

[16] M. Başarır, M. Öztürk, On some generalized B^m-difference Riesz sequence spaces and uniform opial property, J. Inequal. Appl. 2011 (2011), Article ID 485730.

[17] C. Bektas, M. Et, R. Çolak, Generalized difference sequence spaces and their dual spaces, J. Math. Anal. Appl. 292 (2004) 423–432.

[18] S. Demiriz, M. İlkhan, E. E. Kara, Almost convergence and Euler totient matrix, Ann. Funct. Anal. 11 (2020) 604-616.

[19] S. Dutta, P. Baliarsingh, A note on paranormed difference sequence spaces of fractional order and their matrix transformations, J. Egypt. Math. Soc. 22 (2014) 249–253.

[20] M. Et, On some difference sequence spaces, Dŏga-Tr. J. Math. 17 (1993) 18–24.

[21] M. Et, M. Başarır, On some new generalized difference sequence spaces, Periodica Math. Hungar. 35 (1997) 169–175.

[22] M. Et, R. Çolak, On generalized difference sequence spaces, Soochow J. Math. 21 (1995) 377–386.

[23] M. Et, A. Esi, On Köthe-Toeplitz duals of generalized difference sequence spaces, Bull. Malays. Math. Sci. Soc. 231 (2000) 25-32.

[24] H. Furkan, On some λ-difference sequence spaces of fractional order, J. Egypt Math. Soc. 25(1) (2017) 37–42.

[25] K. G. Grosse-Erdmann, Matrix transformation between the sequence spaces of Maddox, J. Math. Anal. Appl. 180 (1993) 223–238.

[26] M. İlkhan, Matrix domain of a regular matrix derived by Euler totient function in the spaces c_0 and c, Mediterr. J. Math. 17 (2020) 27.

[27] M. İlkhan, Certain geometric properties and matrix transformations on a newly introduced Banach space, Fundam. J. Math. Appl. 3(1) (2020) 45–51.

[28] M. İlkhan, S. Demiriz, E. E. Kara, A new paranormed sequence space defined by Euler totient matrix, Karaelmas Sci. Eng. J. 9(2) (2019) 277–282.

[29] M. İlkhan, E.E. Kara, A new Banach space defined by Euler totient matrix operator, Oper. Matrices 13(2) (2019) 527–544.

[30] M. İlkhan, N. Şimşek, E. E. Kara, A new regular infinite matrix defined by Jordan totient function and its matrix domain in ℓ_p, Math. Methods. Appl. Sci. 44 (2021) 7622–7633.

[31] B. B. Jena, S. K. Paikray, S. A. Mohiuddine, V. N. Mishra, Relatively equi-statistical convergence via deferred Nörlund mean based on difference operator of fractional-order and related approximation theorems, AIMS Math. 5(1) (2020) 650–672.

[32] U. Kadak, S. A. Mohiuddine, Generalized statistically almost convergence based on the difference operator which includes the (p, q)-Gamma function and related approximation theorems, Results Math. 73(1) (2018), Article 9.

[33] E. E. Kara, M. Öztürk, M. Başarır, Some topological and geometric properties of generalized sequence space, Math. Slovaca 60(3) (2010) 385–398.

[34] U. Kadak, Generalized statistical convergence based on fractional order difference operator and its applications to approximation theorems, Iran. J. Sci. Technol. Trans. Sci. 43 (2019) 225–237.

[35] U. Kadak, Generalized lacunary statistical difference sequence spaces of fractional order, Int. J. Math. Math. Sci. 2015 (2015) 984283.

[36] U. Kadak, Generalized weighted invariant mean based on fractional difference operator with applications to approximation theorems for functions of two variables, Results Math. 72(3) (2017) 1181–1202.

[37] U. Kadak, P. Baliarsingh, On certain Euler difference sequence spaces of fractional order and related dual properties, J. Nonlinear Sci. Appl. 8 (2015) 997–1004.

[38] H. Kızmaz, On certain sequence spaces, Can. Math. Bull. 24 (1981) 169–176.

[39] C. G. Lascarides, I. J. Maddox, Matrix transformations between some classes of sequences, Proc. Camb. Phil. Soc. 68 (1970) 99–104.

[40] I. J. Maddox, Spaces of strongly summable sequences, Quart. J. Math. Oxford 18 (1967) 345–355.

[41] E. Malkowsky, Recent results in the theory of matrix transformations in sequence spaces, Mat. Vesnik 49 (1997) 187–196.

[42] S. A. Mohiuddine, A. Asiri, B. Hazarika, Weighted statistical convergence through difference operator of sequences of fuzzy numbers with application to fuzzy approximation theorems, Int. J. Gen. Syst. 48(5) (2019) 492–506.

[43] S. A. Mohiuddine, B. Hazarika, Some classes of ideal convergent sequences and generalized difference matrix operator, Filomat 31(6) (2017) 1827-1834.

[44] S. A. Mohiuddine, K. Raj, M. Mursaleen, A. Alotaibi, Linear isomorphic spaces of fractional-order difference operators, Alexandria Eng. J. 60 (2021) 1155–1164

[45] E. Malkowsky, S. D. Parashar, Matrix transformations in spaces of bounded and convergent difference sequences of order m, Analysis 17 (1997) 87–97.

[46] J. Meng, L. Mei, Binomial difference sequence spaces of fractional order, J. Inequal. Appl. 2018 (2018), Article 274.

[47] M. Mursaleen, Generalized spaces of difference sequences, J. Anal. Math. Appl. 203 (1996) 738–745.

[48] L. Nayak, M. Et, P. Baliarsingh, On certain generalized weighted mean fractional difference sequence spaces, Proc. Natl. Acad. Sci., India, Sect. A Phys. Sci. 89 (2019) 163–170.

[49] H. Nergiz, F. Başar, Some geometric properties of the domain of double sequential band matrix $B(\tilde{r}, \tilde{s})$ in the sequence space $\ell(p)$, Abstr. Appl. Anal. 2013 (2013), Article ID 421031.

[50] F. Özger, Characterisations of compact operators on l_p-type fractional sets of sequences, Demonstr. Math. 52 (2019) 105–115.

[51] F. Özger, Some geometric characterisations of a fractional Banach set, Commun. Fac. Sci. Univ. Ank. Ser. A1 Math. Stat. 68(1) (2019) 546–558.

[52] H. Roopei, Norm of Hilbert operator on sequence spaces, J. Inequal. Appl. 2020 (2020), Article 117.

[53] H. Roopei, A study on Copson operator and its associated sequence space, J. Inequal. Appl. 2020 (2020), Article 120.

[54] M. Sengönül, K. Kayaduman, On the Riesz almost convergent sequence space, Abstr. Appl. Anal. 2012 (2012), Article ID 691694.

[55] T. Yaying, Domain of generalized Riesz difference operator of fractional order in Maddox's space $\ell(p)$, AIP Conference Proceedings 2325 (2021) 020053.

[56] T. Yaying, On the paranormed Nörlund difference sequence space of fractional order and geometric properties, Math. Slovaca 71 (1) (2021) 155–170.

[57] T. Yaying, On a new class of generalized difference sequence spaces of fractional order, Proyecciones 38 (3) (2019) 485–497.

[58] T. Yaying, Paranormed Riesz difference sequence spaces of fractional order, Kragujevac J. Math. 46 (2) (2022) 175–191.

[59] T. Yaying, A. Das, B. Hazarika, P. Baliarsingh, Compactness of binomial difference sequence spaces of fractional order and sequence spaces, Rend. Circ. Mat. Palermo, II Ser. 68 (3) (2019) 459–476.

[60] T. Yaying, B. Hazarika, On sequence spaces generated by binomial difference operator of fractional order, Math. Slovaca 69(4) (2019) 901–918.

[61] T. Yaying, B. Hazarika, On sequence spaces defined by the domain of a regular tribonacci matrix, Math. Slovaca 70(3) (2020) 697-706.

[62] T. Yaying, B. Hazarika, A. Esi, Geometric properties and compact operator on fractional Riesz difference space, Kragujevac J. Math. 47(4) (2023) 545–566.

[63] T. Yaying, B. Hazarika, S. A. Mohiuddine, M. Mursaleen, K. J. Ansari, Sequence spaces derived by the triple band generalized Fibonacci difference operator, Adv. Difference Equ. 2020 (2020), Article 639.

[64] M. Yeşilkayagil, F. Başar, On the paranormed Nörlund sequence space of non absolute type, Abstr. Appl. Anal. 2014 (2014), Article ID 858704.

Index

Note: Locators in *italics* represent figures.